Developments in Hydrobiology 212

Series editor

K. Martens

For other titles published in this series, go to
www.springer.com/series/5842

Jellyfish Blooms: New Problems and Solutions

Jellyfish Blooms: New Problems and Solutions

Editors

Jennifer E. Purcell[1,2] & Dror L. Angel[3]

[1]*Western Washington University, Shannon Point Marine Center, 1900 Shannon Point Rd, Anacortes, WA 98221, USA*
[2]*University College Cork, Coastal and Marine Resources Centre, Naval Base, Haulbowline Island, Cobh, Co. Cork, Ireland*
[3]*Department of Maritime Civilizations, Leon Recanati Institute of Maritime Studies, Haifa University, Mt Carmel, Haifa 31905, Israel*

Previously published in *Hydrobiologia,* Volume 645, 2010

Springer

Editors
Jennifer E. Purcell
Western Washington University,
Shannon Point Marine Center,
1900 Shannon Point Rd,
Anacortes, WA 98221,
USA;
University College Cork,
Coastal and Marine Resources Centre,
Naval Base, Haulbowline Island,
Cobh, Co. Cork,
Ireland

Dror L. Angel
Department of Maritime Civilizations,
Leon Recanati Institute of Maritime
Studies, Haifa University,
Mt Carmel, Haifa 31905,
Israel

ISBN 978-90-481-9540-4

Springer Dordrecht Heidelberg London New York

Library of Congress Control Number: 2010928696

Cover illustration: A bloom of *Pelagia noctiluca* jellyfish off the coast of Galicia, Spain. Photo: Alejandro Olariaga.

Printed on acid-free paper.

Springer is part of Springer Science+Business Media (www.springer.com)

Contents

Hydrobiologia (2010) 645:1–2
DOI 10.1007/s10750-010-0220-0

JELLYFISH BLOOMS

Preface

Jennifer E. Purcell · Dror Angel

Published online: 17 March 2010

This third volume in the Jellyfish Blooms series was inspired by the large aggregation of jellyfish scientists that presented papers at the Aquatic Sciences Meeting: A Cruise through Nice Waters, sponsored by ASLO (Advancing the Science of Limnology and Oceanography) in Nice, France, in January, 2009. More than 60 presentations were given in three sessions: *New Approaches to the Study of Gelatinous Zooplankton*, organized and chaired by Drs. Christofer Troedsson and José Luis Acuña; *Harmful Jelly Blooms (HJBs)—What are we doing to understand and mitigate these?*, organized and chaired by Drs. Dror Angel, Alenka Malej, and Ahmet Kideys, and *Implications of Gelatinous Zooplankton Blooms on Food-web Dynamics and Nutrient Cycling*, organized and chaired by Drs. Robert H. Condon, Gaby Gorsky, and William M. Graham.

The character of each Jellyfish Blooms volume has reflected the location of the meeting. *Jellyfish Blooms: Ecological and Societal Importance*, from the First International Conference on Jellyfish Blooms in Gulf Shores, Alabama, USA, had predominantly North American contributions. Similarly, *Jellyfish Blooms: Causes, Consequences, and Recent Advances*, from the Second International Conference on Jellyfish Blooms, in Gold Coast, Australia, had many Australian and Asian contributions. Many scientists at the ASLO sessions in Nice presented the latest European research on gelatinous species. The current volume contains a subset of those presentations, but retains the European character.

New blooms of gelatinous predators (including scyphomedusae, hydromedusae, siphonophores, and ctenophores), and human problems related to these, have increased in recent years. Two papers in this volume directly address this topic. A key for the ephyrae of 18 species will permit early detection of scyphomedusan blooms (Straeler-Pohl & Jarms). The invasive ctenophore, *Mnemiopsis leidyi*, bloomed across the Mediterranean Sea for the first time in 2009 (Fuentes et al.).

Several of the contributions herein reflect concerns about how climate change will affect jellyfish blooms. Acidification of the oceans could affect statolith formation in medusae (Winans & Purcell); lowered

Guest editors: J. E. Purcell & Dror Angel / Jellyfish Blooms: New Problems and Solutions

J. E. Purcell (✉)
Shannon Point Marine Center, Western Washington University, 1900 Shannon Point Rd, Anacortes, WA 98221, USA
e-mail: Jennifer.Purcell@wwu.edu

J. E. Purcell
Coastal and Marine Resources Centre, University College Cork, Naval Base, Haulbowline Island, Cobh, Co., Cork, Ireland

D. Angel
Department of Maritime Civilizations, Leon Recanati Institute of Maritime Studies, Haifa University, Mt Carmel, Haifa 31905, Israel

salinities could affect the distributions of many species (Holst & Jarms); environmental variations cause fluctuations of jellyfish in diverse habitats (Quiñones et al., Kogovšek et al.). On smaller scales, water column characteristics, such as thin layers and mixing affect the behaviors and feeding of jellyfish and ctenophores (Frost et al., Mianzan et al.).

Jellyfish and ctenophores are voracious predators of zooplankton and ichthyoplankton. They potentially affect fisheries by competition with and predation on fish, as addressed for the widespread jellyfish, *Pelagia noctiluca*, in Sabatés et al. A single regression can be used to estimate respiration rates and thereby minimum ingestion rates of scyphozoan medusae (Purcell et al.). Møller et al. illustrates the importance of growth in bioenergetic estimates in the ctenophore, *Pleurobrachia pileus*. The amount of carbon contributed by jellyfish carcasses, and the effects on the microbial web of decomposing scyphomedusae, are presented by Sexton et al. and Tinta et al., respectively.

A critically-important characteristic of jellyfish is their ability to sting, which is important to the health of humans and penned fish, as well as for identification of taxa. Baxter et al. presents use of nematocysts for identification of the jellyfish, *Pelagia noctiluca*, from Continuous Plankton Recorder samples. Wiebring et al. and Lassen et al. use novel methods to separate different types of nematocysts to study their different toxins from the scyphomedusa, *Cyanea capillata*. The toxicities of the stinging tentacles of *C. capillata* and *Aurelia aurita* medusae are tested with a fish gill-cell assay (Helmholz et al.).

This is an exciting time in jellyfish research. Awareness of the roles of jellyfish in the health of the oceans has increased due to their recent proliferations and interference with human commercial enterprises, especially fisheries and tourism. A new project of high importance to the scientific community began in February 2010. This project, Global Expansion of Jellyfish Blooms: Magnitude, Causes and Consequences, is led by Drs. Robert Condon, Monty Graham, and Carlos Duarte and funded by the National Center for Ecological Analysis and Synthesis (NCEAS). The objectives of the project are development of a global jellyfish database and analysis of the global long-term patterns of jellyfish occurrence. The Third International Jellyfish Blooms Symposium will be held on 13–16 July 2010 in Mar del Plata, Argentina, hosted by the Argentinean Association of Marine Sciences and the National Institute for Fishery Research and Development, and organized by Drs. Hermes Mianzan, Gabriel Genzano, Agustín Schiariti, and Marcelo Acha. This meeting promises to be an important venue for researchers of gelatinous and fish species alike.

We would like to thank all the people who organized the ASLO sessions, the presenters, and, most of all, those who submitted papers to be included in this volume and the reviewers who helped all of us to present our hard work in the best possible way. We hope the readers will enjoy the following papers, which we believe represent important contributions to our understanding of jellyfish and their blooms.

Hydrobiologia (2010) 645:3–21
DOI 10.1007/s10750-010-0226-7

Identification key for young ephyrae: a first step for early detection of jellyfish blooms

I. Straehler-Pohl · G. Jarms

Published online: 21 March 2010

Abstract Although jellyfish blooms are a focus of recent research, the roles that the developmental stages of species play are underestimated. Planulae, polyps and ephyrae are inconspicuous and often overlooked. The importance of production of ephyrae from the sessile polyps has become more apparent. Our objective was to establish an identification system for early ephyrae of scyphozoan species in plankton samples. We studied ephyrae of 18 species. Standard measurements were introduced and the variability of marginal lappets analysed. Characters differentiating the 18 species are described. Photographs and drawings of each species are presented as a catalogue of ephyrae of these species. We developed a key for identification of the 18 species.

Keywords Scyphozoa · Ephyrae · Aurelia · Cyanea · Cassiopea · Rhizostoma

Guest editors: J. E. Purcell & Dror Angel / Jellyfish Blooms: New Problems and Solutions

I. Straehler-Pohl (✉)
Zoologisches Institut und Zoologisches Museum, Universität Hamburg, Martin-Luther-King-Platz 3, 20146 Hamburg, Germany
e-mail: I.Straehler-Pohl@web.de

G. Jarms
Zoologisches Institut und Zoologisches Museum, Universität Hamburg, Martin-Luther-King-Platz 3, 20146 Hamburg, Germany
e-mail: gerhard.jarms@zoologie.uni-hamburg.de

Introduction

Jellyfish blooms are amongst the most conspicuous events in the oceans worldwide (Mills, 2001). These mass occurrences affect swimmers, fisheries, aquaculture and industries (Purcell et al., 2007). Many jellyfish species are blooming. A reliable early warning system for tourism and fish farms that includes information about harmful jellyfish species that may bloom in the forthcoming season does not exist yet. One step in establishing an early warning system is recognition of ephyrae in plankton samples. In order to better understand the effects of jellyfish blooms, early medusa stages of species need to be identified because some species like *Pelagia noctiluca* Forsskål, 1775 (Maretič et al., 1991) or *Cyanea capillata* Linnaeus, 1758 (in Williams et al., 1996) are harmful whilst others do not sting as badly, for example *Aurelia aurita* Linnaeus, 1746 but which clog fishing nets and sea water intakes. At first glance, all ephyrae look alike because of which morphological structures are not easily distinguishable (Russell, 1970). Therefore, we have compared living ephyrae of 18 species.

Materials and methods

For the identification list of ephyrae, we used species in our laboratory in the Zoological Institute in Hamburg (Table 1). The ephyrae were bred and

Table 1 List of species used for the ephyra key

Order Coronatae	Order Semaeostomeae	Superorder Rhizostomoidea (former order: Rhizostomeae)
		Suborder Cepheida
Family Nausithoidae	Family Pelagiidae	Family Kolpophora(e)
Nausithoe werneri Jarms, 1990	*Chrysaora hysoscella* Linnaeus, 1767	*Cephea cephea* (Forskål, 1775)
	Sanderia malayensis Goette, 1886	*Netrostoma setouchianum* (Kishinouye, 1902)
		Cotylorhiza tuberculata Macri, 1778
		Mastigias papua (Lesson, 1830)
		Phyllorhiza punctata von Lendenfeld, 1884
		Suborder Cassiopeida
	Family Cyaneidae	Family Cassiopeidae
	Cyanea lamarckii Haeckel, 1880	*Cassiopea andromeda* Eschscholtz, 1829
	Cyanea capillata Linnaeus, 1758	
		Suborder Rhizostomida
	Family Ulmaridae	Family Dactyliophora(e)
	Aurelia aurita[a] Linnaeus, 1746	*Rhizostoma octopus* (Macri, 1778)
	Aurelia limbata Chamisso & Eysenhardt, 1821	*Rhizostoma pulmo* (Macri, 1778)
		Catostylus mosaicus Quoy & Gaimard, 1824
	Phacellophora camtschatica Brandt, 1835	*Rhopilema esculentum* Kishinouye, 1891

[a] Polyps collected in the harbour of Helgoland (2005) by T. Kaminski

reared in different temperature conditions appropriate to their originating habitats in 250-ml glass dishes or in aerated aquaria both filled with natural seawater (Table 2).

In order to compare the ephyrae of the different species, all the specimens were photographed using a digital camera (ColorView, Soft Imaging System GmbH) under the same conditions. Newly released ephyrae (measured within 24 h after detachment) were transferred into the depression of object slides. The depressions were filled with the same water as the culturing dishes. Adaptation time was about 2–5 min. As soon as the ephyrae were fully expanded in a dorsal position (manubrium facing the objective lens), the depressions were covered with coverslips and the pictures were taken. Afterwards, the ephyrae were transferred back into their culturing dishes.

The ephyrae in the pictures were measured, edited and evaluated using the programme analySIS® (Soft Imaging System GmbH). The following measuring points were defined (see also Fig. 1):

- Total Body Diameter (TBD): 2× total length of marginal lappet + diameter of central disc
- Central Disc Diameter (CDD): adradial diameter of the central disc
- Total Marginal Lappet Length (TMLL): length of the lappet stem + length of rhopalial lappet
- Lappet Stem Length (LStL): length from lappet base (line between the bases of two marginal lappet clefts) to base of rhopalial niche (base of cleft between two rhopalial lappets)
- Rhopalial Lappet Length (RLL): length from rhopalial niche (labelled as 'sense niche' in Fig. 2) base to level of rhopalial lappet tips
- Manubrium Length (ML): length between base and rim of manubrium

Next to rhopalial lappets, some scyphozoan and all rhizostome species develop velar lappets that have no rhopalium. They develop between the rhopalial lappets either before detachment, as in Cassiopeida, or a few days after detachment in Aureliinae and in all Rhizostomeae. Velar lappets are also an important character for identification of *Cassiopea andromeda* (Table 3), and therefore, we measured also their lengths.

- Velar Lappet Length (VLL): length from velar lappet base (line between the bases of two lappets) to velar lappet tip

In order to compare the body dimensions and proportions, the following proportions (%) were used:

Table 2 Temperature and culturing conditions of the observed scyphozoan species

Species	Temperature (°C)		
	Polyps	Strobilae	Ephyrae
Nausithoe werneri	15[a]	15[a]	20–23[a]
Chrysaora hysoscella	15[a]	5–10[a]	10–15[a]
Sanderia malayensis	15[a]	20–25[a, c]	20–25[b]
Phacellophora camtschatica	15[a]	15[a]	15[a]
Cyanea lamarckii	15[a]	10[a]	10–15[a]
Cyanea capillata	10[a]	5[a]	5–10[a]
Aurelia aurita	10–15[a]	5–10[a]	15–25[b]
Aurelia limbata	10–15[a]	5–10[a]	15–25[b]
Rhizostoma octopus	10[a]	5–10[a]	15–25[b]
Rhizostoma pulmo	15[a]	10[a]	15–25[b]
Catostylus mosaicus,	15–20[a]	20[a]	20[a]
Rhopilema esculentum	15–20[a]	20[a]	20[a]
Cephea cephea	20–25[c]	25–28[c]	25[c]
Cotylorhiza tuberculata	20–25[c]	25–28[c]	25[c]
Netrostoma setouchianum	20–25[c]	25–28[c]	25[c]
Mastigias papua	20–25[c]	25–28[c]	25[c]
Phyllorhiza punctata	20–25[c]	25–28[c]	25[c]
Cassiopea andromeda	20–25[c]	25–28[c]	25[c]

[a] Glass dishes (250 ml) in incubator without daylight

[b] Aerated aquarium with daylight

[c] Glass dishes (250 ml) on laboratory bench with daylight

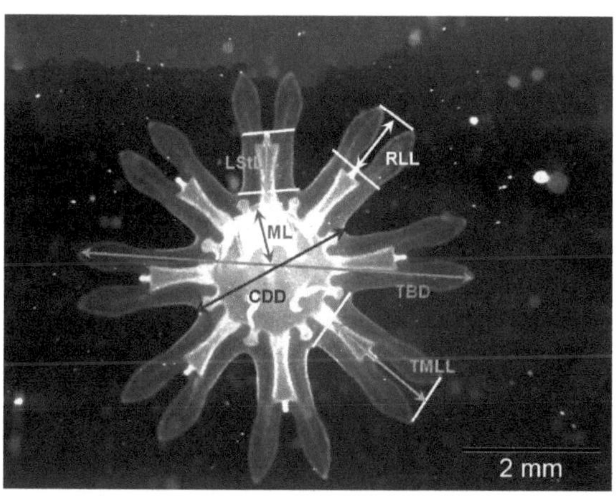

Fig. 1 Measuring points and measurements defined and taken in a newly released ephyra of *Aurelia aurita*. The opacity of the gastric system was increased with food particles. CDD: central disc diameter, LStL: lappet stem length, ML: manubrium length, RLL: rhopalial lappet length, TBD: total body diameter, TMLL: total marginal lappet length

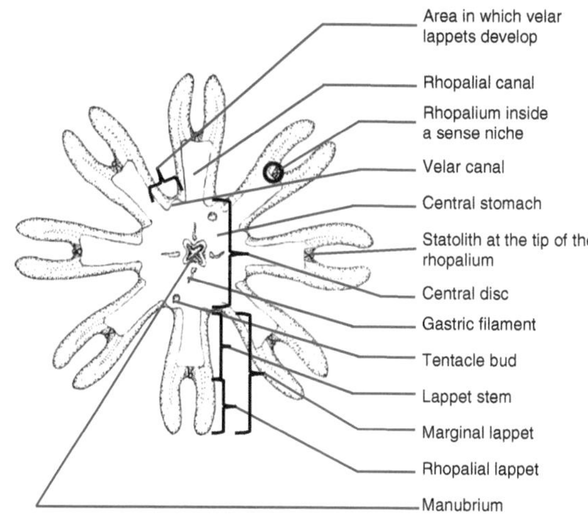

Fig. 2 Anatomy of a scyphozoan ephyra

Measurements compared to body diameter:

- CDD/TBD × 100
- TMLL/TBD × 100
- VLL/TBD × 100

Measurements compared to lappet length:

- RLL/TMLL × 100
- LStL/TMLL × 100

Measurements compared to central disc:

- ML/CDD × 100

Statistics: discriminant function analysis

Six individuals per species were measured (2 ephyrae per strobila from polydisc strobilating species and 1 ephyra per strobila from monodisc strobilating species). We examined the classification of individuals (species) based on body dimensions and proportions and verified the data statistically by using a multivariate process (discriminant function analysis) performed by the programme SPSS 16.0 (Wanker & Fischer, 2001; Backhaus et al., 2006). The correct classification was given as (%) by the programme.

The following variables were examined:

- 'absolute body dimensions' (total body diameter, total marginal lappet length, central stomach diameter, length of manubrium, distance of opposite rhopalia, central disc diameter, number of tentacle buds, number of gastric filaments,

Table 3 Ephyrae: itemized list of morphological characters and proportions

Species/ Distribution	Morphology	Species/ Distribution	Morphology
Nausithoe werneri Mediterranean Sea	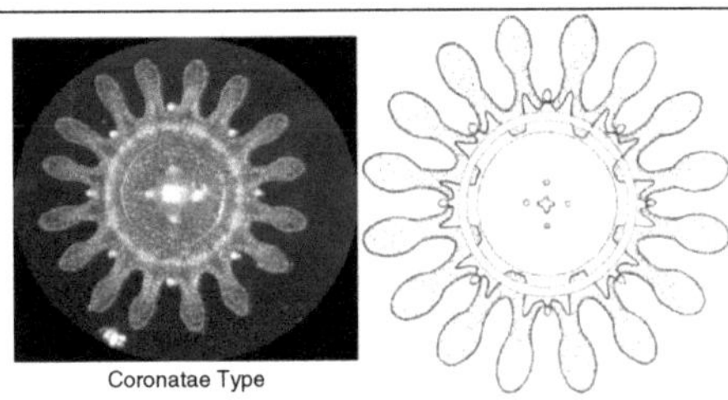 Coronatae Type Anatomy: • 16 round spoon-like marginal lappets without lappet stems • 4 Gastric filament sockets without gastric filaments • white muscle ring between stomach and umbrella rim observable on subumbrella • Rhopalium is directly attached to the central disc • **No marginal tentacles or tentacle bulbs** • Very short manubrium, approx 11% of the CDD Gastric system: • Very short gastric canal like protuberances of the central stomach which reach the bases of the marginal lappets Body proportions: • MLL ≈ 21% of the TBD • CDD ≈ 57% of the TBD	***Phacellophora camtschatica*** Kamtschatca to Peru	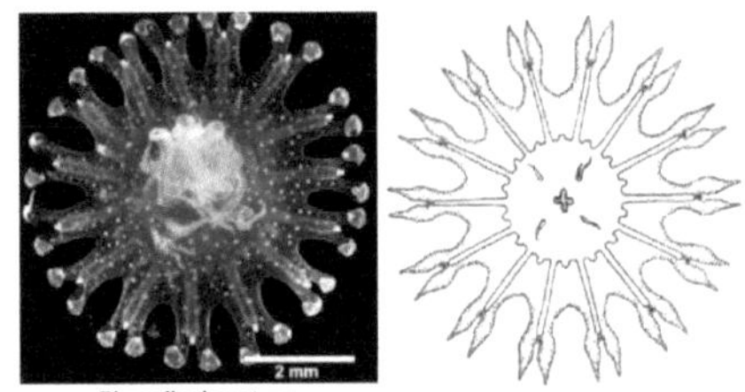 *Phacellophora* type Anatomy: • 14-16 lappet stems • 28-32 flame-like rhopalial lappets • **4-5 gastric filaments** stomach • **0-2 marginal tentacle bulbs** • Medium ML, approx. 28% of the CDD Gastric system: • Rhopalial canals are arrow-like • Velar canals arrow-like Body proportions: • RLL ≈ 11% of the TBD • LStL ≈ 10% of the TBD • CDD ≈ 57% of the TBD • RLL ≈ 53% of the TMLL • LStL ≈ 47% of the TMLL

CDD: Central Disk Diameter LStL: Lappet Stem Length MLL: Marginal Lappet Length RLL: Rhopalial Lappet Length TBD: Total Body Diameter TMLL: Total Marginal Lappet Length

Species/ Distribution	Morphology	Species/ Distribution	Morphology
Chrysaora. hysoscella Atlantic Coasts of France, UK, The Netherlands, Germany, Helgoland, Northern Sea to entrance of Baltic Sea	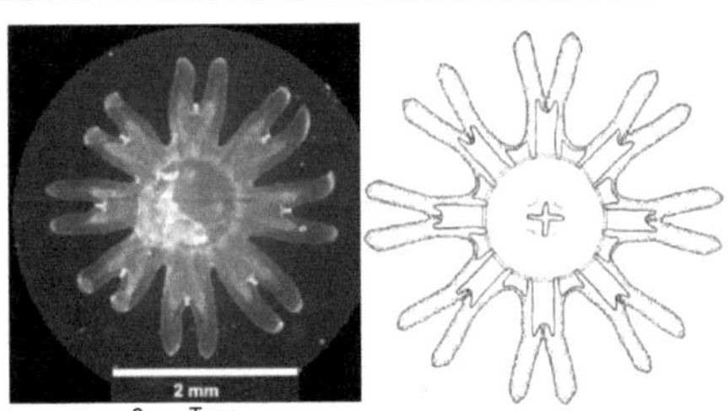 8-ray Type Anatomy: • 8 lappet stems • 16 round spatula-like rholalial lappets • 4 gastric filament sockets without gastric filaments • **No marginal tentacles or tentacle bulbs** • long manubrium, approx 44% of the CDD Gastric system: • Rhopalial canals are forked: Fork arms overtop the rhopalium and reach the bases of the rhopalial lappets, fork tips are sharp • Velar canals are forked: Fork arms reach the lower third of the lappet stems, fork tips are sharp Body proportions: • RLL ≈ 16% of the TBD • LStL ≈ 13% of theTBD • CDD ≈ 42% of the TBD • RLL ≈ 55% of the TMLL • LStL ≈ 45% of the TMLL	***Sanderia malayensis*** Indo-Pacific, Japan, Suez-Canal, South China Sea, Malay Archipelago	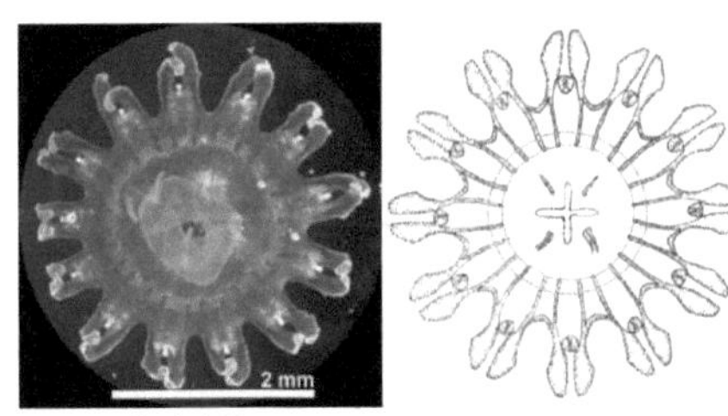 *Sanderia* Type Anatomy: • 13-16 lappet stems • 26-32 bread knife-like rhopalial lappets • 1 -2 gastric filaments per socket • **No marginal tentacles or tentacle bulbs** • long manubrium, approx. 33% of the CDD Gastric system: • Rhopalial canals are club shaped and slightly forked: Fork arms reach the bases of the rhopalial lappets but **do not** overtop the rhopalium, fork tips are sharp • Velar canals are slightly forked: Fork arms reach the lower third of the lappet stems, fork tips are sharp Body proportions: • RLL ≈ 12% of the TBD • LStL ≈ 8% of the TBD • CDD ≈ 60% of the TBD • RLL ≈ 60% of the TMLL • LStL ≈ 40% of the TMLL

CDD: Central Disk Diameter LStL: Lappet Stem Length RLL: Rhopalial Lappet Length TBD: Total Body Diameter TMLL: Total Marginal Lappet Length

Table 3 continued

Species/ Distribution	Morphology	Species/ Distribution	Morphology

Cyanea lamarckii

North Atlantic, English Channel, Northern Sea, Coasts of Norway and Germany, Helgoland, entrance of Baltic Sea

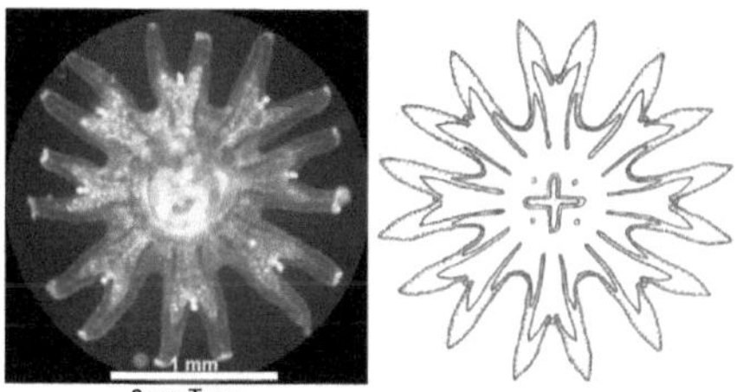

8-ray Type

Anatomy:
- 8 lappet stems
- 16 sword-like rhopalial lappets
- 4 gastric filament sockets without gastric filaments
- **No marginal tentacles or tentacle bulbs**
- Medium ML, approx 21% of the CDD

Gastric system:
- Rhopalial canals are forked:
  Fork arms overtop the rhopalium and reach the lower third of the rhopalial lappets, fork tips are sharp
- Velar canals are forked: Fork arms level with the base of the Rhopalium in the upper third of the lappet stems, fork tips are sharp

Body proportions:
- RLL ≈ 15% of the TBD
- LStL ≈ 12% of the TBD
- CDD ≈ 45% of the TBD
- RLL ≈ 56% of the TMLL
- LStL ≈ 44% of the TMLL

Cyanea capillata

North Atlantic, France, English Channel, North Sea, Helgoland, entrance of Baltic Sea

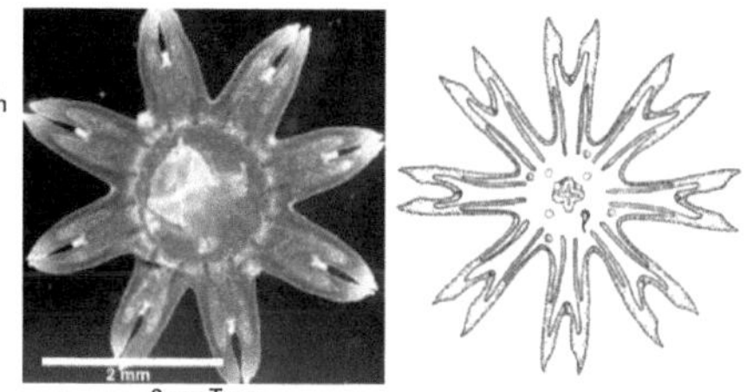

8-ray Type

Anatomy:
- 8 lappet stems
- 16 lancet-like rhopalial lappets
- **0-1 gastric filaments** per quadrant
- **With 0-4 marginal tentacle bulbs**
- Very long manubrium, approx. 41% of the adradial average

Gastric system:
- Rhopalial canals are forked:
  Fork arms overtop the rhopalium and reach the upper half of the rhopalial lappets, fork tips are sharp
- Velar canals are forked:
- Fork arms level with the tip of the Rhopalium in the lower third of the rhopalial lappets, fork tips are sharp

Body proportions:
- RLL ≈ 17% of theTBD
- LStL ≈ 13% of the TBD
- CDD ≈ 41% of the TBD
- RLL ≈ 57% of theTMLL
- LStL ≈ 43% of the TMLL

CDD: Central Disk Diameter LStL: Lappet Stem Length RLL: Rhopalial Lappet Length TBD: Total Body Diameter TMLL: Total Marginal Lappet Length

Species/ Distribution	Morphology	Species/ Distribution	Morphology

***Aurelia aurita* (Helgoland, Polyps found in the Harbour)**

North Sea, Helgoland

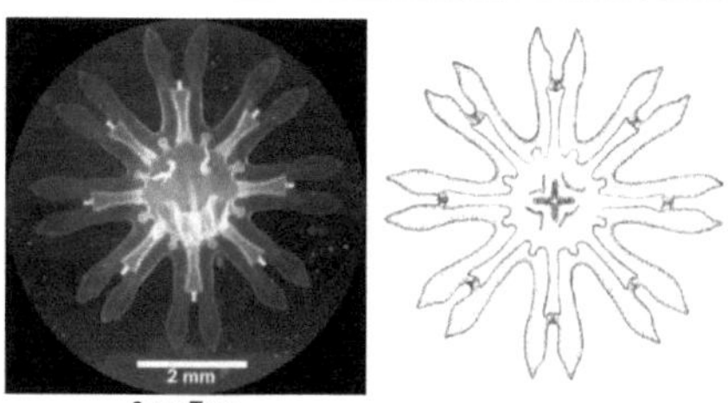

8-ray Type

Anatomy:
- 8 lappet stems
- 16 lancet-like rhopalial lappets
- **0-2 gastric filaments** per quadrant
- **No marginal tentacles or tentacle bulbs**
- Medium ML, approx. 22% of the CDD

Gastric system:
- Rhopalial canals are spade-like or slightly forked:
  Fork arms overtop slightly the base of the rhopalium, fork tips are rounded
- Velar canals are flat rhombical

Body proportions:
- RLL ≈ 14% of the TBD
- LStL ≈ 16% of the TBD
- CDD ≈ 40% of the TBD
- RLL ≈ 48% of theTMLL
- RLL ≈ 52% of theTMLL

Aurelia limbata

Alaska, Labrador, West Greenland, North West Pacific, Japan, Sea of Kamtchatka

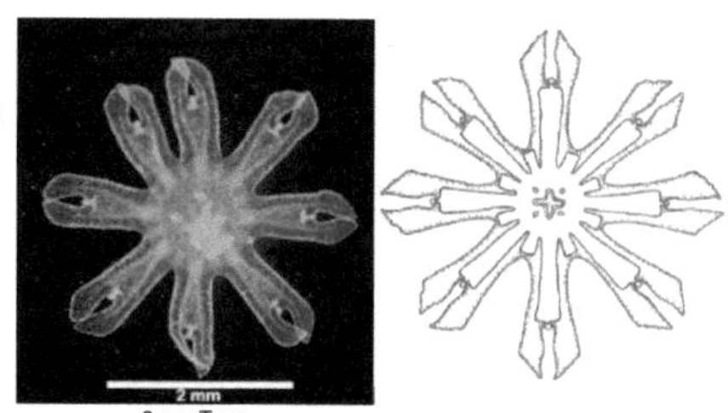

8-ray Type

Anatomy:
- 8 lappet stems
- 16 bread knife-like rhopalial lappets
- **1 gastric filament** per quadrant
- **No marginal tentacles or tentacle bulbs**
- long manubrium, approx. 34% of the adradial average

Gastric system:
- Rhopalial canals are spade-like to minimally forked
- Velar canals are spade-like

Body proportions:
- RLL ≈ 14% of the TBD
- LStL ≈ 19% of the TBD
- CDD ≈ 35% of the TBD
- RLL ≈ 42% of the TMLL
- LStL ≈ 58% of the TMLL

CDD: Central Disk Diameter LStL: Lappet Stem Length RLL: Rhopalial Lappet Length TBD: Total Body Diameter TMLL: Total Marginal Lappet Length

Table 3 continued

Species/ Distribution	Morphology	Species/ Distribution	Morphology
Cephea cephea Malay Archipelago, Maledive Islands, Indian Ocean	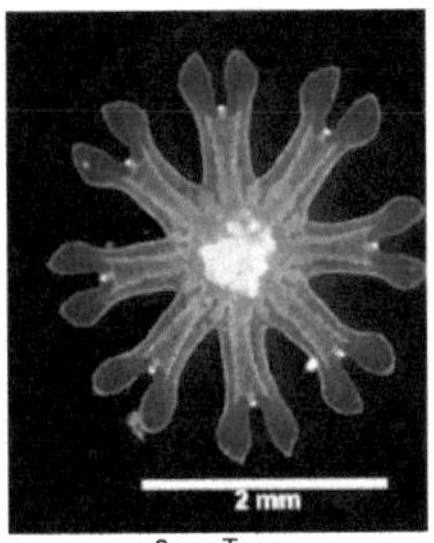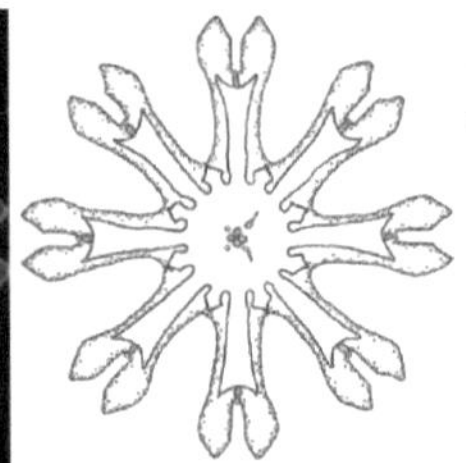 8-ray Type Anatomy: • 8 lappet stems • 16 oval spoon-like to bread knife-like rhopalial lappets • **0-1 gastric filament** per quadrant • Medium ML, approx. 24% of the CDD • Zooxanthellae mark the form of the gastric system Gastric system: • Rhopalial canals are slightly forked: Fork arms overtop the base of the rhopalium, fork tips are sharp • Velar canals are rhombical • **Rhopalia tip level with lappet rim** Body proportions: • RLL ≈ 13% of the TBD • LStL ≈ 18% of the TBD • CDD ≈ 37% of the TBD • RLL ≈ 41% of the TMLL • LStL ≈ 59% of the TMLL	***Netrostoma setouchianum*** Japan, Inland Sea, Misaki, Senzaki	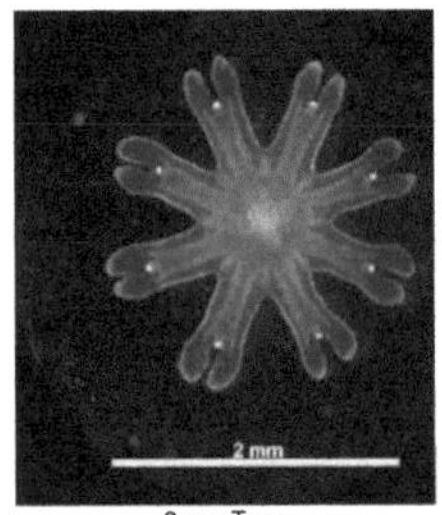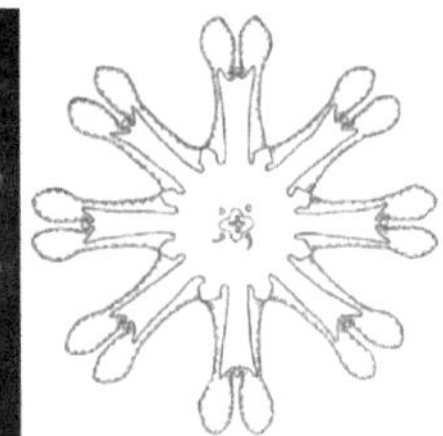 8-ray Type Anatomy: • 8 lappet stems • 16 round to oval spoon-like rhopalial lappets • **0-1 gastric filament** per quadrant • Short manubrium, approx. 17% of the adradial average • Zooxanthellae mark the form of the gastric system Gastric system: • Rhopalial canals are forked: Fork arms overtop the mid section of the rhopalium and reach the lower third of the rhopalial lappets, fork tips are sharp • Velar canals are spade-like to slightly rhombical • **Rhopalia (Statoliths) overtop the lappet rim** Body proportions: • RLL ≈ 12% of the TBD • LStL ≈ 17% of the TBD • CDD ≈ 41% of the TBD • RLL ≈ 41% of the TMLL • LStL ≈ 59% of the TMLL

CDD: Central Disk Diameter LStL: Lappet Stem Length RLL: Rhopalial Lappet Length TBD: Total Body Diameter TMLL: Total Marginal Lappet Length

Species/ Distribution	Morphology	Species/ Distribution	Morphology
Cotylorhiza tuberculata Mediterranean Sea, Red Sea, Suez Canal, Atlantic Coast of Spain, Mar Menor (Spain), Canary Islands	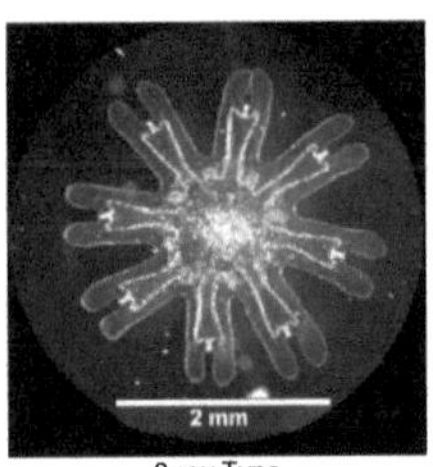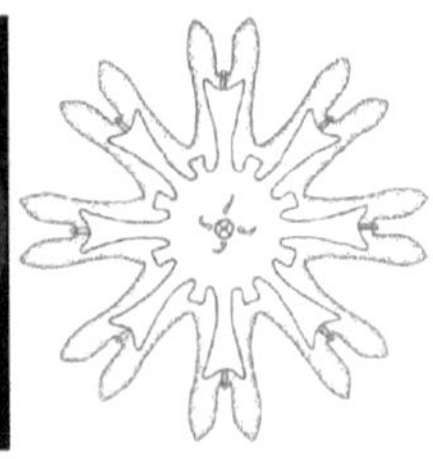 8-ray Type Anatomy: • 8 lappet stems • 16 round spoon-like rhopalial lappets • **1 gastric filament** per quadrant • Medium ML, approx. 21% of the CDD • Zooxanthellae mark the form of the gastric system Gastric system: • Rhopalial canals are slightly forked: Fork arms overtop the base of the rhopalium, fork tips are rounded • Velar canals are spade-like to slightly rhombical Body proportions: • RLL ≈ 11% of the TBD • LStL ≈ 17% of the TBD • CDD ≈ 42% of the TBD • RLL ≈ 39% of the TMLL • LStL ≈ 61% of the TMLL	***Cassiopea andromeda*** Suez Canal, Red Sea to Sumatra, Indian Ocean, East Africa, Malay Archipelago	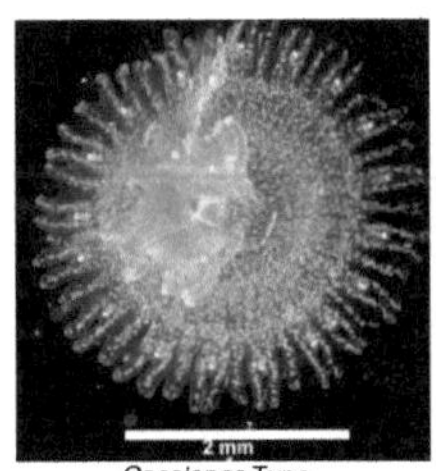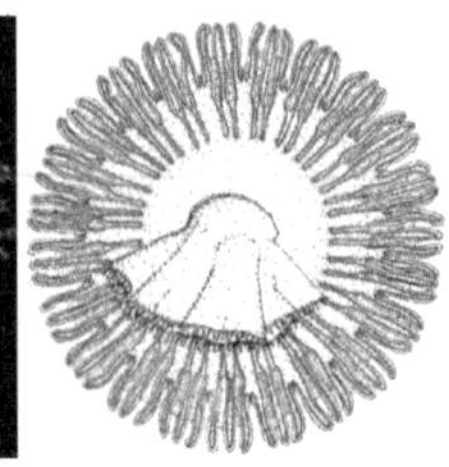 *Cassiopea* Type Anatomy: • 17-23 lappet stems • 34- 46 with outward bent tip rhopalial lappets • 16-21 round spatula-like single velar lappets • **0 gastric filament** per quadrant • Very long manubrium with oral tentacles, approx. 45% of the CDD • Zooxanthellae mark the gastric system Gastric system: • Rhopalial canals are long forked: Fork arms overtop the rhopalium till reaching the tip of the rhopalial lappets, fork tips are rounded • Velar canals are tongue-like and reach the tips of the velar lappets, tips are rounded Body proportions: • RLL ≈ 8%of the TBD • LStL ≈ 8% of the TBD • VLL ≈ 16% of the TBD • CDD ≈ 68% of the TBD • RLL ≈ 52% of the TMLL • LStL ≈ 48% of the TMLL

CDD: Central Disk Diameter LStL: Lappet Stem Length RLL: Rhopalial Lappet Length TBD: Total Body Diameter TMLL: Total Marginal Lappet Length
VLL: Velar Lappet Length

Table 3 continued

Species/ Distribution	Morphology	Species/ Distribution	Morphology
Mastigias papua East Coast of Africa, Indian Ocean, Malay Archipelago, China Sea, Philippines Islands, Japan	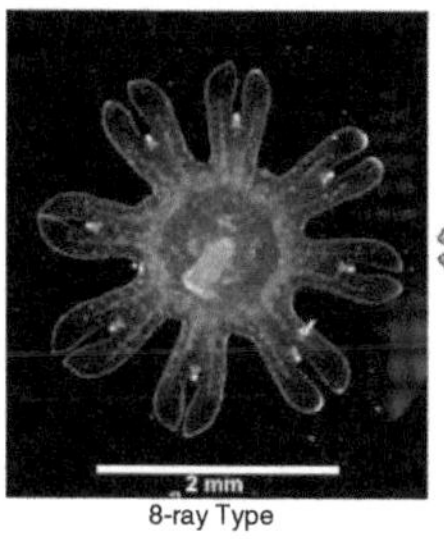 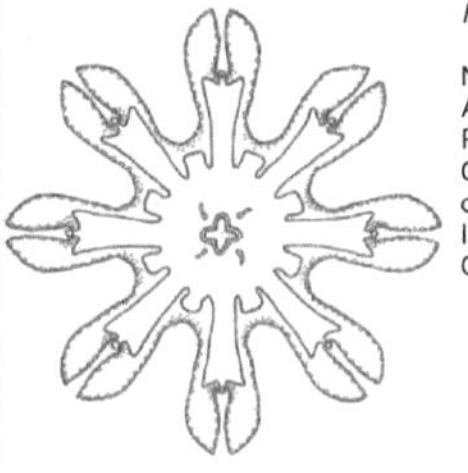 8-ray Type Anatomy: • 8 lappet stems • 16 pointed broad oval spoon-like rhopalial lappets • **1 gastric filament** per quadrant • Long manubrium, approx. 34% of the adradial average • Zooxanthellae mark the gastric system Gastric system: • Rhopalial canals are slightly forked: Fork arms overtop slightly the base of the rhopalium, fork tips are rounded • Velar canals are spade-like to slightly rhombical Body proportions: • RLL ≈ 14% of the TBD • LStL ≈ 14% of the TBD • CDD ≈ 43% of the TBD • RLL ≈ 49% of the TMLL • LStL ≈ 51% of the TMLL	*Phyllorhiza punctata* New South Wales, Australia southwestern Pacific, North Carolina Coast, Caribbean, Gulf of Mexico, Hawaiian Islands, Southern California	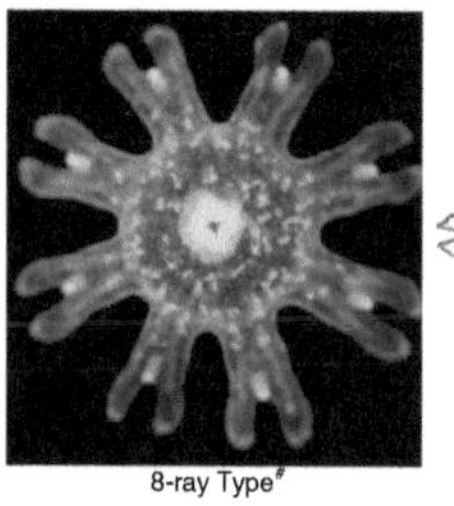 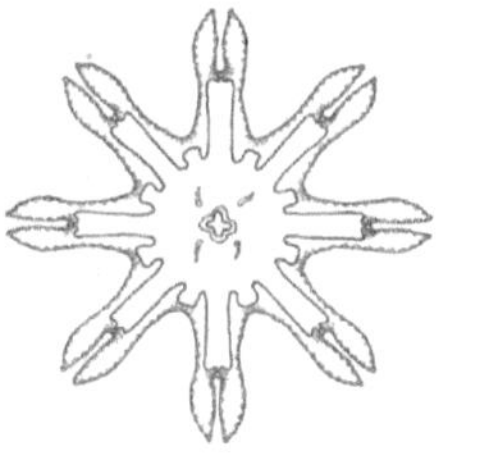 8-ray Type# Anatomy: • 8 lappet stems • 16 lancet-like rhopalial lappets • **1-2 gastric filaments** per quadrant • Long manubrium, approx. 34% of the CDD • Zooxanthellae mark the gastric system Gastric system: • Rhopalial canals are slightly forked: Fork arms overtop slightly the base of the rhopalium, fork tips are rounded • Velar canals are spade-like to slightly rhombical Body proportions: • RLL ≈ 14% of the TBD • LStL ≈ 14% of the TBD • CDD ≈ 43% of the TBD • RLL ≈ 50% of the TMLL • LStL ≈ 50% of the TMLL

CDD: Central Disk Diameter LStL: Lappet Stem Length RLL: Rhopalial Lappet Length TBD: Total Body Diameter TMLL: Total Marginal Lappet Length
#: Photo was taken by A. C. Morandini

Species/ Distribution	Morphology	Species/ Distribution	Morphology
Rhizostoma octopus Atlantic Coasts of Europe, Irish Sea, Northern Sea, Helgoland	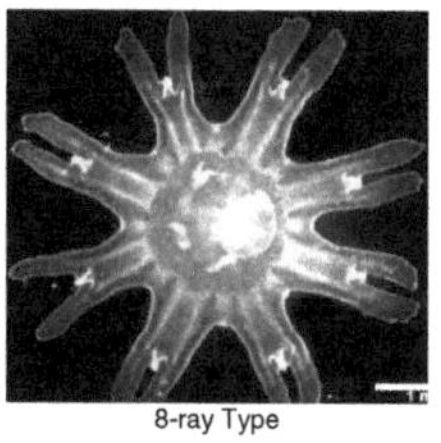 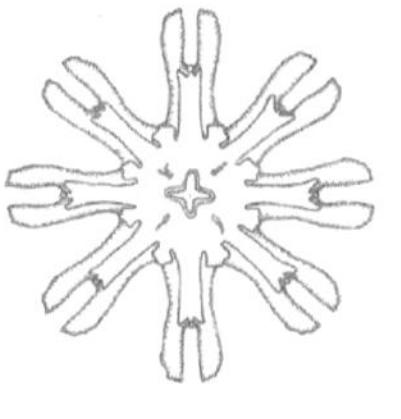 8-ray Type Anatomy: • 8 lappet stems • 16 bread knife-like rhopalial lappets • **2-3 gastric filaments** per quadrant • Long manubrium, approx. 34% of the CDD Gastric system: • Rhopalial canals are slightly forked: Fork arms level the base of the statolith, fork tips are rounded • Velar canals are flat rhombical Body proportions: • RLL ≈ 15% of the TBD • LStL ≈ 15% of the TBD • CDD ≈ 41% of the TBD • RLL ≈ 50% of the TMLL • LStL ≈ 50% of the TMLL	***Rhizostoma pulmo*** Mediterranean Sea, Red Sea	 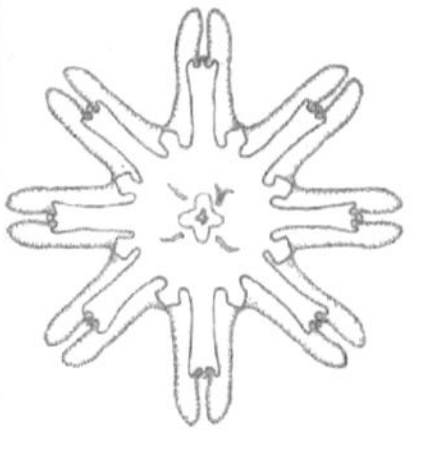 8-ray Type# Anatomy: • 8 lappet stems • 16 round spatula-like rhopalial lappets • **1-3 gastric filaments** per quadrant • Long manubrium, approx. 41% of the CDD Gastric system: • Rhopalial canals are slightly forked: Fork arms overtop the base of the rhopalium slightly, fork tips are rounded • Velar canals are rhombical Body proportions: • RLL ≈ 10% of the TBD • LStL ≈ 15% of the TBD • CDD ≈ 51% of the TBD • RLL ≈ 40% of the TMLL • LStL ≈ 60% of the TMLL

CDD: Central Disk Diameter LStL: Lappet Stem Length RLL: Rhopalial Lappet Length TBD: Total Body Diameter TMLL: Total Marginal Lappet Length

Table 3 continued

Species/ Distribution	Morphology	Species/ Distribution	Morphology
Catostylus mosaicus Australian coasts from Melbourne to Brisbane	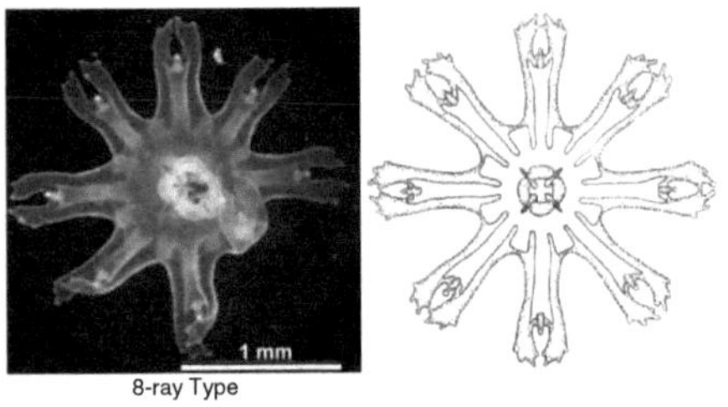 8-ray Type Anatomy: • 8 lappet stems • 16 antler palm-like rhopalial lappets • **1-2 gastric filaments** per quadrant • Long manubrium, approx. 40% of the CDD Gastric system: • Rhopalial canals are slightly forked: Fork arms level the base of the statolith, fork tips are sharp • Velar canals are spade-like Body proportions: • RLL ≈ 10% of the TBD • LStL ≈ 15% of theTBD • CDD ≈ 44% of the TBD • RLL ≈ 44% of the TMLL • LStL ≈ 56% of the TMLL	***Rhopilema esculentum*** Inland Sea of Japan, Indian Sea	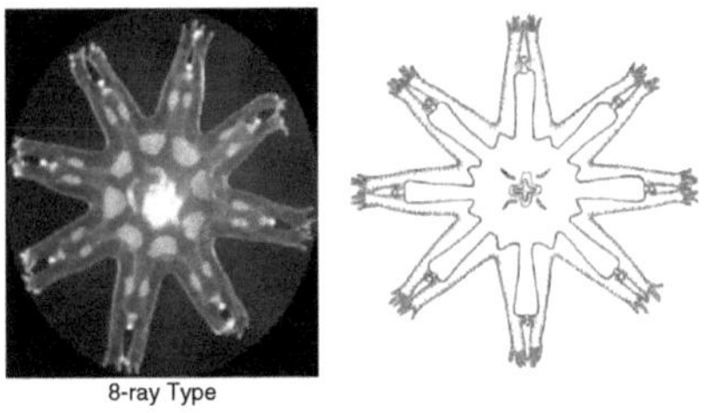 8-ray Type Anatomy: • 8 lappet stems • 16 rhopalial lappets • **1 gastric filament** per quadrant • Manubium was not measured, • Characteristic nematocyst pattern on the exumbrella: half circle-like white nemathocyst clusters in adradial position above the outer rim of the central stomach • Characteristic nematocyst pattern on the lappet stems: one pair each of white oval shaped nematocyst clusters in the upper half of each lappet stem Gastric system: • Rhopalial canals are spatula-like: • Velar canals are arrow tip-like Body proportions: • RLL ≈ 11% of the TBD • LStL ≈ 17% of the TBD • CDD ≈ 43% of the TBD • RLL ≈ 40% of the TMLL • LStL ≈ 60% of the TMLL

CDD: Central Disk Diameter LStL: Lappet Stem Length RLL: Rhopalial Lappet Length TBD: Total Body Diameter TMLL: Total Marginal Lappet Length

number of marginal lappets, number of velar lappets and length of velar lappets)
- 'body dimension proportions' (central stomach diameter, manubrium length, distance of opposite rhopalia and central disc diameter as proportions of total body diameter)
- 'central disc proportions' (central stomach diameter and manubrium length as proportions of central disc diameter)
- 'body length proportions' (total marginal lappet breadth, lappet stem length, rhopalial lappet length and velar lappet length as proportions of total body diameter)
- 'marginal lappet proportions' (lappet stem length and rhopalial lappet length as proportions of total marginal lappet length)

Results

The first part of identification consisted of measurement of the sizes of the ephyrae.

Measurements of the newly released ephyrae of 17 species were grouped by discriminant function analysis into the following 17 groups (=species) and analysed concerning the group memberships.

The following groups (species) were chosen:

1 = *Nausithoe werneri*, 2 = *Chrysaora hysoscella*, 3 = *Sanderia malayensis*, 4 = *Cyanea lamarckii*, 5 = *Cyanea capillata*, 6 = *Aurelia aurita*, 7 = *Aurelia limbata*, 8 = *Phacellophora camtschatica*, 9 = *Cephea cephea*, 10 = *Netrostoma setouchianum*, 11 = *Cotylorhiza tuberculata*, 12 = *Phyllorhiza punctata*, 13 = *Mastigias papua*, 14 = *Cassiopea andromeda*, 15 = *Rhizostoma octopus*, 16 = *Rhizostoma pulmo* and 17 = *Catostylus mosaicus*

The ephyrae of *Rhopilema esculentum* were excluded from the calculation by the SPSS® programme due to the absence of some measurement data. The discriminant function analysis showed that 97.8% of the originally grouped cases based on the variables 'absolute body dimensions', 'body dimension proportions', 'central disc proportions', 'body length proportions' and 'marginal lappet proportions' were correctly classified (Fig. 3).

The ephyrae of groups 2, 4–7, 9–12++ and 15–17, despite clear classification (species), were similar in size and form. Therefore, those groups formed a cluster in the middle of the graphic. Exceptions were

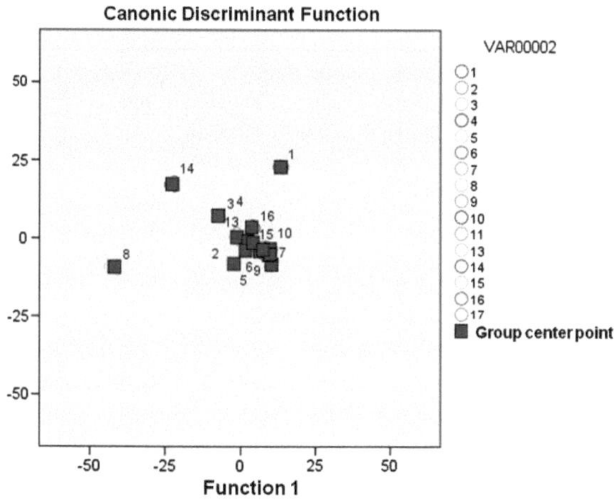

Fig. 3 Results of the discriminant function analysis of 17 Groups (species) of scyphozoan ephyrae based on the variables 'absolute body dimensions', 'body dimension proportions', 'central disc proportions', 'body length proportions' and 'marginal lappet proportions'. The coloured circles (1 colour per group) mark the values of the individuals which differ from the group mean values. Most of these circles are hidden by the group centre points as the observed proportions of the individuals were very similar to the group mean values. The numbers refer to the species as follows: 1 = *Nausithoe werneri*, 2 = *Chrysaora hysoscella*, 3 = *Sanderia malayensis*, 4 = *Cyanea lamarckii*, 5 = *Cyanea capillata*, 6 = *Aurelia aurita*, 7 = *Aurelia limbata*, 8 = *Phacellophora camtschatica*, 9 = *Cephea cephea*, 10 = *Netrostoma setouchianum*, 11 = *Cotylorhiza tuberculata*, 12 = *Phyllorhiza punctata*, 13 = *Mastigias papua*, 14 = *Cassiopea andromeda*, 15 = *Rhizostoma octopus*, 16 = *Rhizostoma pulmo* and 17 = *Catostylus mosaicus*

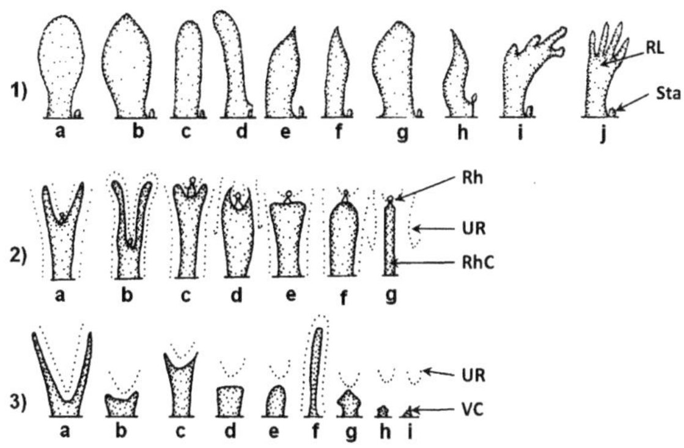

Fig. 4 Rhopalial lappet and gastric canal forms of scyphozoan ephyrae in this study. (1) Rhopalial lappet forms (i.e. left lappet): (a) round spoon-like, (b) pointed spoon-like, (c) round spatula-like, (d) spatula-like with outward bent tip, (e) lancet-like, (f) sword-like, (g) bread knife-like, (h) flame-like, (i) antler palm-like and (j) hand-like (2) Rhopalial canal forms: (a) forked, sharp points, (b) tuning-fork-like, rounded points, (c) slightly forked, rounded points, (d) club-shaped, forked, sharp points, (e) spade-like, (f) spatula-like and (g) arrow-like (3) Velar canal forms: (a) forked, sharp points, (b) slightly forked, rounded points, (c) slightly forked, sharp points, (d) spade-like, (e) spatula-like, (f) tongue-like, (g) rhombic, (h) arrow-like and (i) arrow-tip-like Rh: rhopalium, RhC: rhopalial canal, RL: rhopalial lappet, Sta: statolith, UR: umbrella rim VC: velar canal

1 (*Nausithoe werneri*), 3 (*Sanderia malayensis*), 8 (*Phacellophora camtschatica*) and 14 (*Cassiopea andromeda*) which showed considerable differences in comparison to all other groups (Fig. 3).

The second part of identification was definition of the forms of the rhopalial lappets and gastric canals (Fig. 4). The following lappet and canal forms were defined:

(1) Rhopalial lappet forms (i.e. left lappet): (a) round spoon-like, (b) pointed spoon-like, (c) round spatula-like, (d) spatula-like with outward bent tip, (e) lancet-like, (f) sword-like, (g) bread knife-like, (h) flame-like, (i) antler palm-like and (j) hand-like
(2) Rhopalial canal forms: (a) forked, sharp points, (b) tuning-fork-like, rounded points, (c) slightly forked, rounded points, (d) club-shaped, forked, sharp points, (e) spade-like, (f) spatula-like and (g) arrow-like
(3) Velar canal forms: (a) forked, sharp points, (b) slightly forked, rounded points, (c) slightly forked, sharp points, (d) spade-like, (e) spatula-like, (f) tongue-like, (g) rhombic, (h) arrow-like and (i) arrow-tip-like

Velar canals are also known as 'adradial canals' but in species with more than 8 marginal lappets (e.g. *Sanderia malayensis*, *Phacellophora camtschatica*, *Cassiopea andromeda*) rhopalial canals also can be described as 'adradial'. We call them velar canals, referring to their position between two rhopalial canals.

The following identification key can be used to distinguish the ephyrae of 18 species detailed in this article (Table 4):

Table 4 Identification key for newly released ephyrae

(1)	Marginal lappets without lappet stem	2
	Marginal lappets with lappet stem	3
(2)	16 round spoon-like marginal lappets without lappet stems (Fig 4: 1a), white ring between stomach and umbrella rim observable on subumbrella, rhopalia are directly attached to the central disc	*Nausithoe werneri*
(3)	An average of 8 marginal lappets	4
	An average of more than 11 marginal lappets	18
(4)	Forked velar canals (Fig. 4: 3a–c)	5
	Unforked velar canals (Fig. 4: 3d–i)	7
(5)	Forked arms of velar canals reach into marginal lappet (Fig 4: 3a)	6
	Forked arms of velar canals do not reach into marginal lappet (Fig. 4: 3b), rhopalial lappets spatula-like (Fig. 4: 1c)	*Chrysaora hysoscella*
(6)	Forked arm tips of velar canals level with in the upper third of the lappet stems, rhopalial lappets sword-like (Fig. 4: 1f)	*Cyanea lamarckii*
	Forked arm tips of velar canals level with the lower third of the rhopalial lappets, rhopalial lappets lancet-like (Fig. 4: 1g)	*Cyanea capillata*
(7)	Gastric system with zooxanthellae	13
	Gastric system without zooxanthellae	8
(8)	Rhopalial lappets with finger-like appendices (Fig. 4: 1i–j)	9
	Rhopalial lappets without finger-like appendices (Fig. 4: 1b–h)	10
(9)	Rhopalial canals forked (Fig 4: 3c), velar canals reach umbrella rim, antler palm-like rholalial lappets (Fig. 4: 1i)	*Catostylus mosaicus*
	Rhopalial canals spatula-like (Fig. 4: 2f), velar canals Arrow-tip-like (Fig. 4: 3i) half way between stomach and umbrella rim, large oval nematocyst clusters on lappet stem, half-moon-shaped nematocyst clusters in adradial position above the outer rim of the central stomach on exumbrella and rhopalial lappets hand-like (Fig. 4: 1j)	*Rhopilema esculentum*
(10)	Velar canal tips reach umbrella rim	11
	Velar canal well below umbrella rim	12
(11)	Rhopalial lappets spatula-shaped (Fig 4: 1c), forked arms of rhopalial canal tips above the base of rhopalia but below statoliths, statoliths above umbrella rim	*Rhizostoma pulmo*
	Rhopalial lappets bread knife-shaped (Fig. 4: 1 g), forked arms of rhopalial canals level or just above base of statoliths, statoliths level with umbrella rim	*Rhizostoma octopus*
(12)	Rhopalial lappets lancet-like (Fig. 4: 1e), statolith tips below/level umbrella rim	*Aurelia aurita*
	Rhopalial lappets bread knife-shaped (Fig. 4: 1g), statoliths above umbrella rim	*Aurelia limbata*
(13)	Velar canals reach umbrella rim	14
	Velar canals are well below umbrella rim	15
(14)	Forked arms of rhopalial canals above the base of the rhopalia, rhopalia tip level with lappet rim, rhopalial lappets oval spoon-like with sharp tips (Fig. 4: 1b)	*Cephea cephea*
	Forked arms of rhopalial canals above the mid section of the rhopalium and reach the lower third of the oval spoon-like rhopalial lappets with rounded tips (Fig. 4: 1a), statoliths above the lappet rim	*Netrostoma setouchianum*
(15)	Lappet stem 50% of total marginal lappet length	17
	Lappet stem 60% of total marginal lappet length	16
(16)	Only rim of gastric system is lined with zooxanthellae, ephyra transparent and rhopalial lappets round spoon-like (Fig. 4: 1a)	*Cotylorhiza tuberculata*
(17)	Rhopalial lappets broad, pointed spoon-like (Fig 4: 1b), lappet stems broaden towards rhopalial lappets	*Mastigias papua*
	Rhopalial lappets slim lancet-like (Fig. 4: 1e), lappet stems become smaller towards rhopalial lappets, exumbrella scattered with tiny white nematocyst warts	*Phyllorhiza punctata*
(18)	Velar lappets present after strobilation	19
	Velar lappets absent after strobilation	20

Table 4 continued

(19)	>16 tuning-fork-like marginal lappets with outward bent tips, >14 spatula-like velar lappets, marginal lappets and velar lappets connected by thin lamella, mouth tentacles present, gastric canals extend to lappet tips (Fig. 4: 2b, 3f) and are lined with zooxanthellae	*Cassiopea andromeda*
(20)	Rhopalial canals arrow like (Fig. 4: 2g), velar canals arrow-tip-like (Fig. 4: 3h), gastric system bright yellowish orange, >12 marginal lappets, rhopalial lappets flame-like (Fig. 4: 1h), exumbrella scattered with small, round, and white nematocyst warts	*Phacellophora camtschatica*
	Rholapar canals club-like and forked (Fig. 4: 2d), velar canals forked (Fig. 4: 3c), ephyra deep orange to orange brown, >11 marginal lappets and rhopalial lappets bread-knife-like (Fig. 4: 1g)	*Sanderia malayensis*

Identification list for ephyrae of selected species

Based on the the proportions and definitions mentioned above next to familiar morphological characters (like the number of marginal lappets and gastric filaments) an itemized list of morphological characters of the young ephyrae (see also Figs. 2 and 4) of the following 18 species was generated as an identification key (Table 1).

Discussion

In order to get a definite classification of species for newly detached ephyrae of the Scyphozoa, all the conspicuous characters like central disc, manubrium, lappet stems and rhopalial lappets were measured. To some extent, ephyrae of different strobilae of the same species differed in size, depending on age and nutrition of their polyps (Russell, 1970). Therefore, the absolute sizes are not reliable characters within one species (see also Table 2). It was necessary to transform the data into constant body proportions (Table 5).

The conspicuous morphological characters are differently shaped marginal lappets and the gastric

Table 5 Summary of characters of the ephyrae of 18 scyphozoan species described by different authors

Species	Culture conditions (Temperature)	TBD (mm)	CDD (mm) and/or proportions	No. of marginal lappets	Lappet proportions: LStL/RLL compared to TMLL	Shape of rhopalial lappet/ distinctions	Gastric system/ zooxanthellae	Colour	Reference
Nausithoe werneri	18°C	2.0–2.2	0.7/32–35% of TBD, 52–59% of TBD^{z}	16 (8 rhopalia)	0%/100% of TMLL	From very slim to broad spatula shaped/0–1 gastric filament bud per quadrant	–	Transparent, white ring below ring muscle	Jarms (1990)
	20–23°C	0.65–0.66	0.37–0.38, mean: 0.38/56–58% of TBD	16 (8 rhopalia)	0%/100% of TMLL	Round spoon-like/0 gastric filaments	Very short gastric canal like protuberances of the central stomach which reach the bases of the marginal lappets/ zooxanthellae	Transparent, white muscle ring	Present study

Table 5 continued

Species	Culture conditions (Temperature)	TBD (mm)	CDD (mm) and/or proportions	No. of marginal lappets	Lappet proportions: LStL/RLL compared to TMLL	Shape of rhopalial lappet/distinctions	Gastric system/zooxanthellae	Colour	Reference
Chrysaora hysoscella	5.6–18.9°C	2.0	–/37% of TBDz	8	60%/40% of TMLLz	Rounded tip/nematocyst clusters typical for *Chrysaora* species	Forked radial canals/–	Rose coloured	Delap (1901)
	15°C	1.62–2.0	–	–	–	–/nematocyst clusters typical for *Chrysaora* species	–/–	Pale orange	Morandini et al. (2004)
	15–22°C	1.7–3.3 mean: 2.4	–/44% of TBDp	8	45%/55% of TMLLp	–/nematocyst clusters typical for *Chrysaora* species	Slightly forked radial canals with rounded tips/-	Milky rose	Holst (2008)
	10–15°C	2.84–3.40, mean: 3.18	1.18–1.42, mean: 1.32/42% of TBD	8	45%/55% of TMLL	Spatula shaped with rounded tip/-	Slightly forked radial canals with rounded tips/none	Rose coloured	Present study
Sanderia malayensis	20°C	2.4–4.3	–/59% of TBD^{P}	12–24, mean: 16	40%/60%of TMLL^{P}	Tapered^{P}/nematocyst clusters typical for *Chrysaora* species at the base of the marginal lappets	Slightly forked radial canals; rhopalial canal tips do not reach the rhopalial lappets, velar canal tips do not reach base of marginal lappets/-	Brownish	Uchida & Sugiura (1975, 1978)
	20–25°C	≈3.0	–/62% of TBD	11–16, mean: 14	38%/62% of TMLL^{P}	Tapered^{P}/nematocyst clusters typical for *Chrysaora* species at the base of the marginal lappets	Slightly forked radial canals^{P}/-	Deep brown-orange	Adler (2008)
	20–25°C	3.60–3.79, mean: 3.67	2.12–2.31, mean: 2.19/60% of TBD	13–16, mean: 15	40%/60%of TMLL	Bread knife shaped/1-2 gastric filaments per quadrant	Slightly forked radial canals; rhopalial canal tips do not reach the rhopalial lappets, velar canal tips do not reach base of marginal lappets/none	Brown-orange	Present study

Table 5 continued

Species	Culture conditions (Temperature)	TBD (mm)	CDD (mm) and/or proportions	No. of marginal lappets	Lappet proportions: LStL/RLL compared to TMLL	Shape of rhopalial lappet/ distinctions	Gastric system/ zooxanthellae	Colour	Reference
Cyanea capillata	–	–	–/46% of TBD[z]	8[z]	50%/50% of TMLL[z]	Pointed/marginal tentacles are clearly visible by the time of release	(Forked) radial canals extend out into marginal lappets/-	Orange	Gröndahl & Hernroth (1987)
	19°C	2.0–4.4	–	8[z]	–	Pointed[z]/newly released ephyrae possess 0-1 tentacles	–/–	–	Higgins et al. (2008)
	5–10°C	5.97–9.54, mean: 8.55	2.43–3.79, mean: 3.34/41% of TBD	8	43%/57% of TMLL	Lancet like/2–8 tentacles buds when newly released	Forked rhopalial canals, tips reach far into the rhopalial lappets, forked velar canals, tips reach to tip level of rhopalia into the marginal lappets/none	Red orange	Present study
Cyanea lamarckii	44–68°F	4.00	–	8	–	Lancet like[z]/0 gastric filaments per quadrant[z]	Forked rhopalial canals, tips reach into rhopalial lappet; forked velar lappets, tips reach into lappet stem[z]/-	White	Delap (1905)
	–	4.00	–	8	–	Lancet like with broad blade[z]/1 gastric filament per quadrant[z]	Forked rhopalial and velar canals[z]/-	White	Russell (1970)
	10–15°C	2.15–2.64, mean: 2.33	0.99–1.16, mean: 1.06/45% of TBD	8	44%/56%of TMLL	Sword shaped/0 gastric filament per quadrant	Forked rhopalial canals, tips reach far into the rhopalial lappets; forked velar canals, tips reach to level of base of rhopalia in margial lappet/none	Bluish	Present study

Table 5 continued

Species	Culture conditions (Temperature)	TBD (mm)	CDD (mm) and/or proportions	No. of marginal lappets	Lappet proportions: LStL/RLL compared to TMLL	Shape of rhopalial lappet/ distinctions	Gastric system/ zooxanthellae	Colour	Reference
Aurelia aurita	–	–	–/45% of TBD^z	8^z	66%/34% of TMLL^z	Rounded tips^z/no marginal tentacles visible after release	Radial canals extend not into the rhopalial lappets; tips rounded^z/-	Light blue transparent	Gröndahl & Hernroth (1987)
	44–68°F	A bit smaller than 4 mm	–/44% of TBD^z	8–12, mean: 8	53%/47% of TMLL^z	Rounded tips/0-1 gastric filaments per quadrant	Forked rhopalial canals, tips reach into rhopalial lappets; unforked velar canals/-	–	Delap (1906)
	15–25°C	3.86-4.5, mean 4.19	1.44-1.81, mean: 1.66/40% of TBD	8	52%/48% of TMLL	Broad lancet like/1-2 gastric filaments per quadrant	Spade like to slightly forked rhopalial canals, rhombic velar canals/ none	Milky to bluish	Present study (Helgoland, polyps found in harbour)
Aurelia limbata	9–13°C	3.0–4.5	/43% of TBD^z	8	Marginal lappet a bit longer than radius of disc—52%/48% of TMLL^z	Rather narrow tips which overlap inside/statocyst is bright yellowish, 1–2 gastral filaments per quadrant	Rather wide radial canals, large adradial out-pockets/–	Pale yellow or greyish yellow	Uchida & Nagao (1963)
	15–25°C	2.62–5.33, mean: 3.66	1.07–1.71, mean: 1.26/35% of TBD	8	58%/42% of TMLL	Bread knife shaped/1 gastric filament per quadrant	Spade like, unforked radial canals/none	Greenish yellow with reddish brown	present study
Phacello-phora camtscha-tica	14°C	3.5–6.2, mean: 5.3	–/60% of TBD	13–18, mean: 16	45%/55% of TMLLp	Pointed/2 gastric filaments per quadrant, rudiments of 4 single primary tentacles budding from the subumbrella at the base of the cleft between the marginal arms	Very slim, pointed radial canalsp/–	Yellow to orange	Widmer (2006)
	15°C	10.0–10.8, mean: 10.23	5.64–6.46, mean: 5.85/57% of TBD	14–16, mean: 15	47%/53% of TMLL	Pointed flame shaped/1-2 gastric filaments per quadrant	Arrow shaped rhopalial canals, arrowhead shaped velar canals/none	Yellowish, gastric system: yellowish orange	Present study

Table 5 continued

Species	Culture conditions (Temperature)	TBD (mm)	CDD (mm) and/or proportions	No. of marginal lappets	Lappet proportions: LStL/RLL compared to TMLL	Shape of rhopalial lappet/ distinctions	Gastric system/ zooxanthellae	Colour	Reference
Rhizostoma octopus	–	3.6–5.4, mean: 4.5	1.5–2.3, mean: 1.9/42% of TBD	8	57%/43% of TMLL^{P}	–/2 to 4 gastric filaments per quadrant	Slightly forked rhopalial canals, unforked velar canals^{P}/-	Milky transparent^{P}	Holst & Jarms (2006), Holst et al. (2007), Holst (2008)
	15–25°C	3.30–5.96, mean: 4.81	1.32–2.24, mean: 1.95/41% of TBD	8	50%/50% of TMLL	Bread knife shaped/1 to 2 gastric filaments per quadrant	Slightly forked rhopalial canals; unforked, rhombic velar canals/ none	Milky transparent	Present study
Rhizostoma pulmo	15–25°C	3.19–3.34, mean: 3.27	1.66–1.67, mean: 1.67/51% of TBD	8	60%/40% of TMLL	Spade like/1–2 gastric filaments per quadrant	Slightly forked rhopalial canals; unforked, rhombic velar canals/ none	Milky transparent	Present study
Catostylus mosaicus	21°C	2.0	–/46% of TBD^{P}	8	58%/42% of TMLL^{P}	Antler palm shaped with finger-like appendages^{P}/1-2 gastric filaments per quadrant^{P}	slightly forked rhopalial canals, unforked velar canals^{P}/-	–	Pitt (2000)
	20°C	1.90–2.26, mean: 2.11	0.88–0.94, mean. 0.93/44% of TBD	8	56%/44% of TMLL	Antler palm shaped with finger-like appendages/1-2 gastric filaments per quadrant	Slightly forked rhopalial canals; unforked, spade shaped velar canals/none	Milky transparent	Present study
Rhopilema esculentum	18–20°C	1.5–3.0	–/39^{z}–42^{P}% of TBD	8	50^{P}–53^{z}%/ 47^{z}–50^{P} % of TMLL	Talon-shaped with 4–6 branches/1 gastric filament per quadrant	Unforked rhopalial canals^{z}, velar canals not visible in drawing/–	–	Ding & Chen (1981) Chen & Ding (1983)
	17-20C	2.0–4.0	–	–	–	–	–	–	You et al. (2007)
	20°C	2.11	9.91/43% of TBD	8	60%/40% of TMLL	Hand shaped with 4-6 finger like appendages/1 gastric filament per quadrant, 8 large, white, halfmoon shaped nematocyst clusters above central stomach on exumbrella, 2 white, elongated nematocyst clusters on lappet stems	Unforked, spatula shaped rhopalial canals, unforked triangular velar canals/none	Milky to transparent	Present study

Table 5 continued

Species	Culture conditions (Temperature)	TBD (mm)	CDD (mm) and/or proportions	No. of marginal lappets	Lappet proportions: LStL/RLL compared to TMLL	Shape of rhopalial lappet/ distinctions	Gastric system/ zooxanthellae	Colour	Reference
Cephea cephea	28°C	1.6–2.1	–	8	–	–/1 gastric filament per interradius	8 sacs radiate/–	Pale yellow or yellowish brown	Sugiura (1966)
	25–28°C	2.36–3.24, mean: 3.02	0.95–1.22, mean 1.12/37% of TBD	8	59%/41% of TMLL	Round spoon shaped/0-1 gastric filament per quadrant	Rhopalial canals slightly forked, velar canals unforked and slightly rhombic/zooxanthellae	Yellowish brown	Present study
Netrostoma setouchianum	25–28°C	2.24–2.44, mean: 2.37	0.96–1.02, mean 0.98/41% of TBD	8	59%/41%/of TMLL	Round spoon shaped/0-1 gastric filament per quadrant	Rhopalial canals slightly forked, velar canals unforked and slightly rhombic/zooxanthellae	Reddish to yellowish brown	Present study
Cotylorhiza tuberculata	–	1.5–2.0	–/40% of TBD	8	66%/34% of TMLL[z]	Rounded/1 gastric filament per quadrant	Rhopalial canals slightly forked, velar canals unforked[z]/zooxanthellae	–	Claus (1884)
	20–24°C	1.5–2.0	–	8	–	–	Rhopalial canals slightly forked, velar canals unforked[z]/zooxanthellae line the gastric system	–	Kikinger (1992)
	25–28°C	2.72–3.25, mean: 2.99	1.10–1.49, mean 1.27/42% of TBD	8	61%/39% of TMLL	Round spoon shaped/1 gastric filament per quadrant	Rhopalial canals slightly forked, velar canals unforked and slightly rhombic/zooxanthellae line the gastric system	Transparent with yellow hemmed gastric system	Present study

Table 5 continued

Species	Culture conditions (Temperature)	TBD (mm)	CDD (mm) and/or proportions	No. of marginal lappets	Lappet proportions: LStL/RLL compared to TMLL	Shape of rhopalial lappet/ distinctions	Gastric system/ zooxanthellae	Colour	Reference
Mastigias papua	–	1.5	–/47% of TBD[z]	8	52%/48% of TMLL[z]	Rounded[z]/1 gastric filament per quadrant	Rhopalial canals slightly forked, velar canals unforked[z]/zooxanthellae	Brown	Uchida (1926)
	25°C	1.5–2.7	–/45% of TBD[z]	8[z]	50%/50% of TMLL[z]	Rounded[z]/1 gastric filament per quadrant	Rhopalial canals slightly forked and reach base of rhopalial lappets, velar canals unforked[z]/ zooxanthellae	–	Sugiura (1963)
	25–28°C	2.64–3.91, mean. 3.23	1.16–1.59, mean: 1.39/43% of TBD	8	51%/49% of TMLL	Tapered, broad spoon shaped/1 gastric filament per quadrant	Rhopalial canals slightly forked, velar canals unforked and slightly rhombic/zooxanthellae	Orange brown	Present study
Phyllorhiza punctata	28°C	1.5–2.5	–/44% of TBD[p]	8	57%/43% of TMLL[p]	Rounded/4 gastric filaments; small, white warts on exumbrella	–/zooxanthellae	Yellowish brown[p]	Tronolone et al. (2002)
	25–28°C	0.46	0.22/48% of TBD	8	67%/33%of TMLL	Pointed spoon shaped/0-1 gastric filaments per quadrand, white spots on exumbrella	Rhopalial canals slightly forked, velar canals unforked and slightly rhombic/zooxanthellae	Ochre	Present study
Cassiopea andromeda	–	–	–	12–18, mean: 16	–	/4 gastric filaments, 11-17 velar lappets	–/zooxanthellae	–	Gohar & Eisawy (1960a, b)
	25–28°C	3.69–3.95, mean: 3.79	2.43–2.69, mean: 2.56/68% of TBD	17–23, mean: 19	48%/52% of TMLL	Spatula like/0 gastric filaments; 16-21 (mean: 18) velar lappets; velar lappets and rhopalial lappets are connected by a thin lamella	Severely forked rhopalial canals with rounded tips, fill nearly the whole the rhopalial lappet; unforked, spatula like velar canals, fill nearly the whole velar lappet/ zooxanthellae	Yellowish green	Present study

CDD central disc diameter, *LStL* lappet stem length, *p* data taken from photos, *RLL* rhopalial lappet length, *TBD* total body diameter, *TMLL* total marginal lappet length, *z* data taken from drawings, – no statement given in the publication systems. The condition of collected ephyrae depends on the sampling gear used. Delicate ephyrae can not be reliably identified from high-speed plankton samplers or from net hauls with high towing speeds. Ephyrae are best collected by a net with non-filtering cod-ends or a scoop bucket. Marginal lappets of ephyrae sometimes are damaged in the sea and/or under laboratory conditions. The shape of the rhopalial lappet is very characteristic and of considerable importance in distinguishing species (Gröndahl & Hernroth, 1987) if the marginal lappets are undamaged. The proportions of different parts to the marginal region and the central disc are sufficient and more precise; thus, mistakes due to deformation of the delicate animals can be avoided.

The pattern of the gastric system is always recognizable. This pattern was the only character that was always the same in normal, newly detached ephyrae: this is the most reliable character for identification in our study and was also mentioned in the chapter of 'Identification of Ephyrae' of Russell (1970).

Our key only works for the earliest stage, i.e. within 24–48 h after detachment.

The proportion of the lappet stems and the undamaged rhopalial lappet to the total marginal lappet length was a reliable character for comparison between genera, but not for comparison between species in the same genus in our study. Nevertheless, the standardisation of measurements and the key to identify ephyrae is a first step in establishing a catalogue of species for early detection of jellyfish blooms.

References

Adler, L., 2008. Vegetative Vermehrung von *Sanderia malayensis* Goette, 1886 (Cnidaria, Scyphozoa, Semaeostomeae). Diplomarbeit, Universität Hamburg: 1–89.

Backhaus, K., B. Erichson, W. Plinke & R. Weiber, 2006. Multivariate Analysemethoden: Eine anwendungsorientierte Einfuehrung. 11. Auflage, Springer. Berlin, Heidelberg: 1–14.

Chen, J. & G. Ding, 1983. Effect of temperature on the strobilation of jellyfish (*Rhopilema esculenta* Kishinouye – Scyphozoa, Rhizostomeae) (in Chinese). Acta Zoologica Sinica 29: 195–206.

Claus, C., 1884. Die Ephyren von *Cotylorhiza* und *Rhizostoma*. Zoologisches Institut Wien 5: 1–10.

Delap, M. J., 1901. Notes on the rearing of *Chrysaora isosceles* in an aquarium. Irish Nature 10: 25–28, pl. 1, 2.

Delap, M. J., 1905. Notes on the rearing, in an aquarium, of *Cyanea lamarcki*, Péron & Lesueur. Report of Sea and Fisheries Ireland (1902–3), pt II, Scientific Investigations 20–22, pl. I–II.

Delap, M. J., 1906. Notes on the rearing, in an aquarium, of *Aurelia aurita*, L. and *Pelagia perla* (Slabber). Report of Sea and Fisheries Ireland (1901–1904), pt II, Scientific Investigations 22–26, pl. I–II.

Ding, G. & J. Chen, 1981. The life history of *Rhopilema esculenta* Kishinouye. Journal of Fisheries of China 5: 93–104.

Gohar, H. A. F. & A. M. Eisawy, 1960a. The biology of *Cassiopea andromeda* (with a note on the Species Problem). Publications Marine Biological Station, Al-Ghardaqa 11: 3–39.

Gohar, H. A. F. & A. M. Eisawy, 1960b. The development of *Cassiopea andromeda* (Scyphomedusae). Publications Marine Biological Station, Al-Ghardaqa 11: 148–190.

Gröndahl, F. & L. Hernroth, 1987. Release and growth of *Cyanea capillata* (L.) ephyrae in the Gullmar Fjord, Western Sweden. Journal of Experimental Marine Biology and Ecology 106: 91–101.

Higgins, J. E., M. D. Ford III & J. H. Costello, 2008. Transitions in morphology, nematocyst distribution, fluid motions, and prey capture during development of the scyphomedusa *Cyanea capillata*. Biological Bulletin 214: 29–41.

Holst, S., 2008. Grundlagen der Populationsentwicklung verschiedener Scyphozoa (Cnidaria) der Deutschen Bucht. Dissertation, Universität Hamburg: 1–154.

Holst, S. & G. Jarms, 2006. Substrate choice and settlement preferences of planula larvae of five Scyphozoa (Cnidaria) from German Bight, North Sea. Marine Biology 151: 863–871.

Holst, S., I. Sötje, H. Tiemann & G. Jarms, 2007. Life cycle of the rhizostome jellyfish *Rhizostoma octopus* (L.) (Scyphozoa, Rhizostomeae), with studies on cnidocysts and statoliths. Marine Biology 151: 1695–1710.

Jarms, G., 1990. Neubeschreibung dreier Arten der Gattung *Nausithoe* (Coronata, Scyphozoa) sowie Wiederbeschreibung der Art *Nausithoe marginata* Kölliker, 1853. Mitteilungen des Hamburger Zoologischen Museums und Instituts 87: 7–39.

Kikinger, R., 1992. Cotylorhiza tuberculata (Cnidaria: Scyphozoa) – life history of a stationary population. Marine Ecology 13: 333–362.

Maretič, Z., D. Matič-Piantanida & J. Ladavač, 1991. The bloom of the jellyfish *Pelagia noctiluca* in the Mediterranean and Adriatic and its impact on human health. Proceedings of the II Workshop of Jellyfish in the Mediterranean Sea, 1–5 September 1987. Mediterranean Action Plan Technical Reports Series No. 47. United Nations Environment Program, Athens: 260–267.

Mills, C. E., 2001. Jellyfish blooms: are populations increasing globally in response to changing ocean conditions? Hydrobiologia 451: 55–68.

Morandini, A. C., F. L. da Silveira & G. Jarms, 2004. The life cycle of *Chrysaora lactea* Eschscholtz, 1829 (Cnidaria, Scyphozoa, Discomedusae, Semaeostomeae, Pelagiidae) with notes on the scyphistoma stage of three other species. Hydrobiologia 530(531): 347–354.

Pitt, K. A., 2000. Life history and settlement preferences of the edible jellyfish *Catostylus mosaicus* (Scyphozoa: Rhizostomeae). Marine Biology 136: 269–279.

Purcell, J. E., S.-i. Uye & W.-T. Lo, 2007. Anthropogenic causes of jellyfish blooms and their direct consequences for humans: a review. Marine Ecology Progress Series 350: 153–174.

Russell, F. S., 1970. II. Pelagic Scyphozoa with a supplement to the first volume on hydromedusae. In Russell, F. S. (ed.), The Medusae of the British Isles. Cambridge University Press, London, New York: 1–283.

Sugiura, Y., 1963. On the life-history of rhizostome medusae. I. *Mastigias papua* L. Agassiz. Annotationes Zoologicae Japonenses 36: 194–202.

Sugiura, Y., 1966. On the life-history of rhizostome medusae IV. *Cephea cephea*. Embryologia 9: 105–122.

Tronolone, V. B., A. C. Morandini & A. E. Migotto, 2002. On the occurrence of Scyphozoa ephyrae (Cnidaria, Scyphozoa, Semaeostomeae and Rhizostomeae) in the Southeastern Brazilian coast. Biota Neotropica 2: 1–18.

Uchida, T., 1926. The anatomy and development of a rhizostome medusa, *Mastigias papua* L. Agassiz, with observations on the phylogeny of Rhizostomae. Journal of the Faculty of Science, Imperial University of Tokyo, Section VI - Zoology 1: 45–95.

Uchida, T. & Z. Nagao, 1963. The metamorphosis of the scyphomedusa, *Aurelia limbata* (Brandt). Annotationes Zoologicae Japonenses 36: 83–91.

Uchida, T. & Y. Sugiura, 1975. On the ephyra and postephyra of a semaeostome medusa, *Sanderia malayensis* Goette. Journal of the Faculty of Science, Hokkaido University, Series VI, Zoology 19: 879–881.

Uchida, T. & Y. Sugiura, 1978. On the polyp of the scyphomedusa, *Sanderia malayensis* and its reproduction. Journal of the Faculty of Science, Hokkaido University, Series VI, Zoology 21: 279–285.

Wanker, R. & J. Fischer, 2001. Intra- and interindividual variation in the contact calls of spectacled parrolets (*Forpus conspicillatus*). Behaviour 138: 709–726.

Widmer, C. L., 2006. Life cycle of *Phacellophora camtschatica* (Cnidaria: Scyphozoa). Invertebrate Biology 125: 83–90.

Williams, J. A., P. J. Fenner, J. W. Burnett & J. F. Rifkin (ed.), 1996. Venomous and Poisonous Marine Animals – A Medical and Biological Handbook. University North South Wales Press, Australia: 63–415.

You, K., C. Ma, H. Gao, F. Li, M. Zhang, Y. Qui & B. Wang, 2007. Research on the jellyfish (*Rhopilema esculentum* Kishinouye) and associated aquaculture techniques in China: current status. International Aquaculture 15: 479–488.

Hydrobiologia (2010) 645:23–37
DOI 10.1007/s10750-010-0205-z

JELLYFISH BLOOMS

Blooms of the invasive ctenophore, *Mnemiopsis leidyi*, span the Mediterranean Sea in 2009

Verónica L. Fuentes · Dror L. Angel · Keith M. Bayha · Dacha Atienza · Dor Edelist · Cesar Bordehore · Josep-Maria Gili · Jennifer E. Purcell

Published online: 21 March 2010

Abstract Blooms of the invasive ctenophore, *Mnemiopsis leidyi*, occurred in 2009 along the Mediterranean Sea coasts of Spain and Israel. This voracious zooplanktivore spread throughout the Black Sea basin after its introduction in the early 1980s, throughout northern European coastal waters, and now occurs throughout the Mediterranean Sea. *M. leidyi* occurred throughout the summer along the entire Catalan Spanish and Israeli coasts in 2009. Those locations had high temperatures (18–26°C) and salinities (37–38) during the blooms. The patterns of abundance of large jellyfish along the Catalan coast were unusual in 2009, with low numbers during July, August, and September when ctenophores were abundant. Small populations of those potential predators and food competitors of *M. leidyi* could have contributed to the ctenophore bloom. The identity of the ctenophores from Spain and Israel was confirmed as *M. leidyi* by molecular analysis based on DNA sequencing of the nuclear internal transcribed spacer (ITS) regions. This is the first molecular confirmation of *M. leidyi* in the Mediterranean Sea. Most ctenophores had an ITS genotype previously found in *M. leidyi* from other invaded regions (the Black, Azov, and Mediterranean seas), as well as native regions in the United States,

Guest editors: J. E. Purcell & Dror Angel / Jellyfish Blooms: New Problems and Solutions

V. L. Fuentes (✉) · D. Atienza · J.-M. Gili
Institut de Ciencies del Mar, CSIC, P. Marítim de la Barceloneta 37-49, 08003 Barcelona, Spain
e-mail: vfuentes@icm.csic.es

D. L. Angel · D. Edelist
Department of Maritime Civilizations, Leon Recanati Institute of Maritime Studies, University of Haifa, Mt Carmel, 31905 Haifa, Israel

K. M. Bayha
University of California Merced, School of Natural Sciences, 5200 North Lake Rd, Merced, CA 95343, USA

Present Address:
K. M. Bayha
Dauphin Island Sea Lab, 101 Bienville Blvd., Dauphin Island, AL 36528, USA

C. Bordehore
Departamento de Ecología, Universidad de Alicante, Ap. 9903080 Alicante, Spain

J. E. Purcell
Western Washington University, Shannon Point Marine Center, 1900 Shannon Point Rd, Anacortes, WA 98221, USA

J. E. Purcell
Coastal and Marine Resources Centre, University College Cork, Naval Base, Haulbowline Island, Cobh, Co. Cork, Ireland

suggesting common ancestry. Based on the circulation patterns of Mediterranean surface waters and shipping activities, we conclude that the spread of *M. leidyi* in the Mediterranean probably resulted from re-introductions by ballast water transport and subsequent distribution by currents. We also conclude that the near-simultaneous blooms in opposite ends of both the Mediterranean basins indicate that *M. leidyi* is resident around the Mediterranean. We discuss environmental conditions, food, and predators of *M. leidyi* in both regions that would influence the future effects of this voracious consumer on the pelagic food web of the Mediterranean Sea.

Keywords Jellyfish · Zooplankton · Climate · Israel · Spain

Introduction

Several years after its accidental introduction into the Black Sea in the early 1980s, the ctenophore, *Mnemiopsis leidyi* A. Agassiz, formed massive blooms that coincided with abrupt decreases in zooplankton, ichthyoplankton, and fisheries (Vinogradov et al., 1989; Vinogradov & Shushkina, 1992; GESAMP, 1997). Since the 1980s, the introduced distribution of *M. leidyi* spread to the Sea of Azov (1988), the Sea of Marmara (1989–1990), and the northern Aegean Sea (1990) (Shiganova et al., 2001). Additional introductions have transported the ctenophore to the Caspian Sea (1999) (Ivanov et al., 2000; Shiganova et al., 2001) and the North and Baltic seas (first observations 2006) (Faasse & Bayha, 2006; Javidpour et al., 2006). Within the Mediterranean Sea, *M. leidyi* rapidly spread from the Aegean Sea to adjacent waters of Turkey and Syria (Kideys & Niermann, 1994). More recently, *M. leidyi* was reported from the Northern Adriatic Sea and France (Shiganova & Malej, 2009), Israel (Galil et al., 2009), and Italy (Boero et al., 2009).

Mnemiopsis leidyi is a very versatile species native to estuaries and coastal regions along the eastern coasts of North and South America (GESAMP, 1997). Within its native range it occurs in coastal waters with temperatures ranging between 2 and 32°C and salinities of <2–38 (Purcell et al., 2001). It has flourished in non-native waters with winter temperatures above 4°C, estuarine salinity, and high productivity (e.g., Shiganova et al., 2001). The extraordinary success of *M. leidyi* in the Black, Azov, and Caspian seas has been attributed to the lack of predators in combination with the deteriorated conditions for fish populations in those waters due to eutrophication, pollution, and over fishing (e.g., Purcell et al., 2001; Oguz, 2005). The northern European coastal waters, where *M. leidyi* is considered to be established (Javidpour et al., 2008), share the above characteristics.

The first report of *M. leidyi* in the Mediterranean Sea was in the Aegean Sea, where gelatinous predators (e.g., *Beroe* spp. ctenophores) occur, and temperatures and salinities are much higher than in other non-native locations; *M. leidyi* appeared to have limited success there, and the establishment of persistent populations was uncertain (Kideys & Niermann, 1993; Shiganova & Malej, 2009). Short-term occurrences in Mediterranean ports were attributed to ballast-water transfers from the Black Sea to those ports (summarized in Galil et al., 2009).

Siapatis et al. (2008) developed a predictive model to identify the potential habitat of *M. leidyi* in the Mediterranean basin. Their model shows potential habitat around the entire Mediterranean coast. In each of 3 years modeled from environmental conditions (2004–2006), the Catalan and Levantine coast of Spain and the Levant (eastern Mediterranean) areas including Israel showed high probabilities of *M. leidyi* occurrence. In 2009 shortly after publication of that prophetic paper, *M. leidyi* was reported along Israeli shores (Galil et al., 2009), in the Adriatic Sea (Faris, 2009), and from the Catalan and Levantine coast in Spain (Fuentes et al., 2009).

Here, we report high abundances of *M. leidyi* in coastal waters of Israel and Spain in winter–summer 2009, in conjunction with hydrographic conditions and the abundances of zooplankton and scyphomedusan species. We compare these factors in 2009 with those in 2008 when *M. leidyi* was not widely reported. We also present the first molecular confirmation of the identity of *M. leidyi* in the Mediterranean Sea, along with possible source region(s).

Materials and methods

Field sampling

Spain

Beginning in 2000, semi-quantitative data were compiled on the occurrences of large gelatinous plankton, predominantly the scyphomedusae, *Pelagia noctiluca* (Forskal), *Rhizostoma pulmo* (Macri), and *Cotylorhiza tuberculata* (Macri), along the Catalan coast. The abundances of jellyfish on the beaches (without distinguishing among species) were classified in three categories: 1 (few, <10 animals per beach), 2 (many, <1 medusa m^{-2}), and 3 (a lot, >1 medusa m^{-2}) (Gili et al., 2007, 2008). This classification system was retained for observations that began in 2007 when the "Medusa Project" began monitoring the presence of individual jellyfish species along the Catalan coast.

The "Medusa Project" monitoring comprised a network of organizations covering the entire Catalan coastline. The Medusa Project is financed by the Catalan Water Agency (ACA) with the main goals of studying and monitoring the jellyfish proliferations along the Catalan coast of the NW Mediterranean Sea. The ACA records the relative abundance (few, many, a lot) of jellyfish daily at more than 300 beaches covering the 69 municipalities of Catalonia. Data are recorded by beach inspectors who observe what is on the sand and in the near-shore water at fixed locations and by inspectors in boats 200-m offshore beyond the swimming area. The project also involves participation of the Emergency Services from 26 of the 69 municipalities that each completes a form daily to report the relative abundance of jellyfish at the beaches (ACA, 2009). The Fisherman Associations of Catalonia also report to the Medusa Project. All the collected information, including daily observations of jellyfish from ACA boats, beach inspectors, and emergency services, are summarized on the ACA web page and are available to the public (ACA, 2009). As part of the observation and monitoring protocol, data on water temperature, transparency, salinity, chlorophyll *a*, and nutrients have been recorded since 2000 along the coast.

As part of a cooperative project, information about jellyfish presence in other parts of Spain is also available. In Denia (Levantine coast of Spain, south of Catalonia), daily visual inspection of the beaches has been conducted by the local Administration (Environmental Department) and by its summer beach emergency services (Red Cross) since summer 2001. The presence of jellyfish is recorded daily as part of this monitoring procedure. Water samples have been collected monthly since 2001 at four locations, separated by 3–5 km, along the coast at 4–5 m depth and 300–700 m from shore. Data on water temperature, O_2, pH, and nutrients were recorded in those samples.

Israel

Jellyfish occurrences in coastal or offshore waters of Israel have been recorded on an ad hoc basis; many of the observations were made by fishermen. Data on the occurrences and abundances of medusae and ctenophores in 2009 were compiled during trawling (fishing) cruises and from interviews with trawl, seine, and inshore fishermen at the ports of Jaffa and the Kishon River (Haifa). Trawlers used 40–48-mm stretched diamond mesh nets with a vertical opening of 2–3 m, trawling at depths of 15–100 m. Inshore and purse-seine fishermen trawled at 0–50 m depth with 22–120-mm (mostly 50–60 mm) mesh gill- and trammel-nets with a 12-m vertical opening in shallow waters and up to 24 m in deep waters. Purse-seiners used 18–24-mm mesh nets.

Presence/absence and the relative abundances of gelatinous animals were recorded. The abundances of gelatinous zooplankton reported were "none," "occasional" (sporadic sightings in surface waters or 1–5 specimens in each trawl), or "abundant" (either swarms in surface waters or >5 specimens in each trawl). Zooplankton abundances in coastal waters were determined between March and May 2009 by means of horizontal net tows at 5-m depth for 3 min at 1 m s^{-1} and vertical tows from the seafloor to the surface at 1 m s^{-1} with 64- and 303-μm plankton nets. Surface seawater temperatures were recorded at Ashdod Port during the past 12 years (CAMERI, 2008) and both surface seawater temperatures and salinities were recorded near Ashkelon (IEC, 2009) during 2009.

Molecular analysis

Ctenophores collected in the field were provisionally identified morphologically as *M. leidyi* based on a typical morphotype, namely that the lobes extended to the region of the statocyst (see Faasse & Bayha, 2006). In order to confirm field identification, a few specimens of this consistent morphotype were collected, preserved, and identified molecularly. Molecular species identification was performed making use of the DNA sequence from the nuclear internal transcribed spacer (ITS) region (ITS-1, 5.8S, ITS-2), a region that has been useful for ctenophore identification in the past (Podar et al., 2001; Bayha et al., 2004; Faasse & Bayha, 2006). A total of three ctenophores were collected and preserved in 75–100% ethanol, including one from the vicinity of Haifa, Israel (HAF-1) and two from the vicinity of Salou, Spain (SAL-1 and SAL-2). DNA extraction was performed using a standard CTAB protocol (Dawson et al., 1998), and polymerase chain reaction (PCR) amplifications were performed using an Applied Biosystems 2720 Thermal Cycler. The entire ITS region was PCR amplified using primers KN-8 (ATTACGTCCCTGCCCTTTGTA) and KN-9 (GCA ATCCCAAACAGTCCGACTCTTC) in conditions consisting of 95°C for 3 min, followed by 38 cycles of 95°C for 45 s, 52°C for 1 min, 72°C for 1.5 h, and a single cycle of 72°C for 10 min (then 4°C). Successful PCR products were treated with Exonuclease I and Shrimp Alkaline Phosphatase (USB Corp.) to prevent interference by unused primers and dNTPs in sequencing reactions. PCR products were then directly cycle sequenced by Northwoods DNA, Inc. (Solway, MN) using primers KN-8 and KN-11 (ATTTGAGCTGTTCCCTGTTC GT) (Bayha, 2005). All sequences were deposited into GenBank. All DNA sequences were assembled using Seqman II (DNAStar, Inc.), and electropherograms were amended by eye for bad sequence calls. Loci that showed clear peaks of two different bases were coded as degenerate (i.e., Y = C and T). Sequence identity was evaluated by performing BLASTN searches against GenBank (Altschul et al., 1997).

Results

Field sampling

Spain

During the summer of 2009, we received many reports from the entire length of the Catalan coast describing the presence of an unusual jellyfish species, which we identified as *M. leidyi* (Fig. 1). Only the observations for which we visually confirmed that the reported jellyfish species was *M. leidyi* are given in Table 1. Many other similar reports were received from several additional locations along the Catalan coast, but we were not always available to make identifications; therefore, *M. leidyi* may have been even more commonly distributed than reported here. The confirmed reports show that large swarms of *M. leidyi* were observed from the

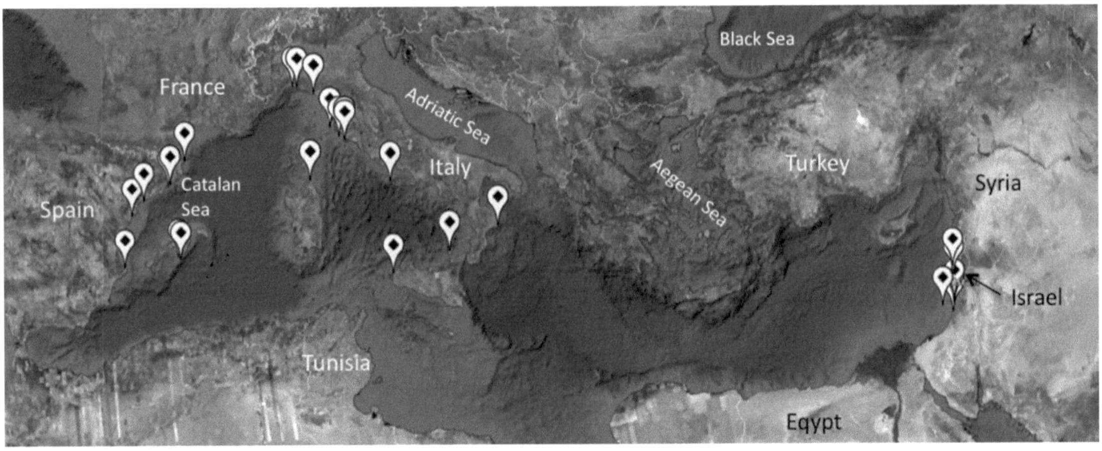

Fig. 1 Reports of blooms of *Mnemiopsis leidyi* in the Mediterranean Sea in 2009 in Spain, Israel, and Italy

Table 1 Records of *Mnemiopsis leidyi* along the Spanish NW Mediterranean coast during summer 2009

Location	Geographic coordinates		Survey date 2009 and (No. of reports)	Source	Water conditions	
	Latitude, N	Longitude, E			Temp. (°C)	Salinity
Catalan coast						
Cap de Creus	42°19′5.78	3°19′31.40	July (2 reports)	ACA, Fishermen	22.7	37.7
Arenys de Mar	41°34′44.6	2°33′10.62	August (1 report)	ACA	23.9	37.6
Mataro port	41°32′1.70	2°27′1.62	August (1 report)	ACA	24.7	37.7
Salou La Pineda beach	41°3′12.28	1°10′8.52	August (reported daily for 1 week)	ACA–Red Cross	23.3	37.9
Cambrils	41°3′46.60	1°10′20.63	August (reported daily for 1 week)	ACA–Red Cross	20.8	37.8
Les Cases d'Alcanar beach	40°33′2.22	0°31′49.37	July–August (several reports)	ACA	25.3	37
Alcanar Del Marjal beach	40°33′15.13	0°32′4.41	July–August (several reports)	ACA	25.3	37
Other Spanish coastal areas						
Cabrera port (Balearic Islands)	39°8′43.13	2°55′49.70	August (1 report)	Red Cross	23	ND
Denia (Alicante) Big port	38°50′57.18	0°6′34.25	July–August (observed frequently)	Red Cross	27.5	ND

ACA Catalan Water Agency, *ND* no data

The temperature and salinity data correspond to the period when *M. leidyi* were observed

beginning of July to the end of September along the entire coast from Cap de Creus in the northern Catalan coast to Alcanar in the south. The presence of *M. leidyi* also was confirmed in the south of Catalonia in Denia (Valencia, Levantine coast of Spain) and Cabrera (Balearic Islands) (Table 1; Fig. 1). Most of the *M. leidyi* observations made by ACA during summer 2009 were in the highest abundance category 3.

The surface coastal waters of the Catalan Sea (NW Mediterranean) during the summer months of 2009, when the presence of *M. leidyi* was confirmed (from July to September), were characterized by salinities between 30.5 in the south, where the delta of the Ebro River is located, and 38.2, and temperatures between 21.5 and 25.8°C. Temperature and salinity (S) of the waters where *M. leidyi* was present ranged from 22.7 to 25.3°C and 37–37.9 S (Table 1).

Along the coast of Denia, a high-density area of *M. leidyi* was detected from 16 to 18 July 2009, with densities ranging from 5 to 100 ind. m^{-2}. The ctenophores were seen only on those days in an area of about 2,500 m^2 at depths of 0–4 m (Table 1). The high-density patch dispersed after 18 July; however, low densities of *M. leidyi* were observed by SCUBA divers between June and August in that area. Temperatures of coastal waters in Denia (2 m depth) ranged from 12.5°C in March to 27.5°C in July. In this area, the effects of river runoff are very important; for example, nitrate values are normally very high, reaching a maximum of 361 $\mu g\ l^{-1}$ in November 2006.

The observations of large jellyfish (scyphomedusae and hydromedusae) were atypical along the Catalan coast during the spring and summer months of 2009. The typical pattern in 2000–2008 showed that the greatest jellyfish abundances occurred during July and August (Gili et al., 2007, 2008); however, in 2009, most jellyfish occurred during May and June (Fig. 2). In July, most of the jellyfish species usually seen then (*P. noctiluca*, *Aequorea forskalea* Péron & Lesueur, and *R. pulmo*) nearly disappeared, which coincided with the period when *M. leidyi* was most abundant and frequently observed (Fig. 2).

Israel

Observations, including relative abundances, of gelatinous zooplankton in Israeli coastal waters were recorded from January to July 2009 with an emphasis

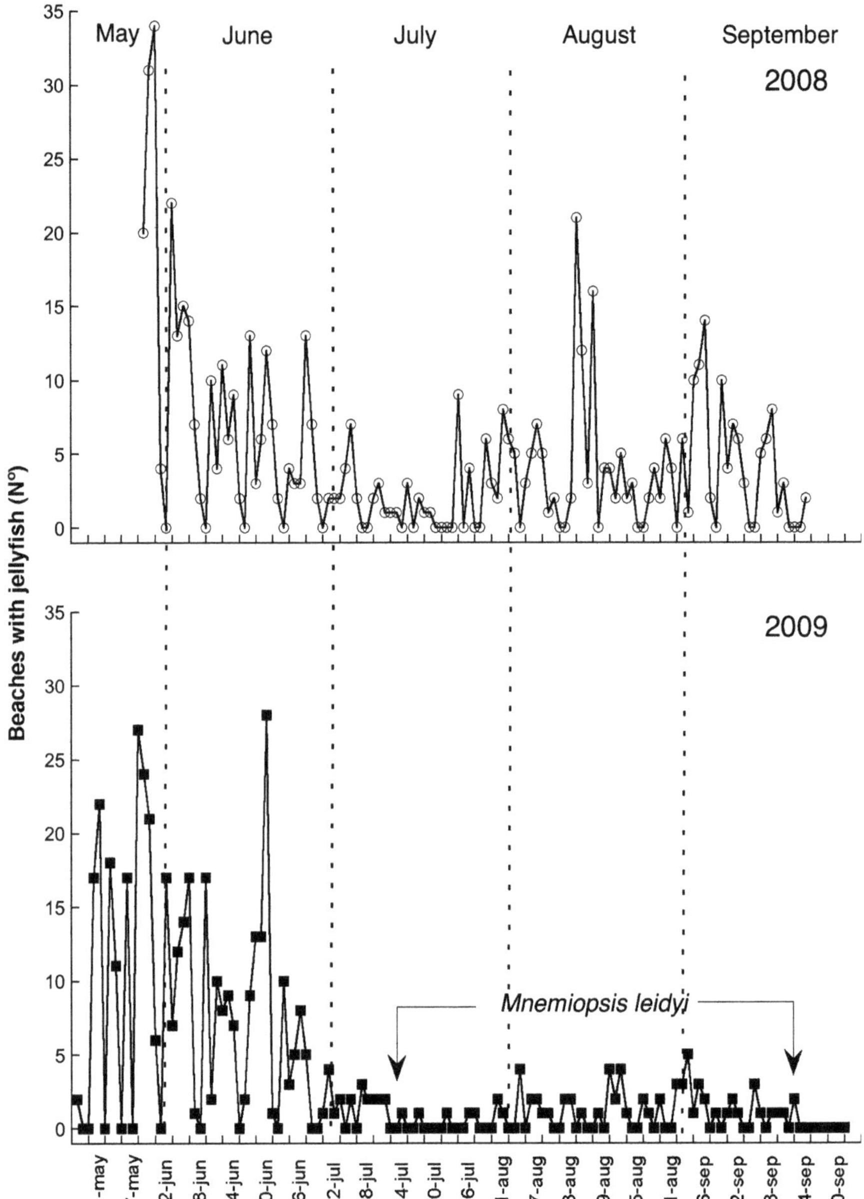

Fig. 2 The numbers of beaches with scyphozoan and hydrozoan jellyfish during the beach season 18 May to 30 September 2008 (*circles*) and 2009 (*squares*). The *arrows* mark the period of time when high numbers of *Mnemiopsis leidyi* ctenophores were observed in 2009 in the Catalan Sea, Spain

on medusae (mainly *Rhopilema nomadica* Galil, Spanier & Ferguson and *R. pulmo*) and ctenophores (Table 2). This record-keeping was stimulated by the appearance in January of a massive bloom of *M. leidyi*, which had not been observed prior to this year in such abundances in these coastal waters. The earliest reports of the ctenophore swarms were from purse seine and trawl fishermen, describing "marmalade" or "medusa jelly" that made it difficult to lift the fishing gear. Continuous onboard observations of the ctenophores led to subsequent visual confirmation during a SCUBA diving expedition to 32°51′02″N–034°56′33″E. All of the ctenophore swarms were observed within 5 nautical miles of shore. Ctenophores were reported to occur either sporadically or abundantly from January to mid-June (Table 2). Most of the medusae observed in 2009 were *R. pulmo*, which was often abundant in January through March, and several patches of *R. nomadica* that occurred in June (Table 2). Routine coastal monitoring did not

Table 2 Observations of ctenophores in coastal waters of Israel during 2009

Location	Month/week	Ctenophore abundance	Medusa abundance	Source	Temp. (°C)
Jaffa Port	January/3rd	Abundant	Abundant	Trawl	18
Jaffa Port	March/1st	Occasional	Occasional	Fishermen	NR
Jaffa Port	March/2nd	Abundant	Abundant	Fishermen	NR
Jaffa Port	March/4th	Abundant	Absent	Trawl + Fishermen	18
Jaffa Port	April/1st	Occasional	Absent	Fishermen	NR
Jaffa Port	April/2nd	Occasional	Absent	Fishermen	NR
Jaffa Port	April/3rd	Abundant	Occasional	Trawl + Fishermen	19
Jaffa Port	April/4th	Abundant	Occasional	Trawl + Fishermen	19
Jaffa Port	May/1st	Abundant	Absent	Fishermen	NR
Jaffa Port	May/2nd	Occasional	Absent	Trawl + Fishermen	21
Jaffa Port	May/4th	Occasional	Absent	Fishermen	NR
Jaffa Port	June/1st	Abundant	Occasional	Fishermen	NR
Jaffa Port	June/2nd	Abundant	Occasional	Fishermen	NR
Jaffa Port	June/3rd	Absent	Occasional	Trawl + Fishermen	26
Haifa Bay	January/3rd	Abundant	NR	Trawl	NR
Haifa Bay	March/4th	Occasional	NR	Trawl	NR
Haifa Bay	April/4th	Abundant	NR	Trawl	NR
Haifa Bay	May/2nd	Occasional	NR	Trawl	NR
Haifa Bay	June/3rd	Absent	NR	Trawl	NR

Observations were made by inshore and purse-seine fishermen and recorded during and following trawl fishing at two fishing ports in Israel. Jaffa (32°03′21.1N and 034°43′53.4E) and Kishon (Haifa: 32°52′44.8N and 034°56′04.4E). Abundance categories were "Occasional" = sporadic sightings of jellyfish (no swarms) or 1–5 specimens in each trawl. "Abundant" = consistent reports of jellyfish in swarms or >5 specimens in each trawl. Ctenophores were not observed or caught by fishermen from either port from the third week of June through July 2009

NR not recorded

reveal unusual patterns in seawater temperatures, salinities (39.1–39.8), or chemistry during spring and early summer 2009 as compared with other years (Herut et al., 2005, 2006, 2008); however, winter 2008/2009 temperatures were higher than average winter temperatures (Fig. 3).

Molecular species identification

The sequences for the nuclear ITS region (ITS-1, 5.8S, and ITS-2) from all three ctenophores sequenced were identical in length (638 base pairs) and nearly identical in sequence to published *M. leidyi* sequences, indicating positive identification as *M. leidyi*. Sequences for the three ctenophores analyzed differed by at most a single degenerate base (T in SAL-1 and HAF-1 and Y [C and T] in SAL-2). All sequences were extremely similar to published *M. leidyi* sequence for ITS (NCBI Accession #AF293700), all being the same length and differing by 3 (SAL-1 and HAF-1) or 4 (SAL-2) bases out of 638 (all degenerate changes, i.e., T vs. Y, etc.), indicative of within-species differentiation. In comparison, the three ctenophores differed markedly from the next closest match (*Bolinopsis* sp.—Accession #BSU65480) in being 2 bp shorter and diverging by 19–20 bp.

Discussion

Arrival and transport of *Mnemiopsis leidyi* in the Mediterranean Sea

Although May 2009 was the first confirmed report of *M. leidyi* on the Spanish coast, this species was present before but remained unrecorded until its populations increased. We located unpublished information about the presence of *M. leidyi* in Cabrera

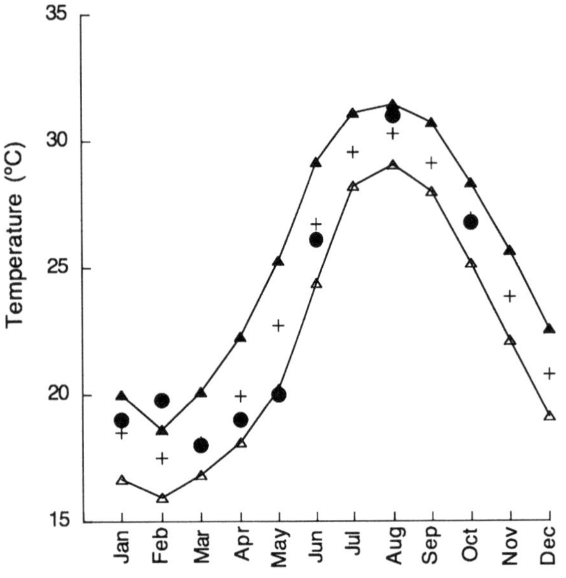

Fig. 3 Sea water temperature (°C) of last 12 years from Palmachim, Israel. *Solid triangles* maxima, *open triangles* minima, *crosses* means. *Solid circles* mark the 2008–2009 winter–spring water temperatures

(Balearic Islands) (OCEANA, 2008); we confirmed its presence there in 2009 (Fig. 1). Earlier occurrences of *M. leidyi* in other Mediterranean areas may be established by examination of formalin-preserved plankton samples, as in Purcell (1988). High numbers of *M. leidyi* along the Mediterranean coastlines of Israel (Galil et al., 2009), Italy (Boero et al., 2009), and Spain during summer 2009, plus the fact that the species was also present in Spanish waters in 2008, strongly suggest that its population is established in the Mediterranean.

Molecular species identification confirmed representative ctenophores as *M. leidyi* through the use of DNA sequence from the nuclear ITS region. Some previous articles on invasive ctenophores in the Baltic Sea apparently have misidentified the ctenophores to the genus *Mnemiopsis* (Kube et al., 2007; Lehtiniemi et al., 2007); the errors only were recognized upon the use of molecular identification tools (Gorokhova et al., 2009).

Although the molecular dataset is small ($n = 3$), it may also give some preliminary indication of the original source region(s) of some of the ctenophores. Two of the sequenced ctenophores (SAL-1 and HAF-1) contained an ITS composite genotype that was previously found in invasive *M. leidyi* from the Black Sea (southwestern Black Sea and Gelendzhik Bay, Russia) and the Sea of Azov (various locations), as well as in native ctenophores from the United States (only from the Gulf of Mexico at Mobile, Alabama and Atlantic Ocean at Miami, Florida) (Bayha, 2005), possibly indicating common recent ancestry for these animals. The third ctenophore (SAL-2) contained an ITS composite genotype not previously found in any geographic region (Bayha, 2005). These results must be interpreted cautiously, however. First, the dataset is extremely small, so the findings are specific only to these individuals and do not reflect anything regarding the greater Mediterranean population (i.e., others may have originated elsewhere). Description of the entire population would require the examination of significantly more Mediterranean individuals. Secondly, while the ITS region historically has been useful at the species level, the ITS genetic region occurs in multiple copies (Arnheim et al., 1980) and its use in population-level studies can be problematic (Vollmer & Palumbi, 2004). Therefore, ITS data ideally should be corroborated with data from additional single-copy genetic marker(s). Nevertheless, our preliminary data are consistent with two of our specimens having a common source with Black and Azov sea invaders, as well as ctenophores from the Gulf of Mexico (Mobile, Alabama) and/or Miami (Florida) in the native range.

The introduction of *M. leidyi* and its distribution throughout the Mediterranean Sea probably resulted from transport by both ballast waters from the Black Sea region and currents. *M. leidyi* was first reported in the Mediterranean Sea in 1990, transported into the Aegean Sea by water flow through the Dardanelles from the Sea of Marmara or by ballast water from ships. The ctenophore was reported in Turkey and Syria soon thereafter (1992 and 1993; Table 3). No other reports were made in the Mediterranean until 2005, when *M. leidyi* appeared in France and 2006 in the Gulf of Trieste (Table 3). The simultaneous blooms of *M. leidyi* all along the Mediterranean coasts of Israel, Spain, and Italy in 2009 (Fig. 1; Boero et al., 2009; Fuentes et al., 2009; Galil et al., 2009) suggest that such wide-spread, abundant ctenophores could not have been distributed by either currents or shipping in 2009, but that the ctenophores already were present and were stimulated to bloom in favorable conditions in 2009.

Table 3 Hydrological characteristics and mesozooplankton biomass (μg C) or abundance (number m^{-3}) of the seas of the Mediterranean basin, and the year when *Mnemiopsis leidyi* was first reported and the seasons when it is known to be active in each location

Location	Temperature (°C)		Salinity	Mesozoo (μg C or # m^{-3})	First report	Active Season
	Winter	Summer				
Black Sea[a]	0–8	24–27	18–22.3	33,000C*	1982	Spr–Aut
Sea of Azov[b]	−0.8 to +1.2	24–30	0–14		1988	Spr–Aut
Sea of Marmara[b]	8–15	24–29	18–29		1989–1990	All year
Caspian Sea[b]	0–11	24–28	0.1–11		1999	Spr–Aut
Aegean Sea[c]	13.3–14.1	24–29	38.7–39.1	450–10,940#	1990	All year
Turkey[d]		~25–26.5	~32.6–33.8		1992	ND
Syria					1993	ND
Gulf of Trieste[e]	<10	20 to >26	32–38	3.6–9C	2005	ND
France[e]		31.5	39.5		2006	ND
Catalan Sea[f]	12–13	25–26	37–39.2	500–8000#*	2008	ND
Italy[g]	13–14	23–26	37.5–37.9	500–4000#*	2009	ND
Israel	16–20	29–32	39.1–39.8	0–2598#	2009	ND

ND no data

* = Before Mnemiopsis, # = number, C = carbon

[a] Purcell et al. (2001)

[b] Shiganova et al. (2001)

[c] Shiganova et al. (2004)

[d] Isinibilir & Tarkan (2002)

[e] Shiganova & Malej (2009)

[f] Calbet et al. (2001)

[g] Kamburska & Fonda-Umani (2009)

Transport of *M. leidyi* in ballast water to Spain could have been to Barcelona, which was listed as the Mediterranean port with greatest ship traffic in 2006 (REMPEC, 2008). Of the Spanish locations with confirmed *M. leidyi*, only Denia represents an important commercial and passenger port; however, because all ships use the container system to transfer cargo, they do not release ballast water; therefore, it is unlikely that *M. leidyi* arrived in Denia by ship transport. The ctenophores found in the southern Catalan Sea as well as in Denia may be explained by circulation patterns in those areas.

The general circulation pattern in the Catalan Sea is characterized mainly by a current that follows the continental slope, in geostrophic equilibrium with a shelf/slope hydrographic front (Font et al., 1988). This current, the Liguro-Provençal-Current (LPC), is directed from the northeast towards the southwest partially governed by the effect of a permanent cyclonic gyre north of the Balearic Islands at the western side of Gulf of Lyon. This current transports the major outside Mediterranean surface waters to the Catalan Sea from the Gulf of Lyon along with considerable runoff from the Rhone River. This mesoscale circulation pattern persists throughout the year but is most intense in winter and spring (Font et al., 1988). In the southern Catalan Sea, the LPC follows the shelf break, entering the Gulf of Valencia and continuing to the Eivissa Channel (Salat, 1995). Water balances in the region show that the volume of water flowing southward through the Eivissa Channel is always smaller than the transport along the shelf break by the LPC and Atlantic waters that penetrate into the eastern Catalan Sea by the Eivissa Channel (García et al., 1994). The southern penetration of Atlantic waters mainly occurs through the channel between Eivissa and Mallorca all year, but this water mass remains in the eastern sector of the region (Monserrat et al., 2008). In years with mild winters, the flow of Atlantic waters through the Eivissa Channel is more intense especially in late spring and summer. These circulation patterns make intrusion of Atlantic waters into the Catalan Sea from the northeast with surface waters more probable than flow through the eastern Balearic Islands and the Gulf of Lyon. However, after mild winters, the intrusions from the south through the Eivissa Channel could had been of great importance incorporating Atlantic waters flowing around the Iberian Peninsula from the Strait of Gibraltar, being this another possibility for the introduction of *M. leidyi* into the southern Catalan Sea waters.

Hamad et al. (2005) analyzed satellite thermal images of Mediterranean surface waters and proposed a revision to the accepted scheme of large-scale flow in the Mediterranean Sea. Hamad et al.'s analysis indicates that there is a direct link between water flowing from the Aegean Sea, where *M. leidyi* has occurred since 1990, to the Levant. Alternatively, because the swarms of *M. leidyi* were observed mainly in the southern half of Israel, between Ashkelon and Netanya and considering that the predominant current in this region is along shore from south to north (Rosentraub & Brenner, 2007; Zviely et al., 2007), it is possible that *M. leidyi* was released in ballast water from ships traveling toward the southern Israeli port of Ashdod or to Egypt.

Factors affecting the success of *Mnemiopsis leidyi* in the Mediterranean Sea

Environmental conditions

Conditions for *M. leidyi* in the Mediterranean are quite different from the conditions in most native and introduced habitats. *M. leidyi* typically inhabits shallow estuaries and coastal waters in its native American waters (reviewed in Kremer, 1994; Purcell et al., 2001). In its northern temperate habitats (Rhode Island and Maryland), temperatures range from 1°C in the winter to 30°C in the summer and salinities range from ≤2 to 32. The annual peak in biomass of northern *M. leidyi* populations is in mid-summer to fall, which have occurred earlier with rising temperatures since 1950 (Sullivan et al., 2001; Costello et al., 2006a). Ctenophore biomass declines in the autumn when temperatures decrease and the ctenophores overwinter in nearshore bottom-water refugia (Costello et al., 2006b). Ctenophores in more southerly native locations (Florida and Texas) experience higher temperatures (winter minima of 7–18°C and summer maxima 31–32°C) and higher salinities (14–45) than in the north. The populations peak in winter when temperatures are lower (~20°C) than the rest of the year, and the population can display several small peaks (20–30 ml m^{-3} live volume) throughout the year.

These same patterns are seen among the locations where *M. leidyi* was introduced earlier. The Black, Caspian, North, and Baltic seas in the northern range are often estuarine, with a summer bloom of *M. leidyi* that subsides in cool temperatures (reviewed in Purcell et al., 2001; Shiganova et al., 2001). The southern introduced habitats in the Sea of Marmara and Mediterranean Sea are similar to the southern U.S. native habitats, with high salinities and temperatures (Shiganova et al., 2001; Shiganova & Malej, 2009). The only study of osmoregulation in ctenophores showed *M. leidyi* to conform to salinity within the range tested (8–23; Foshtomi et al., 2007). Clearly, the habitats of *M. leidyi* seem unrestricted by most physical conditions, with the exception of temperatures <4°C in high latitudes.

Water conditions along the Catalan coast are suitable for *M. leidyi* throughout the year (Fig. 4), because they would not be limited by cold winter temperatures as in northern latitudes. *M. leidyi* is present all year in the Aegean Sea (Shiganova et al., 2004), but its seasonal distributions in other Mediterranean locations are unknown.

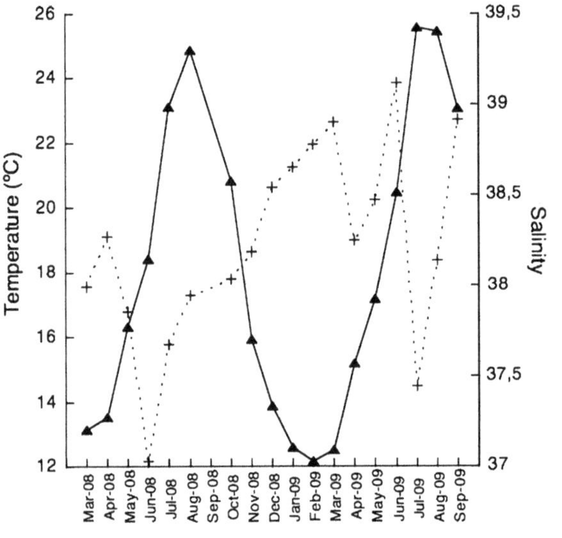

Fig. 4 Monthly measurements of sea temperature (°C) and salinity in the surface waters of the Catalan Sea from March 2008 to September 2009. *Solid triangles* salinity, *crosses* temperature

Food resources

Zooplankton abundance directly affects ctenophore abundance. *M. leidyi* was found in waters with high zooplankton stocks and did not occur where zooplankton biomass was <3 mg C m^{-3} (reviewed by Kremer, 1994; Purcell et al., 2001). In native locations, peak zooplankton biomass ($\sim$100 mg C m^{-3}) was higher in northern than in southern latitudes ($\sim$25 to 50 mg C m^{-3}).

In contrast to rich environments where *M. leidyi* is found, the north western Mediterranean is considered to be an oligotrophic sea, with low nutrient concentrations but relatively moderate levels of pelagic primary production (100–150 g C m^{-2} y^{-1}; reviewed by Estrada, 1996). In particular, the Catalan Sea is a highly dynamic oligotrophic environment, subjected to important forcing from the alternation of stratification and mixing periods and from the strength of mesoscale singularities (Saiz et al., 2007). Two features of special importance for productivity in the Mediterranean are the deep chlorophyll maximum (DCM) at the base of the pycnocline during most of the stratification period, and the occurrence of two phytoplankton blooms (Estrada, 1996), which are followed by two peaks of zooplankton in late winter–spring and in autumn. During stratification, the DCM is an oasis for zooplankton feeding and accentuates the difference between the poor surface layer (0.2–0.3 μg chl l^{-1}) and richer waters at depth (1–2 μg chl l^{-1}) (Estrada, 1996). During the stratified period, the trophic web based on the microbial loop could assume great importance; microzooplankton is responsible for most of the secondary production in warm waters (Fernández de Puelles et al., 2007; Saiz et al., 2007). In the Mediterranean Sea, there appears to be strong coupling between the microbial food webs and the upper trophic levels (Cushing, 1989). During the warm months, the diverse mesozooplankton community is predominated by filter-feeding organisms that efficiently consume the small fractions of the food web (Calbet et al., 2001).

Mnemiopsis leidyi is very versatile in the types of prey it consumes, which range from microplankton to mesozooplankton and ichthyoplankton; *M. leidyi* larvae ≤5 mm diameter contained diatoms, dinoflagellates, and ciliates, and larger larvae also contained copepod nauplii and copepodites (reviewed in Purcell et al., 2001; see also Sullivan & Gifford, 2004; Rapoza et al., 2005). Several studies report inverse correlations of *M. leidyi* and zooplankton abundances (reviewed in Purcell, 1988); when *M. leidyi* were abundant, clearance rates of the ctenophores were sufficient to reduce zooplankton populations (reviewed in Purcell et al., 2001; Shiganova et al., 2001; see also Purcell & Decker, 2005; Riisgård et al., 2007). In addition, *M. leidyi* had higher clearance rates on copepods than did scyphomedusae of equal biomass (Purcell & Decker, 2005). The wide variety of prey types and high feeding rates could have allowed this species to establish in the oligotrophic conditions of the NW Mediterranean.

Other important food resources for *M. leidyi* are fish eggs and larvae. In the NW Mediterranean, ichthyoplankton has a pronounced seasonal variability because most fishes (neritic, small pelagics, and migratory species) spawn during spring and summer. The most abundant pelagic species are European sardine and anchovy, which have non-overlapping spawning periods, autumn–winter and spring–summer, respectively (Sabatés et al., 2007). Consumption of ichthyoplankton by *M. leidyi* can be very important in native and non-native environments (reviewed in Purcell et al., 2001; Shiganova et al., 2001); however, the importance of *M. leidyi* predation on zooplankton and ichthyoplankton in the recently invaded Mediterranean regions are not yet known.

In addition to the wide spectrum of prey that this ctenophore can exploit, the success of introduced *M. leidyi* also could be attributed to reduced competition from over-fished populations of zooplanktivorous fish (Oguz et al., 2008; Siapatis et al., 2008). In the NW Mediterranean Sea, over-fishing is believed to be one reason that gelatinous zooplankton seem to be increasing in abundance in recent decades (Gili & Pagés, 2005). Two important features of the fishing activity in the Mediterranean Sea are the multi-specificity of catches and the absence of large single stocks, especially in the demersal regime, as compared with those that inhabit other seas (Coll et al., 2006). The development of fishing technologies and over-capitalization, with an increasing demand for marine resources, is placing intensive pressure on marine resources in the western basin, and the general assessment suggests that most demersal stocks are fully exploited, while some pelagic stocks are even overexploited (Bas et al., 2003; Leonard & Maynou, 2003; Coll et al., 2006).

It is not clear what lead to the bloom of this ctenophore in the Levant but it may be related to the dietary similarities of larval fish, zooplanktivorous fish, and *M. leidyi*. The fishing pressure along Israeli coasts continues to be very high (Pisanty & Grofit, 1991) and may have caused an untimely imbalance in favor of the invasive ctenophore, enabling it to flourish, despite the relatively low abundances of micro- and mesozooplankton detected in the coastal waters of Israel during the period that *M. leidyi* was present. Zooplankton abundances in the eastern Mediterranean Sea off the coast of Israel generally are very low, and median values ranged from 256 to 1,261 m^{-3} for microzooplankton and 13–196 m^{-3} for mesozooplankton, as recorded from March to May 2009.

Predators and competitors

Interannual variation in *M. leidyi* abundance is strongly related to predator abundance in their native waters (reviewed in Purcell et al., 2001; Purcell & Decker, 2005). A variety of predators occur, including the ctenophore, *Beroe ovata* Brugière, the scyphomedusae, *Chrysaora quinquecirrha* Desor and *Cyanea capillata* (Linnaeus), harvestfish, *Peprilus alepidotus* (Linnaeus), and butterfish *P. triacanthus* (Peck), but their effects on *M. leidyi* populations have rarely been studied. Low and high salinity waters can serve as a refuge for *M. leidyi* from predators, as in Chesapeake Bay. Purcell et al. (2001) concluded that the key difference between native U.S. habitats of *M. leidyi* and the Black, Azov, and Caspian seas was the lack of predators there, perhaps enabling the devastating blooms of the ctenophore in those non-native habitats. Subsequently, *B. ovata* arrived in the Black and Azov seas and has limited *M. leidyi* populations since 1999.

The NW Mediterranean Sea has a diversity of potential predators of *M. leidyi*. The ctenophores *Beroe cucumis* Fabricius and *Beroe forskalii* Milne Edwards are common (Shiganova & Malej, 2009); however, *Beroe* spp. normally are abundant only in spring and early summer along the Catalan coast (F. Pagès, pers. com.), suggesting that they were not abundant there when *M. leidyi* occurred in high numbers (July and August 2009). Nevertheless, at the beginning of July 2009, both *M. leidyi* and *P. noctiluca* were present in the coastal waters and *P. noctiluca* ate *M. leidyi* in the laboratory (V. Fuentes, pers. obs.). In late July 2009, *P. noctiluca* almost disappeared from the Catalan coast; their absence coincided with high abundances of *M. leidyi* (Fig. 2). Another abundant potential predator is the hydromedusa, *A. forskalea*; its congeners are voracious predators of gelatinous species (Purcell, 1991).

In addition to the potential predators, many potential competitors for zooplankton foods occur in the NW Mediterranean, including the large jellyfish, *P. noctiluca*, *C. tuberculata*, *R. pulmo*, and *A. forskalea*, and many species of small hydromedusae and siphonophores (Sabatés et al., this volume), as well as several species of zooplanktivorous fishes (Leonard & Maynou, 2003; Sabatés et al., 2007).

Jellyfish observations in Israel indicate that in addition to the persistent swarms of *R. nomadica* and *R. pulmo* that have been documented along the Mediterranean shores since the 1980s (Galil, 2007), there are occasional blooms of novel exotic gelatinous species that had not been previously recorded. A comprehensive list of scyphomedusae of the Mediterranean coast of Israel includes *P. noctiluca*, *Aurelia aurita*, *C. tuberculata*, *Phyllorhiza punctata* von Lendenfeld, *Cassiopea andromeda* (Forsskal), *R. pulmo*, and *R. nomadica* (in Galil et al., 1990). Despite annual outbreaks of stinging jellyfish, few attempts have been made to quantify these swarms (Spanier & Galil, 1991; Lotan et al., 1992) or to follow their spatio-temporal dynamics. As in the NW Mediterranean, all of these species are potential competitors with *M. leidyi* for zooplankton prey, and *P. noctiluca* is a likely predator of *M. leidyi*. The importance of these various species as predators and competitors of *M. leidyi* in Mediterranean waters remains to be investigated.

The lack of basic information on these exotic invasives throughout the Mediterranean led to the 2008 CIESM initiative, "JellyWatch" (http://www.ciesm.org/marine/programs/jellywatch.htm) and the "Medusa Project" (ACA, 2009) created to document the frequency and distribution of jellyfish blooms in the Mediterranean Sea. Those efforts need to be expanded to include other gelatinous groups in addition to large jellyfish. For example, the records for ctenophores in the Mediterranean Sea are scarce; *M. leidyi* may have been resident previously but was not noticed before the spectacular 2009 blooms.

Conclusions

Nearly simultaneous blooms of the invasive ctenophore, *M. leidyi*, occurred throughout the Mediterranean Sea in 2009. This voracious zooplanktivore has caused wide-spread ecosystem disruption and damage to fisheries following its accidental introduction to the Black Sea basin in the early 1980s. It subsequently invaded waters of the North and Baltic seas, where its spread and abundance may be restricted by cold temperatures and native gelatinous predators. Conditions in the Mediterranean Sea are very different (comparatively warm temperatures, high salinities, and low productivity) from those in the northern habitats invaded by *M. leidyi*; therefore, *M. leidyi* is unlikely to be constrained by physical conditions in the Mediterranean Sea. On the other hand, a variety of gelatinous species are potential predators and competitors of *M. leidyi* that may constrain its populations in the Mediterranean. Whether or not *M. leidyi* flourishes, the presence of this species will change the Mediterranean ecosystems and poses a threat to fisheries.

Acknowledgments Thanks are due to the Catalan Water Agency for funding the Medusa Project. We are indebted to all institutions collaborating with the reporting network. We thank the Haifa University Maritime Workshop for their help in collecting *M. leidyi* at sea; G. Yahel and the School of Marine Sciences and Marine Environment, Ruppin Academic Center, Michmoret Israel for provision of zooplankton abundance data; N. Drimmer at CAMERI - Coastal and Marine Engineering Research Institute Ltd. for the Israel Ports Development & Assets Company Ltd. for the sea-surface temperature data, and A. Glazer at the Israel Electric Corporation for sea-surface temperature and salinity data. All molecular work was performed in the lab of Dr. Michael N Dawson at the University of California Merced.

References

ACA, 2009. Catalan water agency "Medusa Project", http://aca-web.gencat.cat:80/aca/appmanager/aca/aca?_nfpb=true&_pageLabel=P2020096158125077724062 5&_nfls=false.

Altschul, S. F., T. L. Madden, A. A. Schäffer, J. Zhang, Z. Zhang, W. Miller & D. J. Lipman, 1997. Gapped BLAST and PSI-BLAST: a new generation of protein database search programs. Nucleic Acids Research 25: 3389–3402.

Arnheim, N., M. Krystal, R. Schmickel, G. Wilson, O. Ryder & E. Zimmer, 1980. Molecular evidence for genetic exchanges among ribosomal genes on nonhomolohous chromosomes in man and apes. Proceedings of the National Academy of Sciences, USA 77: 7323–7327.

Bas, C., F. Maynou, F. Sardà & J. Leonard, 2003. Variacions demogràfiques a les poblacions d'espècies demersals explotades: es darrers quaranta anys a Blanes i Barcelona. Institut d'Estudis Catalans. Secció de Ciències Biològiques Vol. 135: 202 pp.

Bayha, K. M., 2005. The molecular systematics and population genetics of four coastal ctenophores and scyphozoan jellyfish of the U.S. Atlantic and Gulf of Mexico. Ph.D. Dissertation, The University of Delaware, Newark.

Bayha, K. M., G. R. Harbison, J. H. McDonald & P. M. Gaffney, 2004. Preliminary investigation on the molecular systematics of the invasive ctenophore *Beroe ovata*. In Dumont, H., T. A. Shiganova & U. Niermann (eds), Aquatic Invasions in the Black, Caspian and Mediterranean Seas. Kluwer Academic Publishers, Dordrecht: 167–175.

Boero, F., M. Putti, E. Trainito, E. Prontera, S. Piraino & T. Shiganova, 2009. Recent changes in Western Mediterranean Sea biodiversity: the establishment of *Mnemiopsis leidyi* (Ctenophora) and the arrival of *Phyllorhiza punctata* (Cnidaria). Aquatic Invasions 4: 675–680.

Calbet, A., S. Garrido, E. Saiz, M. Alcaraz & C. M. Duarte, 2001. Annual zooplankton succession in coastal NW Mediterranean waters: the importance of the smaller size fractions. Journal of Plankton Research 23: 319–331.

CAMERI (Coastal and Marine Engineering Institute of the Technion), 2008. Annual report for the Israel Ports Development & Assets Company Ltd, http://eng.israports.co.il/.

Coll, M., I. Palomera, S. Tudela & F. Sardà, 2006. Trophic flows, ecosystem structure and fishing impacts in the South Catalan Sea, Northwestern Mediterranean. Journal of Marine Systems 59: 63–96.

Costello, J. H., B. K. Sullivan & D. J. Gifford, 2006a. A physical–biological interaction underlying variable phonological responses to climate change by coastal zooplankton. Journal of Plankton Research 28: 1099–1105.

Costello, J. H., B. K. Sullivan, D. Van Keuren & D. J. Gifford, 2006b. Seasonal refugia, shoreward thermal amplification, and metapopulation dynamics of the ctenophore *Mnemiopsis leidyi* in Narragansett Bay, Rhode Island. Limnology and Oceanography 51: 1819–1831.

Cushing, D. H., 1989. A difference in structure between ecosystems in strongly stratified waters and in those that are only weakly stratified. Journal of Plankton Research 11: 1–13.

Dawson, M. N., K. A. Raskoff & D. K. Jacobs, 1998. Preservation of marine invertebrate tissues for DNA analyses. Molecular Marine Biology and Biotechnology 7: 145–152.

Estrada, M., 1996. Primary production in the northwestern Mediterranean. Scientia Marina 60(Suppl. 2): 55–64.

Faasse, M. A. & K. M. Bayha, 2006. The ctenophore *Mnemiopsis leidyi* A. Agassiz 1865 in coastal waters of the Netherlands: an unrecognized invasion? Aquatic Invasions 1: 270–277.

Faris, S., 2009. A gelatinous invasion. Time 174: 47–48.

Fernández de Puelles, M. L., F. Alemany & J. Jansá, 2007. Zooplankton time-series in the Balearic Sea (Western Mediterranean): variability during the decade 1994–2003. Progress in Oceanography 74: 329–354.

Font, J., J. Salat & J. Tintoré, 1988. Permanent features of the circulation in the Catalan Sea. Oceanologica Acta 1988: 51–57.

Foshtomi, M. Y., B. Abtahi, A. E. Sari & M. Taheri, 2007. Ion composition and osmolarity of Caspian Sea ctenophore, *Mnemiopsis leidyi*, in different salinities. Journal of Experimental Marine Biology and Ecology 352: 28–34.

Fuentes, V. L., D. Atienza, J.-M. Gili & J. E. Purcell, 2009. First records of *Mnemiopsis leidyi* A. Agassiz 1865 off the NW Mediterranean coast of Spain. Aquatic Invasions 4: 671–674.

Galil, B. S., 2007. Seeing red: alien species along the Mediterranean coast of Israel. Aquatic Invasions 2: 281–312.

Galil, B. S., E. Spanier & W. W. Ferguson, 1990. The scyphomedusae of the Mediterranean coast of Israel, including two Lessepsian migrants new to the Mediterranean. Zoologische Mededelingen 64: 95–105.

Galil, B. S., N. Kress & T. A. Shiganova, 2009. First record of *Mnemiopsis leidyi* A. Agassiz, 1865 (Ctenophora Lobata Mnemiidae) off the Mediterranean coast of Israel. Aquatic Invasions 4: 356–362.

García, E., J. Tintoré, J. Pinot, J. Font & M. Manríquez, 1994. Surface circulation and dynamics of the Balearic Sea. Coastal and Estuarine Studies 46: 73–91.

GESAMP (IMO/FAO/UNESCO-IOC/WMO/WHO/IAEA/UN/UNEP Joint Group of Experts on the Scientific Aspects of Marine Environmental Protection), 1997. Opportunistic settlers and the problem of the ctenophore *Mnemiopsis leidyi* invasion in the Black Sea. Report and Studies GESAMP 58: 84 pp.

Gili, J. M. & F. Pagés, 2005. Les proliferacions de meduses. Bolletí de la Societat d'Història Natural de les Balears 48: 9–22.

Gili, J. M., V. Fuentes & D. Atienza, 2007. Medusa project: data year 2007. Technical report No. 1: 98 pp.

Gili, J. M., V. Fuentes & D. Atienza, 2008. Medusa project: data year 2008. Technical report No. 2: 198 pp.

Gorokhova, E., M. Lehtiniemi, S. Viitasalo-Frosen & S. H. D. Haddock, 2009. Molecular evidence for the occurrence of ctenophore *Mertensia ovum* in the northern Baltic Sea and implications for the status of the *Mnemiopsis leidyi* invasion. Limnology and Oceanography 54: 2025–2033.

Hamad, N., C. Millot & I. Taupier-Letage, 2005. A new hypothesis about the surface circulation in the eastern basin of the Mediterranean Sea. Progress in Oceanography 66: 287–298.

Herut, B., E. Shefer & Y. Cohen, 2005. Environmental quality of Israel's Mediterranean coastal waters in 2004. IOLR Report H34/2005.

Herut, B., Y. Cohen, E. Shefer, N. Gordon, B. Galil, G. Tibor & M. Tom, 2006. Environmental quality of Israel's Mediterranean coastal waters in 2005. IOLR Report H25/2006.

Herut, B., E. Shefer, N. Gordon, B. Galil, G. Tibor & M. Tom, 2008. Environmental quality of Israel's Mediterranean coastal waters in 2007. IOLR Report H52/2008.

IEC (Israel Electric Corporation), 2009. Environmental report 2009 (in preparation).

Isinibilir, M. & A. N. Tarkan, 2002. Distribution of the invasive ctenophore *Mnemiopsis leidyi* in the north-eastern Aegean Sea in August 1998. Turkish Journal of Fisheries and Aquatic Sciences 2: 129–132.

Ivanov, V. P., A. M. Kamakin, V. B. Ushivtsev, T. A. Shiganova, O. P. Zhukova, N. Aladin, S. I. Wilson, G. R. Harbison & H. J. Dumont, 2000. Invasion of the Caspian Sea by the comb jellyfish *Mnemiopsis leidyi* (Ctenophora). Biological Invasions 2: 255–258.

Javidpour, J., U. Sommer & T. A. Shiganova, 2006. First record of *Mnemiopsis leidyi* A. Agassiz 1865 in the Baltic Sea. Aquatic Invasions 1: 299–302.

Javidpour, J., J. Molinero & J. Peschutter, 2008. Seasonal changes and population dynamics of the ctenophore *Mnemiopsis leidyi* after its first year of invasion in the Kiel Fjord, Western Baltic sea. Biological Invasions. doi:10.1007/s10530-008-9300-8.

Kamburska, L. & S. Fonda-Umani, 2009. From seasonal to decadal inter-annual variability of mesozooplankton biomass in the Northern Adriatic Sea (Gulf of Trieste). Journal of Marine Systems 78: 490–504.

Kideys, A. E. & U. Niermann, 1993. Intrusion of *Mnemiopsis mccradyi* (Ctenophora; Lobata) into the Mediterranean Sea. Senckenbergiana Maritima 23: 43–47.

Kideys, A. E. & U. Niermann, 1994. Occurrence of *Mnemiopsis* along the Turkish coasts (from northeastern Mediterranean to Istanbul). ICES Journal of Marine Science 51: 423–427.

Kremer, P., 1994. Patterns of abundance for *Mnemiopsis* in US coastal waters: a comparative overview. ICES Journal of Marine Science 51: 347–354.

Kube, S., L. Postel, C. Honnef & C. B. Augustin, 2007. *Mnemiopsis leidyi* in the Baltic Sea – distribution and overwintering between autumn 2006 and spring 2007. Aquaic Invasions 2: 137–145.

Lehtiniemi, M., J. P. Pääkkönen, J. Flinkman, T. Katajisto, E. Gorokhova, M. Karajalanein, S. Viitasalo & H. Björk, 2007. Distribution and abundance of the American comb jelly (*Mnemiopsis leidyi*) – a rapid invasion to the northern Baltic Sea during 2007. Aquatic Invasions 2: 445–449.

Leonard, J. & F. Maynou, 2003. Fish stocks assessment in the Mediterranean: state of the art. In Ulltang, O. & G. Blom (eds), Fish Stocks Assessments and Predictions: Integrating Relevant Knowledge. Scientia Marina 67 (Suppl. 1): 37–49.

Lotan, A., R. Ben-Hillel & Y. Loya, 1992. Life cycle of *Rhopilema nomadica*: a new immigrant scyphomedusan in the Mediterranean. Marine Biology 112: 237–242.

Monserrat, S., J. L. López-Jurado & M. Marcos, 2008. A mesoscale index to describe the regional circulation around the Balearic Islands. Journal of Marine Systems 71: 413–420.

OCEANA, 2008. Illes Balears: propuesta para la gestión de habitat amenazados y la pezca, http://oceana.org/file admin/oceana/uploads/europe/reports/Illes_Balears_OCEANA.pdf.

Oguz, T., 2005. Long-term impacts of anthropogenic forcing on the Black Sea ecosystem. Oceanography 18: 112–121.

Oguz, T., B. Fach & B. Salihoglu, 2008. Invasion dynamics of the alien ctenophore *Mnemiopsis leidyi* and its impact on anchovy collapse in the Black Sea. Journal of Plankton Research 30: 1385–1397.

Pisanty, S. & E. Grofit, 1991. Limiting effort in the Israeli trawl fishery. Fisheries and Fishbreeding in Israel 24: 100–133. (in Hebrew with English abstract).

Podar, M., S. H. D. Haddock, M. L. Sogin & G. R. Harbison, 2001. A molecular genetic framework for the phylum Ctenophora using 18s rRNA genes. Molecular Phylogenetics and Evolution 21: 218–230.

Purcell, J. E., 1988. Quantification of *Mnemiopsis leidyi* (Ctenophora, Lobata) from formalin-preserved plankton samples. Marine Ecology Progress Series 45: 197–200.

Purcell, J. E., 1991. Predation by *Aequorea victoria* on other species of potentially competing pelagic hydrozoans. Marine Ecology Progress Series 72: 255–260.

Purcell, J. E. & M. B. Decker, 2005. Effects of climate on relative predation by scyphomedusae and ctenophores on copepods in Chesapeake Bay during 1987–2000. Limnology and Oceanography 50: 376–387. http://aslo.org/lo/toc/vol_50/issue_1/0376.pdf.

Purcell, J. E., T. A. Shiganova, M. B. Decker & E. D. Houde, 2001. The ctenophore *Mnemiopsis* in native and exotic habitats: U.S. estuaries versus the Black Sea basin. Hydrobiologia 451: 145–176.

Rapoza, R., D. Novak & J. H. Costello, 2005. Life-stage dependent, in situ dietary patterns of the lobate ctenophore *Mnemiopsis leidyi* Agassiz 1865. Journal of Plankton Research 27: 951–956.

REMPEC (Regional Marine Pollution Emergency Response Centre for the Mediterranean Sea), 2008. Study of maritime traffic flows in the Mediterranean Sea. Final report, Unrestricted Version, EU-Funded MEDA Regional Project MED 2005/109-573.

Riisgård, H. U., L. Bøttiger, C. V. Madsen & J. E. Purcell, 2007. Invasive ctenophore *Mnemiopsis leidyi* in Limfjorden (Denmark) in late summer 2007 – assessment of abundance and predation effects. Aquatic Invasions 2: 395–401.

Rosentraub, Z. & S. Brenner, 2007. Circulation over the southeastern continental shelf and slope of the Mediterranean Sea: direct current measurements, winds, and numerical model simulations. Journal of Geophysical Research: Oceans 112(C11001): 1–21.

Sabatés, A., M. P. Olivar, J. Salat, I. Palomera & F. Alemany, 2007. Physical and biological processes controlling the distribution of fish larvae in the NW Mediterranean. Progress in Oceanography 74: 355–376.

Sabatés, A., F. Pagès, D. Atienza, V. Fuentes, J. E. Purcell & J.-M. Gili, this volume. Planktonic cnidarian distribution and feeding of *Pelagia noctiluca* from near shore to open sea in the NW Mediterranean. Hydrobiologia. doi: 10.1007/s10750-010-0221-z.

Saiz, E., A. Calbet, D. Atienza & M. Alcaraz, 2007. Feeding and production of zooplankton in the Catalan Sea (NW Mediterranean). Progress in Oceanography 74: 313–328.

Salat, J., 1995. The circulation between the Catalan and Balearic currents in the southern Catalan Sea. Oceanologica Acta 18: 227–234.

Shiganova, T. A. & A. Malej, 2009. Native and non-native ctenophores in the Gulf of Trieste, Northern Adriatic Sea. Journal of Plankton Research 31: 61–71.

Shiganova, T. A., Z. A. Mirzoyan, E. A. Studenikina, S. P. Volovik, I. Siokou-Frangou, S. Zervoudaki, E. D. Christou, A. Y. Skirta & H. J. Dumont, 2001. Population development of the invader ctenophore *Mnemiopsis leidyi* in the Black Sea and other seas of the Mediterranean basin. Marine Biology 139: 431–445.

Shiganova, T. A., E. D. Christou, J. V. Bulgakova, I. Siokou-Frangou, S. Zervoudaki & A. Siapatis, 2004. Study on the distribution and biology of the invader *M. leidyi* in the northern Aegean Sea, comparison with indigenous species Bolinopsis vitrea. In Dumont, H., T. A. Shiganova & U. Niermann (eds), The Ctenophore *Mnemiopsis leidyi* in the Black, Caspian and Mediterranean Seas and Other Aquatic Invasions. NATO Science Series: IV. Earth and Environmental Sciences, Vol. 35. Kluwer Academic Publishers, Dordrecht: 113–135.

Siapatis, A., M. Giannoulaki, V. D. Valavanis, A. Palialexis, E. Schismenou, A. Machias & S. Somarakis, 2008. Modelling potential habitat of the invasive ctenophore *Mnemiopsis leidyi* in Aegean Sea. Hydrobiologia 612: 281–295.

Spanier, E., & B. S. Galil, 1991. Lessepsian migration: a continuous biogeographical process. Endeavour 15: 102–106.

Sullivan, L. J., & D. J. Gifford, 2004. Diet of the larval ctenophore *Mnemiopsis leidyi* A. Agassiz (Ctenophora, Lobata). Journal of Plankton Research 26: 417–431.

Sullivan, B. K., D. Van Keuren & M. Claucy, 2001. Timing and size of blooms of the ctenophore *Mnemiopsis leidyi* in relation to temperature in Narragansett Bay, RI. Hydrobiologia 451: 113–120.

Vinogradov, M. E. & E. A. Shushkina, 1992. Temporal changes in community structure in the open Black Sea. Oceanology 32: 485–491.

Vinogradov, M. E., E. A. Shushkina, E. I. Musayeva & P. Y. Sorokin, 1989. A newly acclimated species in the Black Sea: the ctenophore *Mnemiopsis leidyi* (Ctenophora: Lobata). Oceanology 29: 220–224.

Vollmer, S. V. & S. R. Palumbi, 2004. Testing the utility of internal transcribed spacer sequences in coral phylogenetics. Molecular Ecology 13: 2763–2772.

Zviely, D., E. Kit & M. Klein, 2007. Longshore sand transport estimates along the Mediterranean coast of Israel in the Holocene. Marine Geology 238: 61–73.

Hydrobiologia (2010) 645:39–52
DOI 10.1007/s10750-010-0224-9

JELLYFISH BLOOMS

Effects of pH on asexual reproduction and statolith formation of the scyphozoan, *Aurelia labiata*

Amanda K. Winans · Jennifer E. Purcell

Published online: 21 March 2010

Abstract Although anthropogenic influences such as global warming, overfishing, and eutrophication may contribute to jellyfish blooms, little is known about the effects of ocean acidification on jellyfish. Most medusae form statoliths of calcium sulfate hemihydrate that are components of their balance organs (statocysts). This study was designed to test the effects of pH (7.9, within the average current range, 7.5, expected by 2100, and 7.2, expected by 2300) combined with two temperatures (9 and 15°C) on asexual reproduction and statolith formation of the moon jellyfish, *Aurelia labiata*. Polyp survival was 100% after 122 d in seawater in all six temperature and pH combinations. Because few polyps at 9°C strobilated, and temperature effects on budding were consistent with published results, we did not analyze data from those three treatments further. At 15°C, there were no significant effects of pH on the numbers of ephyrae or buds produced per polyp or on the numbers of statoliths per statocyst; however, statolith size was significantly smaller in ephyrae released from polyps reared at low pH. Our results indicate that *A. labiata* polyps are quite tolerant of low pH, surviving and reproducing asexually even at the lowest tested pH; however, the effects of small statoliths on ephyra fitness are unknown. Future research on the behavior of ephyrae with small statoliths would further our understanding of how ocean acidification may affect jellyfish survival in nature.

Keywords Jellyfish · Ocean acidification · Statolith · pH · Reproduction

Guest editors: J. E. Purcell & Dror Angel / Jellyfish Blooms: New Problems and Solutions

A. K. Winans
Biology Department, Western Washington University, 516 High Street, Bellingham, WA 98225, USA

A. K. Winans (✉) · J. E. Purcell
Shannon Point Marine Center, Western Washington University, 1900 Shannon Point Rd, Anacortes, WA 98221, USA
e-mail: amandawinans@gmail.com

J. E. Purcell
Coastal and Marine Resources Centre, University College Cork, Naval Base, Haulbowline Island, Cobh, Co. Cork, Ireland

Introduction

Research on jellyfish and their possible responses to human influences, such as global climate change, has increased in recent years because of negative effects of jellyfish on human enterprises. Jellyfish can bloom in very high densities and seriously affect their prey populations, which include zooplankton, larval fish, and fish eggs (Purcell, 1997). As well as being jellyfish prey, the diets of zooplanktivorous fish, including anchovies, herring, pollock, sandlance, and sardines, overlap with jellyfish diets and competition for food may occur (Purcell & Grover, 1990; Purcell & Sturdevant, 2001; Brodeur et al., 2008b). Many anthropogenic influences may be contributing to recent jellyfish blooms, including global warming, reduction of fish stocks, and eutrophication (reviewed in Purcell et al., 2007). Because jellyfish play key ecological and economic roles, it is important to determine how such changes may affect them.

One of the greatest anthropogenic causes of global climate change is the burning of fossil fuels. Since before the industrial revolution (about 250 years ago), the concentration of CO_2 in the earth's atmosphere has increased by about 36% (Forster et al., 2007). About 30% of this CO_2 has been taken up by the ocean (Feely et al., 2004). When CO_2 dissolves in seawater it forms carbonic acid, which then can dissociate into bicarbonate (HCO_3^-), carbonate (CO_3^{2-}), and hydrogen (H^+) ions; increased hydrogen ions lower the pH. The current average sea surface pH is about 7.9–8.3 (Bindoff et al., 2007), a decrease of about 0.1 pH units since before the industrial revolution (Royal Society, 2005). Because pH is on a logarithmic scale, a difference of one pH unit is a tenfold change. Thus, the predicted decrease in ocean pH of 0.3–0.4 units by the end of this century would be a 150% increase in the concentration of H^+ ions (Orr et al., 2005).

As CO_2 is taken up into the ocean, carbonate chemistry and pH changes will affect marine life in several ways (Doney et al., 2009): (1) reduced calcification rates of shell-forming organisms in response to supersaturation of CO_2; (2) undersaturation of aragonite, a more soluble form of $CaCO_3$ in the oceans than the more abundant calcite, which also is important for some calcifying organisms; and (3) biological effects of decreased pH other than on calcification. With the first threat, as more CO_2 is taken up in seawater, more carbonate ions are converted to bicarbonate and hydrogen ions, lowering the saturation state of calcium carbonate (Kleypas et al., 1999), which leaves less carbonate available for shell and skeleton formation (Doney et al., 2009). Calcification rates decreased in various marine organisms grown in seawater with elevated CO_2, including mussels (Michaelidis et al., 2005), gastropods and sea urchins (Shirayama & Thornton, 2005), corals (Gattuso et al., 1998; Kleypas et al., 1999; Langdon et al., 2000), and coccolithophores (Riebesell et al., 2000). With the second threat, in lowered pH, calcification decreased in certain species of corals (Gattuso et al., 1998) and pteropods, whose shells dissolved (Fabry et al., 2008). In the third threat, in addition to reduced calcification, decreased pH can cause acidosis (lowered pH in tissues and body fluids) as well as decreased metabolic, growth, and reproduction rates (Portner et al., 2005). Decreased metabolic and growth rates occurred in mussels exposed to low pH (Michaelidis et al., 2005), and warmer temperatures may exacerbate those effects (Portner et al., 2005; Metzger et al., 2007).

Because of the obvious issues related to changing carbonate chemistry, most ocean acidification research focuses on calcifying organisms. How ocean acidification may affect pelagic cnidarians, such as jellyfish, is poorly studied, and the results of existing studies are inconsistent. Attrill et al. (2007) showed a negative correlation of jellyfish frequency and pH (averaging from 8.0 to 8.35) in the North Sea, and suggested increasing jellyfish with ocean acidification. However, Richardson & Gibbons (2008) looked at a larger area and found no significant correlation of jellyfish frequency and pH from in the North Sea and North Atlantic. Activity and feeding of the jellyfish, *Rhopilema esculenta* Kishinouye, decreased below pH 7.4 in the laboratory (Gu et al., 2005).

Scyphozoans have two main stages in their life cycles, the benthic polyps and pelagic jellyfish. Scyphozoan polyps reproduce asexually by budding polyps and through the process of strobilation, in which ephyrae (juvenile jellyfish) are produced by transverse fission. Asexual production rates of many scyphozoan jellyfish are affected by water chemistry variables, such as temperature, salinity (Purcell et al., 1999; Purcell, 2007), and dissolved oxygen (DO) (Condon et al., 2001). Polyps of temperate scyphozoans, including *Aurelia labiata* Chamisso & Eysenhardt, produce more ephyrae but fewer buds in warm temperatures (Purcell et al., 1999; Purcell, 2007).

Like many other marine invertebrates, jellyfish have statocysts, balance organs that enable them to sense gravity. Scyphozoan jellyfish have statocysts inside their rhopalia, the club-like structures around the bell margin (Arai, 1997). Inside these statocysts are numerous statoliths of trigonal crystals of calcium sulfate hemihydrate ($CaSO_4 \cdot 0.5H_2O$) that are formed during strobilation (Spangenberg & Beck, 1968; Spangenberg, 1976; Becker et al., 2005).

The number and size of statoliths formed are influenced by the presence or absence of certain nutrients, minerals, and hormones. Spangenberg (1984) found that polyps grown in the presence of thyroxine, a hormone produced during strobilation in *Aurelia aurita* Linnaeus, had significantly fewer statoliths than polyps grown without thyroxine addition. Statoliths also were fewer and smaller in ephyrae from polyps grown in high concentrations of tetracycline HCl; "minor" pH changes (not reported) also occurred with the addition of tetracycline (Spangenberg & Beck, 1972). Starved polyps produced ephyrae with fewer statoliths than did fed polyps (Spangenberg, 1984). Statolith formation was almost completely inhibited in sulfate-deficient water, and ephyrae with no statoliths were irregular and weak swimmers (Spangenberg, 1968). Calcium sulfate behaves similarly to calcium carbonate, and can dissolve in the presence of hydrogen ions.

In our study, we raised polyps of *Aurelia labiata*, the local species of the cosmopolitan genus *Aurelia*, in combinations of different pH levels (7.2, 7.5, and 7.9) and temperatures (9 and 15°C) to determine the effects on asexual production of buds and ephyrae and on statolith formation (number and size). Because few polyps strobilated at 9°C, we were only able to test the null hypothesis that temperature and pH do not affect the numbers of buds produced. For 15°C, we tested the null hypotheses that pH does not affect the numbers of buds or ephyrae produced by the polyps, the time until strobilation, the number of strobilations, or the sizes and numbers of statoliths in the ephyrae.

Methods

Experimental protocols

Aurelia labiata polyps produced by medusae collected from Dyes Inlet, Washington were placed in 1.5-l glass finger bowls with filtered seawater (5 μm) pumped from Puget Sound into the Shannon Point Marine Center (SPMC) in Anacortes, Washington, USA on December 28, 2007. The polyps were kept in an incubator at 9°C in the dark without food for 78 days to inhibit them from strobilating until stimulated at the beginning of the experiment; feeding after a period of fasting, temperature increase after cooling, and light after darkness can trigger strobilation (Arai, 1997; Purcell et al., 1999; Lucas, 2001; Purcell, 2007).

Each of 108 arbitrarily selected polyps was placed in an individual 125-ml glass jar and allowed a few days for re-attachment. Eighteen polyps were arbitrarily assigned to each of the six combinations of two temperatures (9 and 15°C) and three pH levels (7.2, 7.5, and 7.9). The lower temperature is near average for waters of Puget Sound in autumn and early spring, and the high temperature occurs in summer. The pH of 8.1 was chosen to represent the current ocean pH mean of 7.9–8.3 (Bindoff et al., 2007). Average pH levels at SPMC range from 7.7 to 8.2. Ocean pH levels are predicted be 0.3–0.4 units lower by the end of the century (Orr et al., 2005) and 7.3 within the next 300 years (Caldeira & Wickett, 2003); therefore, 7.7 and 7.3 were chosen as intermediate and extreme pH levels. Because of changes in pH levels during the experiment, the actual pH levels in the treatments averaged 7.2, 7.5, and 7.9.

Every 2–4 days, renewed seawater solutions were prepared to replace the water in each of the six treatments, the polyp observations and maintenance were performed, and the water from some of the jars re-tested. To prepare the treatments, 5-μm-filtered seawater from SPMC was stored in six 8-l covered plastic containers in each incubator at the experimental temperatures. One container for each of the six treatments was placed on a magnetic stirrer and the pH tested with an Accumet® Basic AB15 pH meter using the NBS scale. The pH was adjusted to the desired level with drops of 1 M HCl or 1 M NaOH; HCl and NaOH have been used to manipulate seawater pH (e.g., Riebesell et al., 2000; Jokiel et al., 2008; Kuffner et al., 2008). Gattuso & Lavigne (2009) describes this as one of the five techniques that is "most useful in the context of ocean acidification," with the drawback that the use of acids and bases to alter pH also changes the alkalinity; low pH levels have low alkalinities (Riebesell et al., 2000; Rost et al., 2008). After the target pH level was obtained, the water was tested for DO with a YSI® 550A meter, salinity with a YSI® 30 meter, and alkalinity with the titration method of Palmer (1992). The pH and DO meters were calibrated at the beginning of each analysis day, using 4.0, 7.0, and 10.0 buffers to calibrate the pH meter. Although this method of using one container to mix each treatment may risk causing a container effect, it was important to ensure that all individuals within each treatment received identical water. The containers were arbitrarily switched so that any one would not always be used for the same treatment.

The jars with polyps initially were kept in four temperature-controlled incubators, two at each temperature. Polyps were arbitrarily assigned to an incubator, with equal numbers from each pH treatment in each incubator. On day 72 of the experiment, the number of incubators available decreased to 1 at each temperature. One temperature logger iButton® was kept in a jar of seawater beside the polyps in each incubator until day 72, then two loggers were kept in each incubator. Temperature in the jars also was measured along with water chemistry. Average temperatures (and standard error) in each incubator were 9.1 (0.007), 9.4 (0.01), 15.4 (0.004), and 15.3 (0.003). Fluorescent lights were placed equidistant from the trays in each incubator and were on for 12 h d^{-1}. Light levels at the trays ranged from 103 to 111 lux.

Polyp maintenance and asexual reproduction

The experiment began on March 4, 2008 and continued for 122 d until July 4, 2008. Every 2–4 days, the chemistry of the incubation water in the jars was measured, polyp data were collected, and the incubation water was replaced with the newly prepared treatment water. First, DO, pH, salinity, and temperature were measured in three arbitrarily chosen jars from each treatment; one of those jars also was tested for alkalinity. One jar was removed from the incubator at a time to minimize temperature changes.

Ephyrae were counted and removed from each of the 108 jars and then preserved in 100% EtOH in microcentrifuge tubes specifically labeled by polyp and date. Then, about three-quarters of the water was poured out and 1–3 day old *Artemia* sp. nauplii were added to the remaining water for at least an hour at amounts sufficient to saturate polyp feeding and to provide equal food to all polyps. Each polyp then was checked for survival, strobilation, and new buds with the aid of a dissecting microscope. Buds were enumerated and removed from the jar when they had separated from the parent polyp. The bottom of each jar was cleaned with foam swabs about once weekly, starting about 3 weeks into the experiment. Finally, each jar was emptied of water, rinsed with 5-μm-filtered seawater, and refilled with the appropriate newly prepared treatment seawater. The lid with parafilm was carefully replaced to prevent air bubbles, and then the jar was returned to its incubator. On the last day of the experiment, all buds attached to the parent polyps and ephyrae that were ready to detach were counted.

Statolith analyses

Ephyrae collected from the second strobilation (days 101–120 of the experiment) were chosen for statolith analysis because the polyps had been in the pH treatments much longer than for the first strobilation. Only polyps from 15°C were used because few polyps strobilated at 9°C. Uncurled ephyrae that could be flattened easily on a microscope slide were chosen to ensure accurate measurement of their diameter and counts and measurements of the statoliths. Each ephyra was pipetted onto a slide, and then a drop of water-free glycerol added and mixed with the alcohol to decrease the viscosity of the glycerol. After gently applying the coverslip and absorbing excess glycerol–alcohol from the edge of the coverslip, the diameter of the ephyra was measured with a compound microscope ocular micrometer at 40× magnification. Then the ephyra was flattened further by gentle pressure on the coverslip to maximize accuracy of the statolith counts and size measurements. The number of statoliths in each rhopalium (8–12 rhopalia per ephyra) was counted. For each ephyra, the length (long axis) and width (short axis) of all statoliths in three rhopalia were measured on a computer screen using Image Pro Plus software at 400× magnification. To test for differences in numbers and sizes of statoliths among ephyrae collected from the same polyp, six preserved ephyrae from one polyp on one collection day were chosen from each of the three pH treatments at 15°C. To test for effects of different pH levels on statolith numbers and sizes among ephyrae, one ephyra from each of six polyps at each pH level at 15°C were chosen and analyzed as above.

Statoliths from one ephyra at pH 7.2 and 15°C were also examined with an scanning electron microscope (SEM). The ephyra was preserved in 100% ethanol and then dissected with needles to isolate the statoliths on a 12-mm-diameter aluminum pin with double-sided carbon tabs. The sample was coated for a minute with a Quorum (Polaron) SC7640 sputter-coater equipped with a gold/palladium target. The SEM used was a Vega 5136MM (Tescan).

Statistical analyses

The water chemistry data for three jars in each combination were averaged for each day. Two-way ANOVAs (analysis of variance) tested for differences in pH, alkalinity, DO, pCO_2, and salinity for "renewed water" (the newly prepared water) and for "incubation water" (water from the jars with polyps) (Table 1). pCO_2 was calculated with CO_2Sys. In order to meet assumptions of the tests, all "pH" and "renewed DO" data were transformed to their base-10 logarithms. The pCO_2 tests did not meet the assumption of homogenous variance, and so α was halved to 0.025 in order to make the test more conservative (Keppel & Wickens, 2004; Gamst et al., 2008). Scheffe post-hoc tests were used to test the differences among the three pH treatments.

ANOVAs were used to test for differences in asexual reproduction among all treatments. Because few polyps strobilated at 9°C, those polyps and temperature effects were tested only for the numbers of buds. For this test, a two-way ANOVA was used to test for differences among pH and temperature treatments. The "number of buds" was transformed to its base-10 logarithm in order to meet the assumption of homogeneous variance. pH treatments at 15°C were tested for differences by one-way ANOVAs for the total number of ephyrae produced by each polyp over 122 d, the numbers of ephyrae produced in the first and second strobilations, the numbers of strobilations per polyp, and the days until the first and second strobilations. The equation: no. of ephyrae/(no. ephyrae + no. buds) $\times$ 100 was used to calculate the percentages that ephyrae contributed to total asexual production of each polyp. These proportions were arcsine transformed and then tested for differences among pHs at 15°C with a one-way ANOVA. They did not meet the assumption for homogeneity of variance, so α was changed to 0.025.

The numbers of statoliths per rhopalium were compared among and within pH treatments in the 15°C treatments using a one-way ANOVA. To test differences in statolith size, statolith volumes were estimated by calculating the volume of a regular hexagonal prism from measurements of the length and width of each statolith. Then, volumes were compared among and within pH treatments with a nested one-way ANOVA. For all statolith volume analyses,

Table 1 Two-way ANOVA results comparing "renewed" and "incubation" water measurements of pH, alkalinity, dissolved oxygen (DO), and salinity in an experiment, in which *Aurelia labiata* polyps were grown in combinations of temperature (9 and 15°C) and pH (7.2, 7.5, and 7.9)

Factor	Temperature		pH	
	Test statistic	*P* value	Test statistic	*P* value
pH				
Renewed	$F_{1,252} = 1.873$	$P = 0.172$	$F_{2,252} = 16904.353$	$\mathbf{P < 0.0001}$
Incubation	$F_{1,252} = 14.532$	$\mathbf{P < 0.0001}$	$F_{2,252} = 687.079$	$\mathbf{P < 0.0001}$
Alkalinity (μequ l^{-1})				
Renewed	$F_{1,240} = 1.368$	$P = 0.243$	$F_{2,240} = 127.153$	$\mathbf{P < 0.0001}$
Incubation	$F_{1,252} = 1.428$	$P = 0.233$	$F_{2,252} = 165.081$	$\mathbf{P < 0.0001}$
DO (mg l^{-1})				
Renewed	$F_{1,246} = 144.437$	$\mathbf{P < 0.0001}$	$F_{2,246} = 0.115$	$P = 0.891$
Incubation	$F_{1,246} = 200.995$	$\mathbf{P < 0.0001}$	$F_{2,246} = 0.948$	$P = 0.389$
Salinity				
Renewed	$F_{1,252} = 2.081$	$P = 0.150$	$F_{2,252} = 0.038$	$P = 0.962$
Incubation	$F_{1,252} = 70.517$	$\mathbf{P < 0.0001}$	$F_{2,252} = 0.011$	$P = 0.989$
pCO_2				
Renewed	$F_{1,240} = 13.785$	$\mathbf{P < 0.0001}$	$F_{2,240} = 456.618$	$\mathbf{P < 0.0001}$
Incubation	$F_{1,252} = 30.05$	$\mathbf{P < 0.0001}$	$F_{2,252} = 487.1$	$\mathbf{P < 0.0001}$

$P < 0.05$ is considered to be significant, except for pCO_2, where $P < 0.025$ is significant

Bold *P* values are significant

individual ephyrae were nested in pH treatment and individual rhopalia were nested in individual ephyra and pH treatment. In this ANOVA, the assumption of homogeneity of dispersion matrices was not met according to Levene's test, so the base-10 logarithms of the data were used. Because of the nested design of this test, we also were able to test differences in statolith volume among individual ephyrae. A one-way ANOVA was used to test for differences in ephyra age and diameters among pHs. This enabled us to determine if other factors, i.e., size or age, could have affected the variation because statoliths grow in size and number with age (Holst et al., 2007). All statistical tests were performed using SPSS 17.0.

Results

Water chemistry

The temperature treatments significantly differed in DO, pH, and salinity. DO was lower in 15°C than in 9°C in both renewed and incubation water (Table 1, Fig. 1). On average, DO decreased between water changes by about 3 mg l^{-1} in the 9°C treatment and by about 4 mg l^{-1} in 15°C. The pH levels in the 15°C treatments decreased significantly more (by 0.03–0.08 pH units) between water changes than in the 9°C treatments (Table 1, Fig. 1). Differences in salinity were not significant in the 9°C treatment; however, in the 15°C treatment, the salinity of the incubation water was 0.2–0.3 lower than the renewed water (Table 1, Fig. 1).

Total alkalinity and pCO_2 differed consistently among pH levels; lower pH levels had lower alkalinities (Table 1, Fig. 1) and higher pCO_2. The alkalinity measurements in renewed water differed by 5–7% among pH treatments, but alkalinities of incubation and renewed water did not differ within any pH level. Although pH decreased between water changes, each of the three pH treatments at both temperatures remained different from the others, as shown by a Scheffe's post-hoc test.

Polyp asexual reproduction

All *Aurelia labiata* polyps, except for seven that were inadvertently discarded during water changes, survived for the duration of the experiment (122 d); therefore, the temperature and pH treatments did not affect survival of the polyps.

Interactions between pH and temperature were not significant in the statistical tests on polyp reproduction (Table 2); therefore, temperature and pH effects were considered separately hereafter.

Bud production by polyps was affected by temperature, but not by pH (Table 2). The mean numbers of buds $polyp^{-1}$ ranged from 25.1 to 29.2 in 9°C treatments and from 11.1 to 15.4 buds $polyp^{-1}$ in 15°C treatments after 122 d. This significant difference matched previous results (Purcell, 2007), and was not considered further here. There were no significant differences in the numbers of buds produced $polyp^{-1}$ among pHs in either temperature.

The numbers of polyps that strobilated differed dramatically between the two temperatures (Table 2). In the 9°C treatments, only 16% of the surviving polyps strobilated, whereas 100% of the polyps at 15°C strobilated from 1 to 3 times. The first strobilation may have been stimulated by the initial change in conditions, as in Purcell (2007). Because so few polyps strobilated (8 of 51) at 9°C, we were unable to test the combined effects of temperature and pH on strobilation, ephyra production, and statolith formation. Therefore, the 9°C treatments and the effects of temperature were not considered further.

No significant effects of pH on strobilation were found in the 15°C treatments (Table 2). The numbers of ephyrae produced per polyp in the first strobilation, second strobilation, and over the entire experiment did not differ significantly among pHs, nor did the numbers of days until first or second strobilation, the numbers of strobilations, or the percentages of ephyrae of the total asexual production.

Statolith formation

Because few polyps strobilated in the 9°C treatments, only ephyrae from the 15°C treatment were used for statolith analyses (Table 3, Fig. 2). There were no significant differences in numbers of statoliths $rhopalium^{-1}$ among pH levels. Statolith volume showed significant differences ($P < 0.0001$) among pH levels. Statoliths were smaller at lower pH levels and larger at higher pH levels (Fig. 2). Statolith volumes also differed significantly among individual ephyrae ($P < 0.0001$) within each pH, but not among rhopalia within each ephyra. The possible ages of each ephyra

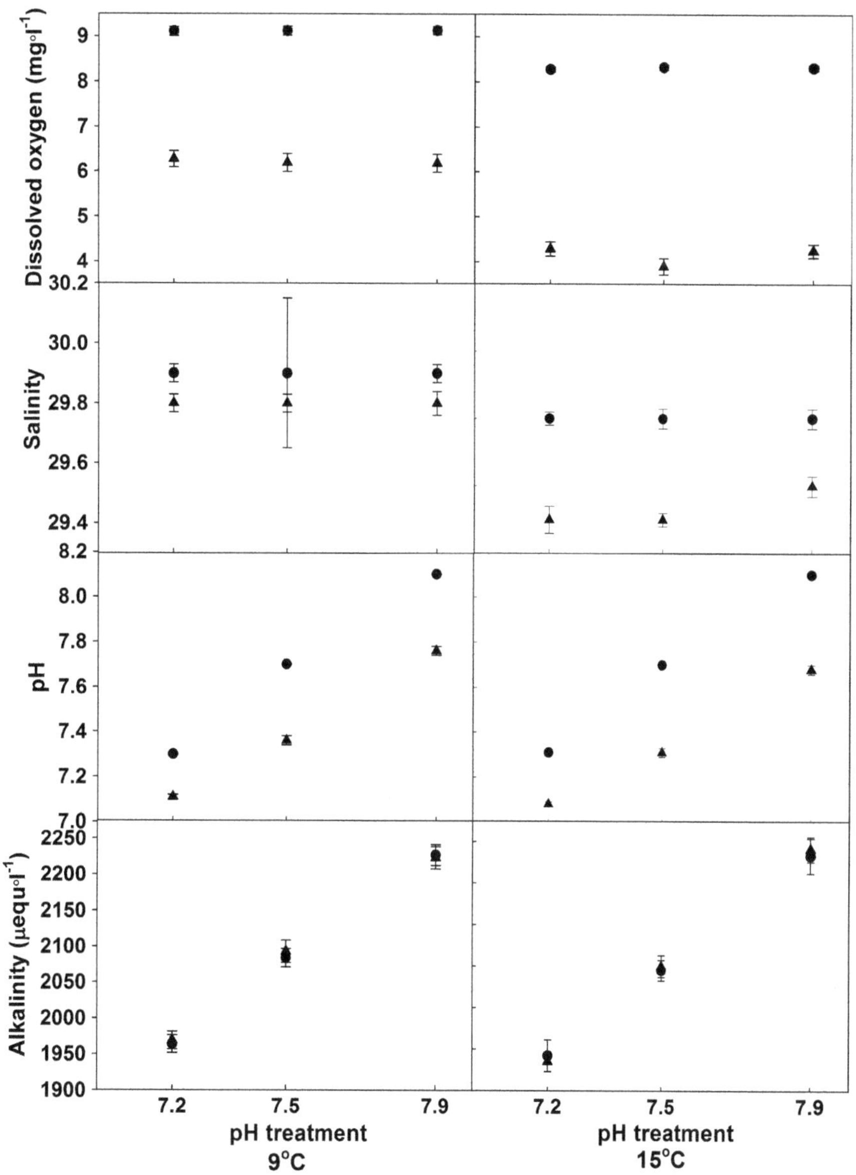

Fig. 1 Water chemistry measurements (mean ± SE) of dissolved oxygen, salinity, pH, and alkalinity during a 122 d experiment in which *Aurelia labiata* polyps were incubated at 9 and 15°C and pH levels of 7.2, 7.5, and 7.9. The symbols represent incubation (*filled circle*) and renewed seawater (*filled triangle*)

ranged from 1 to 4 d due to the collection schedule, which might have affected statolith numbers or volumes; however, the ages and diameters of the ephyrae did not different significantly among pHs (Table 3). Therefore, they should not have affected statolith numbers or volumes.

Statolith volume differences among genetically identical ephyrae collected from the same jar and sample collection also were significant (Table 4). The possible ages of ephyrae ranged from 1 to 3 d in 7.2 and 7.5 pH and 1–4 d in 7.9 pH. Differences in statolith volumes among rhopalia of individual ephyrae were not significant at any pH.

SEM showed that the statoliths of an ephyra raised in the 7.2 pH and 15°C treatment varied in size and shape (Fig. 3), as in *Aurelia aurita* (in Becker et al., 2005). No pitting or dissolution of the statoliths was apparent in SEM at 1,550× magnification.

Discussion

Effects of temperature on asexual reproduction

The effects of temperature on asexual reproduction were consistent with previous studies on

Table 2 Survival, production of buds, and strobilation of *Aurelia labiata* polyps grown in combinations of temperature (9 and 15°C) and pH (7.2, 7.5, and 7.9) over 122 d

pH	Temperature		Effects	Test statistic	*P* value
	9°C	15°C			
Strobilating/surviving polyps					
7.2	1/18	15/15			
7.5	3/16	18/18			
7.9	4/17	17/17			
	Mean (±SE)	Mean (±SE)			
Buds polyp^{-1}					
7.2	25.6 (1.62)	11.1 (0.71)	*T*	$F_{1,95} = 103.066$	**$P < 0.0001$**
7.5	25.1 (1.28)	14.8 (1.91)	pH	$F_{2,95} = 2.771$	$P = 0.068$
7.9	29.2 (2.24)	15.4 (1.64)	$T \times$ pH	$F_{2,95} = 0.792$	$P = 0.456$
Days until first strobilation					
7.2	41 (0)	28.9 (3.40)			
7.5	24.8 (4.53)	33.6 (4.36)	pH	$F_{2,51} = 0.485$	$P = 0.619$
7.9	12.3 (2.14)	34.4 (4.76)			
Days until second strobilation, 15°C					
7.2	NA	80.2 (4.66)			
7.5	NA	89.8 (5.19)	pH	$F_{2,41} = 1.048$	$P = 0.360$
7.9	NA	88.9 (4.85)			
Strobilations polyp^{-1}, 15°C					
7.2	0.06 (0.06)	2.13 (0.17)			
7.5	0.19 (0.10)	1.94 (0.10)	pH	$F_{2,47} = 0.560$	$P = 0.575$
7.9	0.24 (0.11)	2.0 (0.12)			
Ephyrae polyp^{-1} first strobilation, 15°C					
7.2	1.7 (1.72)	22.4 (3.68)			
7.5	2.3 (1.33)	23.9 (3.85)	pH	$F_{2,47} = 49.443$	$P = 0.811$
7.9	1.2 (0.59)	20.5 (3.66)			
Ephyrae polyp^{-1} second strobilation, 15°C					
7.2	NA	21.7 (3.89)			
7.5	NA	24.9 (3.31)	pH	$F_{2,47} = 0.558$	$P = 0.576$
7.9	NA	27.9 (4.86)			
Total ephyrae polyp^{-1}, 15°C					
7.2	1.7 (1.72)	46.8 (3.48)			
7.5	2.3 (1.33)	49.4 (4.67)	pH	$F_{2,47} = 0.098$	$P = 0.907$
7.9	1.2 (0.59)	48.5 (3.91)			
Mean no. ephyrae/(mean no. ephyrae + mean no. buds) × 100, 15°C					
7.2	4.0 (4.0)	79.4 (2.2)			
7.5	6.9 (4.0)	74.6 (3.4)	pH	$F_{2,51} = 0.852$	$P = 0.432$
7.9	5.4 (2.7)	75.2 (2.3)			

As few polyps strobilated at 9°C, those polyps and temperature effects were tested only for bud production

NA not applicable

Bold *P* values are significant

Table 3 Effects of pH on statolith sizes and numbers in *Aurelia labiata* ephyrae

pH 7.2	pH 7.5	pH 7.5	Effects	Test statistic	*P* value
No. statoliths analyzed					
346	314	363			
Mean maximum age of ephyrae in days					
3.7	3	3.5	pH	$F_{2,14} = 1.316$	$P = 0.300$
Mean ephyra diameter (mm)					
1.8	1.9	2.1	pH	$F_{2,14} = 1.410$	$P = 0.277$
Mean statolith length in μm (SE)					
11.39 (0.25)	13.75 (0.24)	15.78 (0.23)			
Mean statolith width in μm (SE)					
5.92 (0.11)	7.10 (0.13)	8.53 (0.14)			
Mean statolith volume in μm^3 (SE)					
304.17 (13.28) a	530.73 (20.24) b	865.92 (29.48) c	pH	$F_{2,14} = 15.857$	**$P < 0.0001$**
			E (pH)	$F_{14,33} = 7.380$	**$P < 0.0001$**
			R (pH * E)	$F_{33,973} = 0.867$	$P = 0.683$
Mean no. statoliths rhopalium^{-1} (SE)					
19.89 (0.69)	21.33 (0.91)	22.94 (1.07)	pH	$F_{2,14} = 0.773$	$P = 0.480$
			E (pH)	$F_{14,134} = 5.967$	**$P < 0.0001$**

Polyps that produced the ephyrae were grown at 15°C and pH levels of 7.2, 7.5, and 7.9. The ephyrae analyzed (5–6 from each pH) were collected from days 101 to 120. One-way ANOVAs were run for all tests. Significantly different groups from a post-hoc Bonferroni test are shown with different letters (a, b, and c)

E ephyra, *R* rhopalium

Bold *P* values are significant

scyphozoans, including *Aurelia labiata* (in Purcell et al., 1999; Purcell, 2007). More ephyrae were produced at 15°C and more buds were produced at 9°C. Higher proportions of medusae relative to buds in warmer temperatures also were seen in other species (Purcell et al., 1999; Ma & Purcell, 2005). Our study, as well as previous laboratory and field studies with other species, reinforces the idea that increasing ocean temperatures could increase the magnitude, frequency, and duration of blooms of many species of jellyfish (reviewed in Purcell, 2005; Purcell et al., 2007). Unfortunately, poor strobilation at 9°C prevented us from testing the combined effects of temperature and pH on strobilation and statolith formation. Conceivably, warm temperatures could exacerbate the effects of low pH. We believe that the poor strobilation at 9°C was due to the lack of temperature change at the beginning of the experiment; therefore, strobilation results at this temperature were considered to be uncharacteristic (see Purcell, 2007).

Effects of pH on asexual reproduction

This is the first published study of the effects of pH on asexual reproduction in a scyphozoan. The pH of seawater had no effect on asexual reproduction, suggesting that a decrease in pH to 7.2 will not affect the production of buds and ephyrae by *A. labiata* polyps. The mesoglea of jellyfish may be able to support diffusion of oxygen to the more metabolically active tissues, giving them a greater tolerance to CO_2 than would be expected (Thuesen et al., 2005; Fabry et al., 2008). Thus, the mesoglea may have aided the polyps in transporting ions and maintaining an acid–base balance in their tissues.

Effects of pH on statolith formation

This is the first experiment to test the effects of pH on statolith formation in a scyphozoan. The pH significantly affected statolith size, with smaller statoliths occurring at lower pH levels. Statolith formation is a

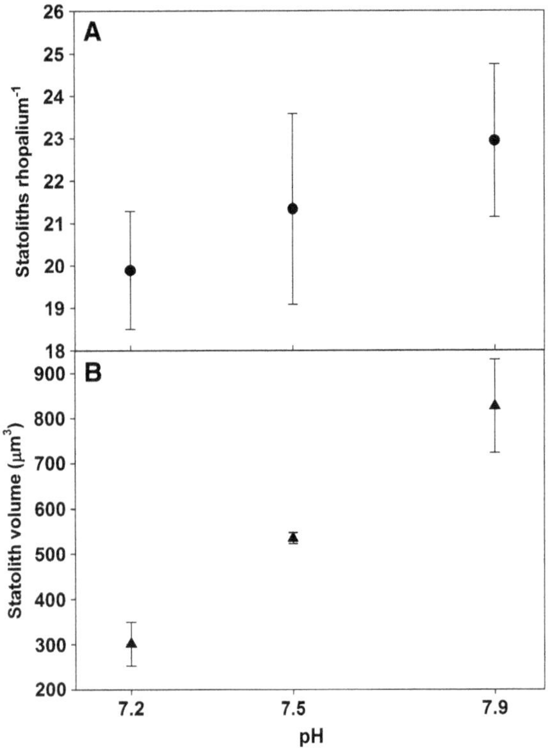

Fig. 2 Effects of pH on numbers of statoliths rhopalium^{-1} (mean ± SE) (**A**) and on statolith lengths and widths (mean ± SE) (**B**) in *Aurelia labiata* ephyrae. Polyps that produced the ephyrae were incubated at 15°C and pH levels of 7.2, 7.5, and 7.9 for 101–120 d. The symbols represent length (*filled circle*) and width (*filled triangle*)

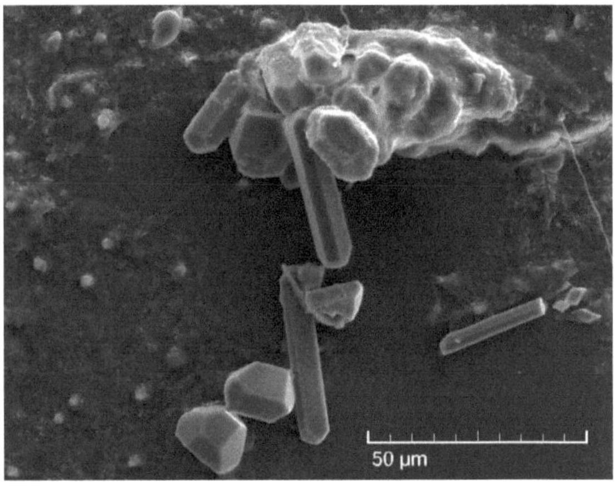

Fig. 3 Scanning electron micrograph of statoliths from an *Aurelia labiata* ephyra produced by a polyp maintained in seawater of 7.2 pH at 15°C for >122 d in the laboratory

complex process, and there are three possible explanations for this result. First, supersaturation of CO_2, which lowers the saturation state of calcium carbonate and limits its precipitation to form shells, would not affect *A. labiata* because its statoliths are composed of calcium sulfate, not calcium carbonate. Spangenberg (1968) found that statolith formation was inhibited in ephyrae from polyps raised in sulfate-deficient water; however, sulfate occurs in high abundance in the ocean. Second, dissolution of exposed calcareous parts, would not affect the statoliths of jellyfish because they are isolated from low pH seawater by their location in the rhopalia. The SEM pictures of *A. labiata* statoliths from an ephyra in our lowest pH (7.2) showed no evidence of dissolution, in contrast to the shells of pteropods exposed to low pH (Fabry et al., 2008). Third, negative effects of acidification on metabolism were

Table 4 Variation in statolith sizes in *Aurelia labiata* ephyrae produced within 1–4 d from individual polyps

		pH 7.2	pH 7.5		pH 7.9	
Age range of ephyrae (d)		1–3	1–3		1–4	
No. statoliths analyzed		308	440		471	
Statolith length (μm)		10.00 (0.17)	12.14 (0.18)		19.08 (0.23)	
Statolith width (μm)		6.45 (0.11)	6.59 (0.11)		8.52 (0.13)	
Statolith volume (μm^3)		309.00 (11.74)	420.47 (16.60)		1051.38 (31.37)	
Effects	Test statistic	*P* value	Test statistic	*P* value	Test statistic	*P* value
E	$F_{5,12} = 3.472$	***P* = 0.035**	$F_{5,12} = 14.917$	***P* < 0.0001**	$F_{5,12} = 3.185$	***P* = 0.045**
R (E)	$F_{12,290} = 0.924$	$P = 0.523$	$F_{12,422} = 0.640$	$P = 0.808$	$F_{12,453} = 0.930$	$P = 0.517$

Polyps (1 per pH) that produced the ephyrae (6 per pH) were grown at 15°C and pH levels of 7.2, 7.5, and 7.9 for 108–115 d. $P < 0.05$ is significant, except for pH 7.5, where $P < 0.025$ is significant. Data are means (±SE)

E ephyra, *R* rhopalium

Bold *P* values are significant

most likely to cause smaller statoliths in *A. labiata* ephyrae in lowered pH. Low pH can cause acidosis, which can suppress metabolism (Portner et al., 2005). When this happens, metabolically expensive processes often are shut down (Fabry et al., 2008). Similarly, reduced ability of the polyps and ephyrae to regulate acids and bases in their tissues could have affected the statolith formation (Fabry et al., 2008). Also, if low pH reduced the polyps' ability to feed, that could reduce their metabolism and statolith formation. All polyps were fed to saturation but ingestion was not quantified and could have differed among treatments. Spangenberg (1984) observed that the number of statoliths produced was reduced by a few days without food, but the sizes of the statoliths were not measured. In our experiment, the numbers of statoliths per rhopalium decreased with pH, but not significantly.

Water chemistry

Changes during polyp incubation occurred in DO, pH, and salinity. pH decreased during incubations, probably due to respiration (CO_2 produced) by the polyps and bacteria; however, the pH levels remained significantly different from each other. DO always decreased between water changes due to respiration of polyps and bacteria. Although DO < 2 mg l^{-1} is considered hypoxic and detrimental to many organisms (Rabalais et al., 2002), excursions into hypoxic conditions (minimum 1.08 mg l^{-1} DO) occurred infrequently in our experiment (in 15°C: five times in 7.2 pH, five times in 7.5 pH, and twice in 7.9 pH) and were brief (<4 d). We believe that hypoxia was unlikely to have affected polyp health because of the brief exposures and because scyphozoan polyps have been shown to be very tolerant of low DO (Condon et al., 2001). Salinity changes (0.2–0.3) in our treatments were small and unlikely to affect the polyps.

Because the pH in our experiment was manipulated with an acid and not CO_2, our results may differ from those in actual ocean acidification. Manipulation of pH with strong acids and bases causes a decrease in alkalinity with pH. Similar alkalinity differences were reported in experiments on pH and coccolithophores by Riebesell et al. (2000), in which the addition of HCl and NaOH caused alkalinity to change by −6.4% and +3.1%, respectively. Similar changes (5–7%) occurred in our experiment. Both acid and CO_2 have been used to manipulate pH to test the effects of acidification on organisms (Gattuso & Lavigne, 2009). Studies that compared these methods found differences between results from the two methods. The Japanese flounder was tested for 48 h in water with a pH of 6.8; they had 100% mortality in water adjusted with CO_2 but no mortality in water adjusted with H_2SO_4 (Hayashi et al., 2004). Mortality rates were much higher in water adjusted with CO_2 than in water adjusted with HCl for another marine fish (20–38 times greater for larvae in pH 5.9 for 24 h and 24–108 times greater for eggs in pH 6.2 for 6 h) (Kikkawa et al., 2004). Portner et al. (2005) explains that such differences may be because CO_2 diffuses into the organism more rapidly than H^+ ions. It is unknown if these pH-manipulation techniques still would have different effects in an experiment as long as ours (122 d). The above comparisons of acid- and CO_2-manipulation of pH suggest that the results of our experiment may underestimate any possible effects of pH on *A. labiata* polyps.

Possible worldwide implications

As climate change progresses, many different variables come into play when considering the future of gelatinous zooplankton. As prior papers have suggested, gelatinous zooplankton seem to have increased in the past few decades in connection with temperature increases and other possible causes (e.g., Mills, 2001; Lynam et al., 2004; Purcell, 2005; Purcell et al., 2007; Brodeur et al., 2008a). Our study was consistent with earlier results on *A. labiata* in which more jellyfish were produced in warm temperatures (Purcell, 2007); thus, warming could increase blooms of this species in nature.

In general, gelatinous zooplankton appear to tolerate, and perhaps benefit from, changes humans have brought to the oceans, such as eutrophication, hypoxia, pollution, and overfishing (Mills, 2001; Purcell, 2005; Purcell et al., 2007). Our study showed that low pH had no significant effects on survival or asexual reproduction of *A. labiata* polyps. This suggests that *A. labiata* polyp populations would tolerate a pH decrease to 7.2; however, decreased pH still may detrimentally affect the jellyfish. At reduced pH levels, significantly smaller statoliths were produced. The effects of small statoliths on survival of ephyrae are unknown. Spangenberg (1968) saw swimming abnormalities in *A. aurita* ephyrae that had fewer than normal statoliths, but statolith sizes were not measured. Gu et al. (2005) found reduced swimming and feeding below pH 7.4 in the swimming and feeding of *Rhopilema esculenta* Kishinouye scyphomedusae.

Recent papers discuss how jellyfish populations may respond to ocean acidification in the North Sea and North Atlantic Ocean. Although Attrill et al. (2007) showed a negative correlation of jellyfish frequency and pH in the North Sea from continuous plankton recorder (CPR) data over 43 years and concluded that jellyfish might increase as an indirect consequence of ocean acidification, Richardson & Gibbons (2008) found no significant correlation of jellyfish frequency and pH from CPR data over 57 years in the North Sea and North Atlantic. The validity of the Attrill et al. (2007) analyses was challenged by Haddock (2008), but Attrill & Edwards (2008) concluded they were robust. Attrill et al. (2007) invoked indirect "bottom-up" mechanisms to explain why jellyfish might increase in lower pH. In our experiment, we tested possible direct effects of pH on survival and reproduction of *Aurelia labiata* polyps and found no significant effects, suggesting that this species is quite tolerant of low pH; however, small statoliths produced by *A. labiata* ephyrae in low pH may affect their survival and ultimately the medusa population size.

Organisms may experience low pH conditions more often than realized. A study in Puget Sound found that pH fluctuates from 7.2 to 9.6 in 2.4-m-deep water over the span of a couple of days (K. L. Van Alstyne, SPMC, pers. comm.). With such large pH fluctuations due to plant photosynthesis during the day and respiration at night, many organisms may be exposed to low pH conditions routinely.

Much more research is needed to determine the fate of jellyfish under the influence of ocean acidification. Different species and populations of jellyfish may have different responses to environmental variables (Liu et al., 2009). Effects of low pH on various organisms have been tested in acute experiments; however, organisms may be able to acclimate or adapt to slowly changing pH conditions. CO_2 should be used to adjust pH and carbonate chemistry to simulate ocean conditions more accurately and because CO_2-adjusted pH had greater toxic effects than did acid-adjusted pH (Hayashi et al., 2004; Kikkawa et al., 2004). The effects of small statoliths on swimming and feeding of the ephyrae need to be determined. If small statoliths negatively affect *A. labiata* ephyra survival, then ocean acidification could be detrimental to this species and counteract the probable positive effects of warming and other factors (reviewed in Purcell, 2005; Purcell et al., 2007). Hopefully our research will encourage further research to determine how ocean acidification will affect jellyfish populations.

Acknowledgments We thank Dr. V. J. Fabry for inviting J. E. P. to the Ocean Carbon and Biogeochemistry Scoping Workshop on Ocean Acidification Research in 2007, and her advice on methodology for this research. We thank Drs. Gisele Muller-Parker, Benjamin Miner, and Brian Bingham for their valuable comments. We are indebted to the following people for their assistance; Rachel Garcia, Ritupreet Virk, Brandy Carpenter, Jen Moore, Dr. Gary Winans, Nathan Schwarck, Gene McKeen, Carissa Haug, Jeannie Gilbert, Peter Thut, Chandler Colahan, Alyssa Gehman, Sylvia Graham, Lauren Chomiczewski, and Carmen Guerra. Jellyfish polyps were provided by the Seattle Aquarium with help from Kathryn Kegel and Roland Anderson. We also thank Dr. K. L. Van Alstyne for sharing her information on diel intertidal pH levels. Financial support was provided by Western Washington University through the RSP Fund for the Enhancement of Graduate Research and the Biology Faculty Fellowship Fund.

References

Arai, M. N., 1997. A Functional Biology of Scyphozoa. Chapman & Hall, London.

Attrill, M. J. & M. Edwards, 2008. Reply to Haddock. S.H.D., reconsidering evidence for potential climate-related increases in jellyfish. Limnology and Oceanography 53: 2763–2766.

Attrill, M. J., J. Wright & M. Edwards, 2007. Climate-related increases in jellyfish frequency suggest a more gelatinous future for the North Sea. Limnology and Oceanography 52: 480–485.

Becker, A., I. Sotje, C. Paulmann, F. Beckmann, T. Donath, R. Boese, O. Prymak, H. Tiemann & M. Epple, 2005. Calcium sulfate hemihydrate is the inorganic mineral in statoliths of scyphozoan medusae (Cnidaria). Dalton Transactions 1545–1550.

Bindoff, N. L., J. Willebrand, V. Artale, A. Cazenave, J. Gregory, S. Gulev, K. Hanawa, C. Le Quéré, S. Levitus, Y. Nojiri, C. K. Shum, L. D. Talley & A. Unnikrishnan, 2007. Observations: oceanic climate change and sea level. In Solomon, S., D. Qin, M. Manning, Z. Chen, M. Marquis, K. B. Averyt, M. Tignor & H. L. Miller (eds), Climate Change 2007: The Physical Science Basis. Contribution of Working Group I to the Fourth Assessment Report of the Intergovernmental Panel on Climate

Change. Cambridge University Press, Cambridge, UK and New York, NY, USA.
Brodeur, R. D., M. B. Decker, L. Ciannelli, J. E. Purcell, N. A. Bond, P. J. Stabeno, E. Acuna & G. L. Hunt Jr, 2008a. Rise and fall of jellyfish in the eastern Bering Sea in relation to climate regime shifts. Progress in Oceanography 77: 103–111.
Brodeur, R. D., C. L. Suchman, D. C. Reese, T. W. Miller & E. A. Daly, 2008b. Spatial overlap and trophic interactions between pelagic fish and large jellyfish in the northern California Current. Marine Biology 154: 649–659.
Caldeira, K. & M. E. Wickett, 2003. Anthropogenic carbon and ocean pH. Nature 425: 365.
Condon, R. H., M. B. Decker & J. E. Purcell, 2001. Effects of low dissolved oxygen on survival and asexual reproduction of scyphozoan polyps (*Chrysaora quinquecirrha*). Hydrobiologia 451: 89–95.
Doney, S. C., V. J. Fabry, R. A. Feely & J. A. Kleypas, 2009. Ocean acidification: the other CO_2 problem. Annual Review of Marine Science 1: 162–192.
Fabry, V. J., B. A. Seibel, R. A. Feely & J. C. Orr, 2008. Impacts of ocean acidification on marine fauna and ecosystem processes. ICES Journal of Marine Science 65: 414–432.
Feely, R. A., C. L. Sabine, K. Lee, W. Berelson, J. Kleypas, V. J. Fabry & F. J. Millero, 2004. Impact of anthropogenic CO_2 on the $CaCO_3$ system in the oceans. Science 305: 362–366.
Forster, P., V. Ramaswamy, P. Artaxo, T. Berntsen, R. Betts, D. W. Fahey, J. Haywood, J. Lean, D. C. Lowe, G. Myhre, J. Nganga, R. Prinn, G. Raga, M. Schulz & R. Van Dorland, 2007. Changes in atmospheric constituents and in radiative forcing. In Solomon, S., D. Qin, M. Manning, Z. Chen, M. Marquis, K. B. Averyt, M. Tignor & H. L. Miller (eds), Climate Change 2007: The Physical Science Basis. Contribution of Working Group I to the Fourth Assessment Report of the Intergovernmental Panel on Climate Change. Cambridge University Press, Cambridge, UK and New York, NY, USA.
Gamst, G., L. S. Meyers & A. J. Guarino, 2008. Analysis of Variance Designs: A Conceptual and Computational Approach with SPSS and SAS. Cambridge University Press, Cambridge.
Gattuso, J.-P. & H. Lavigne, 2009. Technical note: approaches and software tools to investigate the impact of ocean acidification. Biogeosciences 6: 2121–2133.
Gattuso, J.-P., M. Frankignoulle, I. Bourge, S. Romaine & R. W. Buddemeier, 1998. Effect of calcium carbonate saturation of seawater on coral calcification. Global and Planetary Change 18: 37–46.
Gu, L., M. Guo, H.-y. Lu, Z.-z. Sun & Y.-z. Lei, 2005. The physical-chemical factors affecting jellyfish *Rhopilema esculenta* breeding. Journal of Dalian Fisheries University 20: 41–44.
Haddock, H. S. D., 2008. Reconsidering evidence for potential climate-related increases in jellyfish. Limnology and Oceanography 53: 2759–2762.
Hayashi, K., J. Kita & A. Ishimatsu, 2004. Comparison of the acid–base responses to CO_2 and acidification in Japanese flounder (*Paralichthys olivaceus*). Marine Pollution Bulletin 49: 1062–1065.
Holst, S., I. Sotje & H. Tiemann, 2007. Life cycle of the rhizostome jellyfish *Rhizostoma octopus* (L.) (Scyphozoa, Rhizostomeae), with studies on cnidocysts and statoliths. Marine Biology 151: 1695–1710.
Jokiel, P. L., K. S. Rodgers, I. B. Kuffner, A. J. Anderson, E. F. Fox & F. T. Mackenzie, 2008. Ocean acidification and calcifying reef organisms: a mesocosm investigation. Coral Reefs 27: 473–483.
Keppel, G. & T. D. Wickens, 2004. Design and Analysis: A Researcher's Handbook, 4th ed. Prentice Hall, Upper Saddle River.
Kikkawa, T., J. Kita & A. Ishimatsu, 2004. Comparison of the lethal effect of CO_2 and acidification on red sea bream (*Pagrus major*) during the early developmental stages. Marine Pollution Bulletin 48: 108–110.
Kleypas, J. A., R. W. Buddemeier, D. Archer, J.-P. Gattuso, C. Langdon & B. N. Opdyke, 1999. Geochemical consequences of increased atmospheric carbon dioxide on coral reefs. Science 284: 118–120.
Kuffner, I. B., A. J. Anderson, P. L. Jokiel, K. S. Rodgers & F. T. Mackenzie, 2008. Decreased abundance of crustose coralline algae due to ocean acidification. Nature Geoscience 1: 114–117.
Langdon, C., T. Takahashi, C. Sweeney, D. Chipman & J. Goddard, 2000. Effect of calcium carbonate saturation state on the calcification rate of an experimental coral reef. Global Biogeochemical Cycles 14: 639–654.
Liu, W.-C., W.-T. Lo, J. E. Purcell & H.-H. Chang, 2009. Effects of temperature and light intensity on asexual reproduction of the scyphozoan, *Aurelia aurita* (L.) in Taiwan. Hydrobiologia 612: 247–258.
Lucas, C. H., 2001. Reproduction and life history strategies of the common jellyfish, *Aurelia aurita*, in relation to its ambient environment. Hydrobiologia 451: 229–246.
Lynam, C. P., S. J. Hay & A. S. Brierley, 2004. Interannual variability in abundance of North Sea jellyfish and links to the North Atlantic Oscillation. Limnology and Oceanography 49: 637–643.
Ma, X. & J. E. Purcell, 2005. Temperature, salinity and prey effects on polyp versus medusa bud production by the invasive hydrozoan *Moerisia lyonsi*. Marine Biology 147: 225–234.
Metzger, R., F. J. Sartoris, M. Langenbuch & H. O. Portner, 2007. Influence of elevated CO_2 concentrations on thermal tolerance of the edible crab *Cancer paguris*. Journal of Thermal Biology 32: 144–151.
Michaelidis, B., C. Ouzounis, A. Paleras & H. O. Portner, 2005. Effects of long-term moderate hypercapnia on acid–base balance and growth rate in marine mussels *Mytilus galloprovincialis*. Marine Ecology and Progress Series 293: 109–118.
Mills, C. E., 2001. Jellyfish blooms: are populations increasing globally in response to changing ocean conditions? Hydrobiologia 451: 55–68.
Orr, J. C., V. J. Fabry, O. Aumont, L. Bopp, S. C. Doney, R. A. Feely, A. Gnanadesikan, N. Gruber, A. Ishida, F. Joos, R. M. Key, K. Lindsay, E. Maier-Reimer, R. Matear, P. Monfray, A. Mouchet, R. G. Najjar, G.-K. Plattner, K. B. Rodgers, C. L. Sabine, J. L. Sarmiento, R. Schlitzer, R. D. Slater, I. J. Totterdell, M.-F. Weirig, Y. Yamanaka & A. Yool, 2005. Anthropogenic ocean acidification over the

twenty-first century and its impact on calcifying organisms. Nature 437: 681–686.

Palmer, M., 1992. Standard Operating Procedure for GLNPO Total Alkalinity Titration. United States Environmental Protection Agency, Chicago.

Portner, H. O., M. Langenbuch & B. Michaelidis, 2005. Synergistic effects of temperature extremes, hypoxia, and increases in CO_2 on marine animals: From Earth history to global change. Journal of Geophysical Research 110: C09S10.

Purcell, J. E., 1997. Pelagic cnidarians and ctenophores as predators: selective predation, feeding rates and effects on prey populations. Annales de l'Institut océanographique, Paris 73: 125–137.

Purcell, J. E., 2005. Climate effects on formation of jellyfish and ctenophore blooms: a review. Journal of the Marine Biological Association of the United Kingdom 85: 461–476.

Purcell, J. E., 2007. Environmental effects on asexual reproduction rates of the scyphozoan *Aurelia labiata.* Marine Ecology Progress Series 348: 183–196.

Purcell, J. E. & J. J. Grover, 1990. Predation and food limitation as causes of mortality in larval herring at a spawning ground in British Columbia. Marine Ecology Progress Series 59: 55–61.

Purcell, J. E. & M. V. Sturdevant, 2001. Prey Selection and dietary overlap among zooplanktivorous jellyfish and juvenile fishes in Prince William Sound, Alaska. Marine Ecology Progress Series 210: 67–83.

Purcell, J. E., J. R. White, D. A. Nemazie & D. A. Wright, 1999. Temperature, salinity and food effects on asexual reproduction and abundance of the scyphozoan *Chrysaora quinquecirrha.* Marine Ecology Progress Series 180: 187–196.

Purcell, J. E., S.-i. Uye & W.-T. Lo, 2007. Anthropogenic causes of jellyfish blooms and their direct consequences for humans: a review. Marine Ecology Progress Series 350: 153–174.

Rabalais, N. N., R. E. Turner & W. J. Wiseman Jr., 2002. Gulf of Mexico hypoxia A.K.A. "the dead zone". Annual Review of Ecology and Systematics 33: 235–263.

Richardson, A. J. & M. J. Gibbons, 2008. Are jellyfish increasing in response to ocean acidification? Limnology and Oceanography 53: 2040–2045.

Riebesell, U., I. Zondervan, B. Rost, P. D. Tortell, R. E. Zeebe & F. M. M. Morel, 2000. Reduced calcification of marine plankton in response to increased atmospheric CO_2. Nature 407: 364–367.

Rost, B., I. Zondervan & D. Wolf-Gladrow, 2008. Sensitivity of phytoplankton to future changes in ocean carbonate chemistry: current knowledge, contradictions, and research directions. Marine Ecology Progress Series 373: 227–237.

Royal Society, 2005. Ocean Acidification Due to Increasing Atmospheric Carbon Dioxide. The Royal Society, London.

Shirayama, Y. & H. Thornton, 2005. Effect of increased atmospheric CO_2 on shallow water marine benthos. Journal of Geophysical Research 110: C09S08.

Spangenberg, D. B., 1968. Statolith differentiation in *Aurelia aurita.* Journal of Experimental Zoology 169: 487–500.

Spangenberg, D. B., 1976. Intracellular statolith synthesis in *Aurelia aurita.* In Watabe, N. & K. Wilbur (eds), The Mechanisms of Mineralization in the Invertebrates and Plants. University of South Carolina Press, Columbia: 231–248.

Spangenberg, D. B., 1984. Effects of exogenous thyroxine on statolith synthesis and resorption in *Aurelia.* American Zoologist 24: 917–923.

Spangenberg, D. B. & C. W. Beck, 1968. Calcium sulfate dehydrate statoliths in *Aurelia.* Transactions of the American Microscopical Society 8: 329–335.

Spangenberg, D. B. & C. W. Beck, 1972. Tetracycline effects on statolith and nematocyst differentiation in *Aurelia.* Calcified Tissue Research 9: 122–133.

Thuesen, E. V., L. D. Rutherford Jr., P. L. Brommer, K. Garrison, M. A. Gutowska & T. Towanda, 2005. Intragel hypoxia promotes hypoxia tolerance of scyphomedusae. The Journal of Experimental Biology 208: 2475–2482.

Hydrobiologia (2010) 645:53–68
DOI 10.1007/s10750-010-0214-y

Effects of low salinity on settlement and strobilation of Scyphozoa (Cnidaria): Is the lion's mane *Cyanea capillata* (L.) able to reproduce in the brackish Baltic Sea?

Sabine Holst · Gerhard Jarms

Published online: 26 March 2010

Abstract Several species of scyphozoan medusae occur in river estuaries and other brackish waters but it is often unknown if the planulae settle and the scyphopolyps reproduce in those low-salinity waters. In the present study, scyphozoan species from the German Bight (North Sea) were tested in laboratory experiments to investigate their tolerance of low salinity. Planula larvae released from medusae in salinity 32 were still active after the salinity was reduced to 10 (*Cyanea capillata*, *Cyanea lamarckii*) and to 7 (*Chrysaora hysoscella*) in laboratory treatments. Planulae did not settle on the undersides of floating substrates when salinity was reduced to <20. By contrast, planulae released from *C. capillata* medusae in Kiel Bight (western Baltic Sea) in salinity 15 developed into polyps in laboratory cultures. Polyps reared from planulae in salinity 36 survived a reduction to 12 (*C. capillata*, *C. lamarckii*) and to 8 (*Aurelia aurita*). Polyps of all tested species strobilated and released young medusae (ephyrae) in salinity 12. These results show a high tolerance of planulae and polyps to low salinity, indicating their possible occurrence in estuaries and brackish waters. In addition to laboratory observations, young *C. capillata* ephyrae were collected in the western Baltic Sea (Kiel Bight) in salinity 15, which indicates that they were probably released by a local polyp population. We suggest that the polyps of the painfully stinging lion's mane, *C. capillata*, may be more widespread in the Baltic Sea than previously assumed and that the occurrence of the medusae may not only depend on inflow of water masses from the North Sea.

Keywords Scyphozoa · Jellyfish · Salinity · *Cyanea capillata* · Settlement · Strobilation

Guest editors: J. E. Purcell & Dror Angel / Jellyfish Blooms: New Problems and Solutions

S. Holst (✉)
Senckenberg am Meer, German Centre for Marine Biodiversity Research, c/o Biozentrum Grindel und Zoologisches Museum, Martin-Luther-King Platz 3, 20146 Hamburg, Germany
e-mail: sabine.holst@senckenberg.de

G. Jarms
Biozentrum Grindel und Zoologisches Museum, Martin-Luther-King Platz 3, 20146 Hamburg, Germany

Introduction

The majority of scyphozoan species have a metagenetic life cycle comprising a conspicuous pelagic medusa generation and an inconspicuous sessile polyp generation. Knowledge about the ecology and distribution of scyphozoan polyps is poor; in most species, the polyp stage never has been found in the field. The complete life cycle is described in <25% of known scyphozoan species (Tronolone et al., 2002).

Water salinity is an important environmental factor limiting the distribution of marine species in brackish waters. Species numbers of marine animals are reduced in brackish waters with a species minimum in salinities 5–8. In this narrow salinity range, marine forms do not usually penetrate into lower salinities and freshwater forms are absent at higher salinities (Khlebovich, 1969). Physiological stress from low salinity reduces growth rates and final sizes in several marine species that immigrate into brackish-water habitats, e.g., mussels, starfish, and polychaetes (Groth & Thede, 1989). The number of Hydrozoa occurring in brackish-water seas and estuaries is low compared with marine habitats (Dumont, 1994). Several species of scyphozoan medusae occur in river estuaries and brackish waters, e.g., in the Baltic Sea (Möller, 1980) and adjacent waters (Kertinge Nor, Olesen et al., 1994), the Elbe estuary (Merck, 1989), Chesapeake Bay (Purcell et al., 1999), the Niantic River (Brewer, 1989; Brewer & Feingold, 1991; Colin & Kremer, 2002), and the Yangtze estuary (Xian et al., 2005), but it often is unknown if establishment and reproduction of the sessile stage occurs at the low salinities in these waters.

The Baltic Sea represents the world's largest brackish-water sea area, with a total surface of 382,000 km^2. The salinity decreases dramatically from the western to the northeastern regions. In the northwest, the Baltic Sea is connected to the North Sea through the Skagerrak and the Kattegat (Fig. 1). In the Skagerrak salinities up to 35 are usual (Danielssen et al., 1996). Salinity of the Kattegat surface water is about 26, being influenced by the Baltic Current, low-salinity coastal currents, and local river inflow (Jakobsen, 1997). In the Bornholm Deep (central Baltic Sea), salinity is $<$18 in bottom waters and eight in surface waters. In the Gulf of Bothnia, salinity decreases to 5–8 in the south and 3–5 in the north (Fonselius & Valderrama, 2003).

Three species of scyphozoan medusae usually occur in the Baltic Sea during the summer months, the moon jellyfish, *Aurelia aurita* (L.), the lion's mane, *Cyanea capillata* (L.), and the blue jellyfish, *Cyanea lamarckii* Péron & Lesueur, 1809. *Chrysaora hysoscella* (Linnaeus, 1766) and *Rhizostoma octopus* (L.) medusae have been recorded infrequently in the Skagerrak, the Kattegat, and the Baltic Sea (Kramp, 1934; Russell, 1970). *Aurelia aurita* medusae occur throughout the Baltic Sea (Möller, 1980). Polyps and ephyrae of this species were documented up to the Gulf of Finland in salinity 5.5 (Wikström, 1932; Palmén, 1954). Therefore, *A. aurita* certainly reproduces in most areas of the Baltic Sea. By contrast, *C. lamarckii* medusae only occur along the west coast of Sweden in most years (Gröndahl, 1988) and sometimes along the Danish coast (Rasmussen, 1973). Settlement and reproduction of the sessile stage of *C. lamarckii* do not occur on the west coast of Sweden, suggesting that its polyps also may not reproduce in the Baltic Sea (Gröndahl, 1988). By contrast, settlement of *C. capillata* planulae was observed on the west coast of Sweden, but small numbers of polyps and ephyrae suggested that this area is the northern limit for reproduction of *C. capillata* (Gröndahl & Hernroth, 1987; Gröndahl, 1988). The optimum salinity to culture *C. capillata* polyps from Chesapeake Bay was 20 and polyps were not found in the field in salinities much below 20 (Cargo, 1984). Thus, others suggested that *C. capillata* polyps could not reproduce in the Baltic Sea in salinities below 20 (Båmstedt et al., 1997; Barz et al., 2006); however, *C. capillata* medusae occur up to the Gulf of Finland in salinities below 10 (Verwey, 1942; Haahtela & Lassig, 1967; Möller, 1980).

Reproduction of scyphozoan species in brackish waters could be limited by the inability of planula larvae to survive and settle or the inability of polyps to develop and strobilate at low salinities. In this study, we investigated the lower salinity limits for the settlement of planulae and for the budding and strobilation activity of scyphozoan polyps from the German Bight (North Sea). In addition, from Kiel Bight (western Baltic Sea), we documented very young *C. capillata* medusae and planulae collected from mature females there were reared at the low salinity of 15 in the laboratory.

Materials and methods

Salinity was measured with a hand refractometer in all samples and experiments.

Effects of reduced salinity on planula settlement

In July and August 2005, medusae were captured from waters of salinity 32 around the island of Helgoland, German Bight, North Sea (Fig. 1). Three species were collected and transported in individual buckets to the Biologische Anstalt Helgoland (part of the Alfred Wegener Institute for Polar and Marine

Fig. 1 Map of the Baltic Sea and the eastern North Sea (created with ArcGis 9, ArcMap 9.2.) with connections through the Skagerrak and the Kattegat, and study areas on scyphozoans in Helgoland (German Bight, the North Sea), and Kiel (western Baltic Sea)

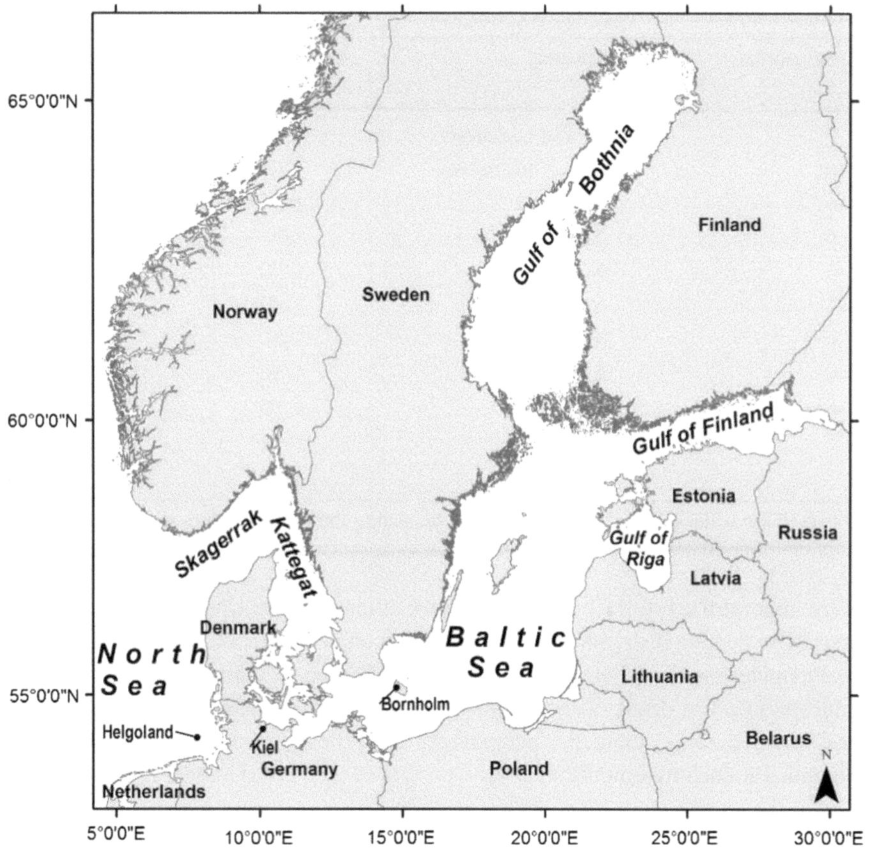

Research). Individual females of *C. lamarckii* and *Ch. hysoscella* were transferred into 2,000-ml glass beakers filled with filtered seawater of salinity 32, and *C. capillata* females were transferred into individual 30-l aquariums. After 1 h, released planulae of each species were collected with pipettes from the container bottoms and transferred into 500-ml beakers. 10 ml of this planula suspension was pipetted into each of 36 plastic 60-ml jars. The planulae were cultured in an incubator at 15°C during the period of salinity reduction.

Salinity treatments for each species were carried out in 10 steps (Table 1) to reach five salinity levels (32, 25, 20, 15, 10, and 5). In treatment 1 (control), the salinity 32 was not changed. In treatments 2–6, the salinity was reduced in steps by adding de-ionized water of 15°C to the seawater in the jars. At each step, equal water volumes were added to all jars, including the control. Changes in the swimming activity, shape, and size of planulae were observed before and after each addition by use of a stereomicroscope; apparent changes were documented photographically.

Settlement experiments were conducted at room temperature (22 ± 3°C) at five salinities, each with six replicates, in 55-mm-diameter polystyrene Petri dishes. Because scyphozoan planulae prefer to settle on the undersides of substrates (e.g., Brewer, 1976, 1984; Holst & Jarms, 2007), Petri dishes were prepared as in Fig. 2. Before beginning the experiments, the lids of all dishes were filled with seawater (10 ml at salinity 32) and the dish bottoms were placed on the water surface to allow bacterial films to grow on the undersides. The water from 18 Petri dish lids (treatments 1–3) was replaced after 48 h. Planulae in seawater (10 ml) from treatments 1–3 (Table 1) were transferred into Petri dish lids and Petri dish bottoms were replaced on the water surface. The same protocol was used for planulae acclimated to salinity 15 (treatment 4) and 10 (treatment 5), which were transferred into Petri dish lids at 72 and 96 h, respectively. Planulae settled on the undersides of the floating Petri dish bottom were visible through the transparent plastic and were counted with the aid of a stereomicroscope. Counts

Table 1 Schedule of steps of salinity reduction in six treatments with scyphozoan planula larvae

Treatment no.	Schedule	Water volume (ml)	Total water volume (ml)	Salinity
1	Start of treatments	0	10	32
2–6	Start of treatments	1.38	11.38	28
2–6	After 14 h	1.31	12.69	25
3–6	After 24 h	1.66	14.35	22
3–6	After 38 h	1.37	15.71	20
4–6	After 48 h	2.62	18.33	17
4–6	After 62 h	2.29	20.63	15
5–6	After 72 h	4.76	25.38	12
5–6	After 86 h	4.62	30.00	10
6	After 96 h	11.25	41.25	7
6	After 110 h	13.75	55.00	5

10 ml of sea water (salinity 32) were diluted by adding the listed volumes of de-ionized water. The total water volumes and the salinities after dilution are given

were repeated every 24 h on days 3 to 6 after the start. The floating dish bottoms were removed on day 6. Planulae remaining in the lids were fixed by the addition of a few drops of formalin (4% in seawater) and counted to calculate the percentages of settled planulae in each treatment.

Effects of reduced salinity on polyp reproduction, mortality, and strobilation

Polyps of the species *A. aurita*, *C. capillata*, and *C. lamarckii* were reared from planulae collected from female medusae in June and July 2004 around the island of Helgoland, as above. Polyps settled on watch glasses or polyethylene plates in laboratory cultures were transported to the Zoological Institute of the University of Hamburg (Biocenter Grindel). These colonized substrates were transferred into 150-ml glass bowls and cultured in filtered sea water in salinity 36 ± 2 during the first year after settlement. Polyps were fed nauplii of *Artemia salina* every 7–10 d. In October and November 2005, polyps were acclimated to reduced salinities. Filtered seawater was diluted with de-ionized water and colonized substrates were transferred into glass bowls filled with the reduced-salinity seawater. The salinity reductions from 36 to 28, 20, 12, and 8 were conducted in a stepwise manner during 1 month (Table 2), during which all cultures were kept at 15°C in an incubator. Visible changes in these polyps, in comparison with control polyps cultured without salinity reduction, were documented photographically for 3 months (mid-November 2005 to mid-February 2006) after the salinity reduction. In cultures with a salinity reduction to 8, the salinity was increased to 12 after 3 months (mid-February 2006). Polyps were counted twice each month to document polyp reproduction and mortality rates at different salinities.

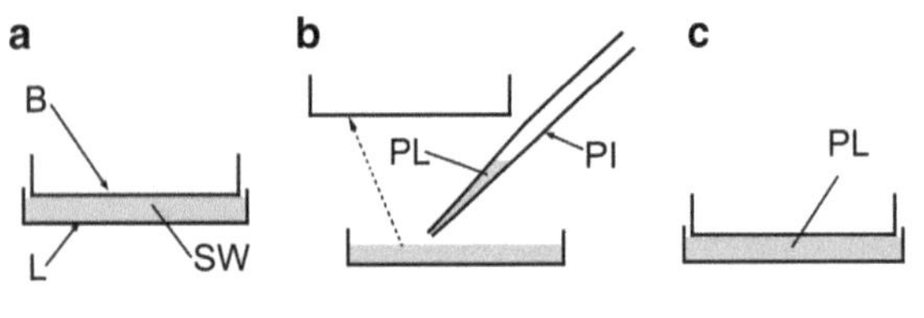

Fig. 2 Design of settlement experiments on scyphozoan planulae. **a** Petri dish lid filled with 10 ml of filtered natural sea water (salinity 32) and dish bottom placed on the water surface; **b** Petri dish bottom removed from the water surface and the seawater exchanged by 10 ml of planulae cultured in seawater of different salinities; **c** Petri dish bottoms replaced on the water surface. *B* petri dish bottom, *L* petri dish lid, *PI* pipette, *PL* planulae in seawater of different salinities, *SW* natural filtered seawater of salinity 32

Strobilation of polyps cultured in salinities 36, 28, 20, and 12 (Table 3) was initiated 1 week after the salinity reduction was completed by reducing the water temperature in the cultures in steps of 2.5°C. Each step of cooling was followed by 2-wk acclimation. The temperature was lowered from 15 to 10°C within 1 month in *A. aurita* and *C. lamarckii* cultures, and from 15 to 5°C within 2 months in *C. capillata* cultures. Strobilating polyps and released ephyrae

Table 2 Schedule of steps of salinity reduction in scyphozoan polyp cultures

	Date of salinity reduction										
	14 Oct	16 Oct	18 Oct	20 Oct	22 Oct	24 Oct	26 Oct	31 Oct	4 Nov	9 Nov	14 Nov
Salinity	33	30	28	25	22	20	18	15	12	10	8
	–	–	33	30	28	25	22	20	18	15	12
	–	–	–	–	–	33	30	28	25	22	20
	–	–	–	–	–	–	–	–	33	30	28

Colonized substrates (watch glasses and polyethylene plates) were transferred into bowls filled with seawater of the listed salinity to reach the final culture salinities of 8, 12, 20, and 28 on November 14

Table 3 Total and mean numbers (±SD of six replicates) of scyphozoan polyps cultured at different salinities to investigate the survival, reproduction, and strobilation during the experiment, which began in November 21, 2005

Species	Salinity	Polyp numbers		Duration (d)
		Tot. no.	Mean ± SD	
Aurelia aurita	36	162	27.0 ± 5.7	170
	28	191	31.8 ± 6.3	
	20	198	33.0 ± 6.0	
	12	205	34.2 ± 6.4	
	8–12	177	29.5 ± 4.8	
Cyanea capillata	36	148	24.7 ± 8.0	229
	28	166	27.7 ± 8.3	
	20	159	26.5 ± 10.0	
	12	152	25.3 ± 10.3	
Cyanea lamarckii	36	112	18.7 ± 9.3	114
	28	125	20.8 ± 11.0	
	20	123	20.5 ± 10.8	
	12	123	20.5 ± 11.0	

Tot. no. Total number

were counted twice each week in the following months to document the effect of reduced salinity on strobilation.

Investigations on *C. capillata* medusae from Kiel Bight (western Baltic Sea)

Young *C. capillata* ephyrae were captured with hand nets in the marina of Strande (Kiel Bight, Fig. 1) in salinity 15 in early summer (June 6, 2006 and May 24, 2009). The ephyrae were measured and photographed in the laboratory (Zoological Institute, University of Hamburg). Female *C. capillata* planulae medusae carrying planulae in their brood pouches were collected with buckets from Strande marina in salinity 15 in August 2008. Planulae released from the medusae were collected with pipettes from the bucket bottoms and transferred into a 250-ml glass bowl filled with seawater of salinity 15 from Strande marina. Four 40-mm-diameter watch glasses, as substrate for planula settlement, were placed in the bowl in a 15°C incubator.

Statistics

The percentages of settled planulae, percentages of polyps strobilating per initial numbers of polyps, and the mean numbers of ephyrae produced per strobila at the different salinities were tested by multiple comparisons. The following null hypotheses were tested. H_1: the percentages of planula settlement were independent of salinity. H_2: the percentages of polyp strobilation were independent of the salinity. H_3: the numbers of ephyrae produced per polyp were independent of salinity. Percentages were arcsine square root transformed before analysis. Data with normal distributions were tested by one-way analysis of variance (ANOVA) followed by a Fisher's Least Significance Difference (LSD) post-hoc test. Data not normally distributed were tested by a Kruskal–Wallis ANOVA on ranks and a Nemenyi post-hoc test.

Results

Planula settlement

Planulae of all species tested had reduced swimming, increased length, and flattened bodies after salinity reduction from 32 to 10 (Fig. 3). In salinity 10, swimming was weak and *C. capillata* and *C.*

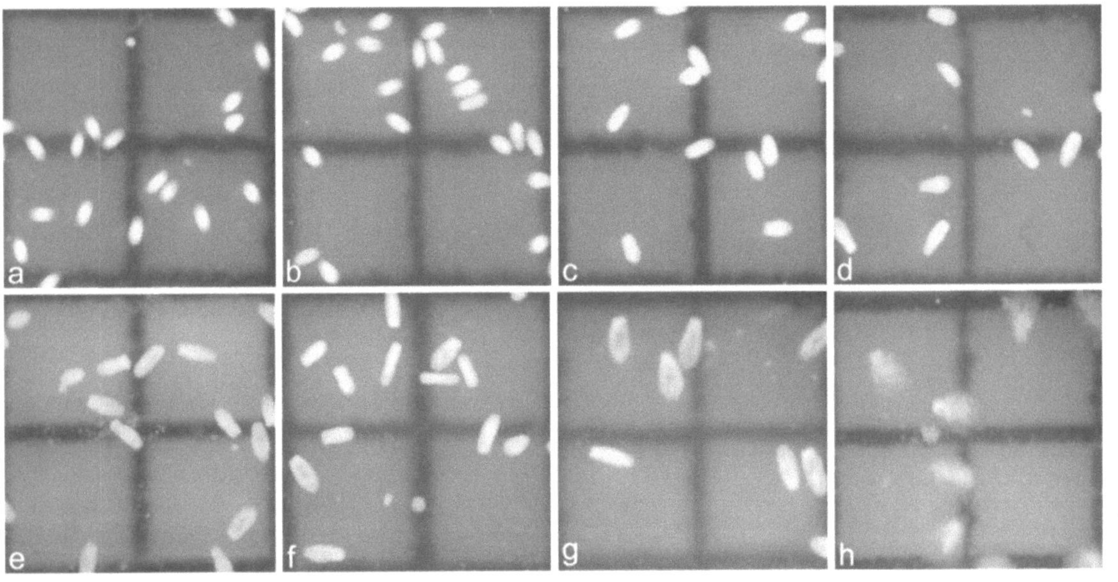

Fig. 3 Changes in shape and size of *Cyanea lamarckii* planulae after stepwise salinity reduction from 32 to 7 within 4 d. **a–f** before each step of salinity reduction as listed in Table 1; salinities in **a** 32, **b** 25, **c** 20, **d** 15, **e** 12, **f** 10. **g** and **h** Disintegrating planulae immediately (**g**) after salinity reduction from 10 to 7 after 30 min (**h**) in salinity 7

lamarckii planulae disintegrated immediately upon further reduction to salinity 7 (Fig. 3g, h). By contrast, planulae of *Ch. hysoscella* still swam very weakly 24 h after salinity reduction from 10 to 7, but disintegrated after a salinity reduction to 5. Planulae settled on the floating Petri dish bottoms in salinities of 32–20 in all tested species (Fig. 4). The percentages settling were smaller for *C. capillata* in all salinities as compared with the other species tested. Planulae of *C. lamarckii* formed cysts after settlement and <10% of cysts directly metamorphosed into polyps in all salinities. Planulae of *C. capillata* and *Ch. hysoscella* did not produce cysts. Planulae of *Ch. hysoscella* were attached to the undersides of the floating substrates in salinity 15 on day 4, but were detached again the next day (Fig. 4). In salinities of 15 and 10, planulae did not settle on the floating substrates and did not develop into polyps, but swimming activity was low until the experiment was terminated (6 d).

The percentages of planulae settling differed significantly among salinities and thus H_1 was rejected for all species tested (Table 4). Post-hoc tests showed that the percentages of settled planulae differed significantly between higher salinities (32 and 25) and 20 for *C. lamarckii*. For *C. capillata*, the percentages of planulae settling differed significantly between the control without a salinity reduction (32) and reduced salinities (25 and 20). For *Ch. hysoscella*, differences were significant among all salinities tested.

Polyp survival

Polyps of all species tested survived salinity reduction from 36 to 12. The polyps were able to catch and feed on *Artemia* nauplii in salinity 12, although the tentacles were shorter than tentacles in salinity 36 (Fig. 5). Polyps of *C. capillata* and *C. lamarckii* died after salinity reduction from 12 to 10. Polyps of *A. aurita*, however, survived a salinity reduction to 8 for 3 months without eating. The tentacles, which were reduced to short stubs in salinity 8 (Fig. 5c), regenerated after salinity was increased to 12 and polyps again were able to feed.

The initial polyp numbers (Table 3) increased by budding or excystment and decreased from mortality following the various salinity changes (Fig. 6). Polyps of *A. aurita* increased by 4–18% in salinities of 12–36; however, their polyps in salinity 8 decreased by 12% after 3 months, and then salinity was increased to 12. Thereafter, little mortality occurred and totaled 14% of polyps at the end of the experiment in May. Polyp reproduction was high in *C. capillata* cultures in salinities of 20–36 where numbers had increased by

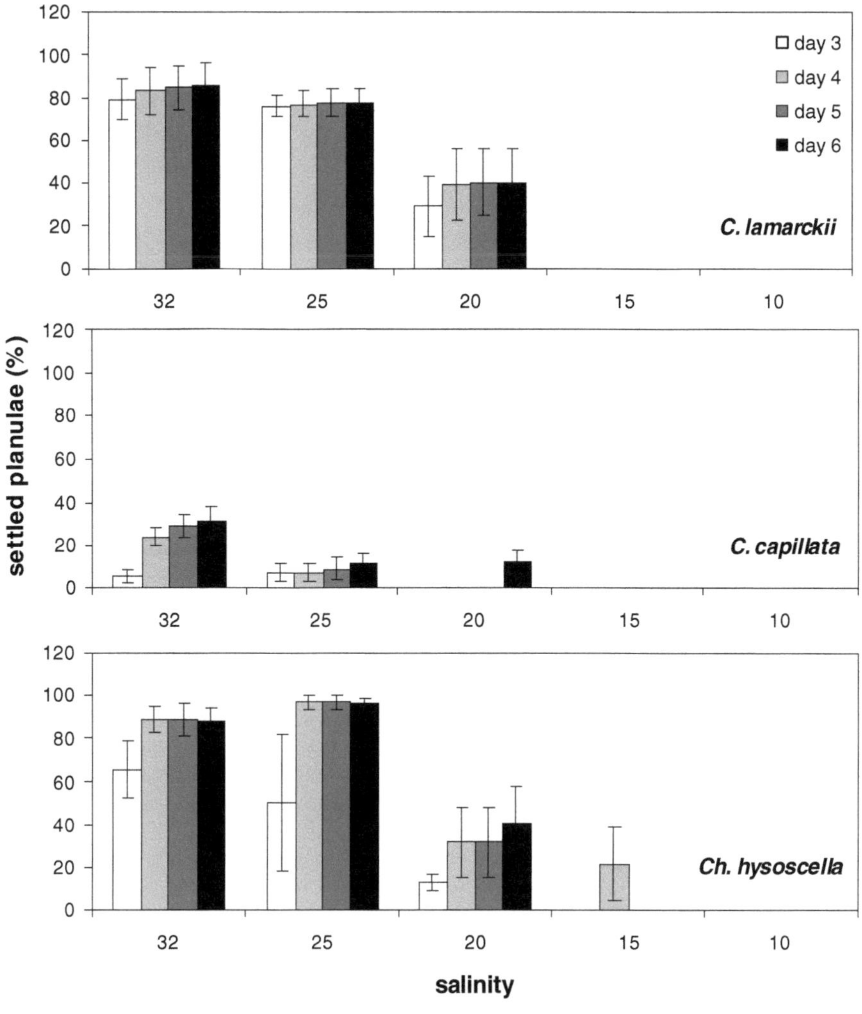

Fig. 4 Percentages of *Cyanea lamarckii*, *Cyanea capillata*, and *Chrysaora hysoscella* planulae settled on the undersides of floating Petri dish bottoms at different salinities

Table 4 Total numbers (tot. no) and percentages (mean ± SD of six replicates) of planulae settled on the undersides of floating Petri dish bottoms in six replicates at different salinities

Species	Salinity	Settled planulae		Test statistic	*P*
		Tot. no.	%, mean ± SD		
Cyanea lamarckii	32	3,266	85 ± 11.02	$F_{2,15} = 23.849$	<0.001
	25	2,987	77.67 ± 6.00		
	20	1,376	10.25 ± 15.59		
Cyanea capillata	32	261	30.63 ± 7.48	$F_{2,15} = 17.387$	<0.001
	25	95	11.21 ± 4.54		
	20	104	11.88 ± 5.95		
Chrysaora hysoscella	32	726	88.01 ± 6.74	$F_{2,15} = 50.982$	<0.001
	25	812	96.45 ± 2.04		
	20	356	40.78 ± 17.59		

Differences among salinities were tested by one-way ANOVA

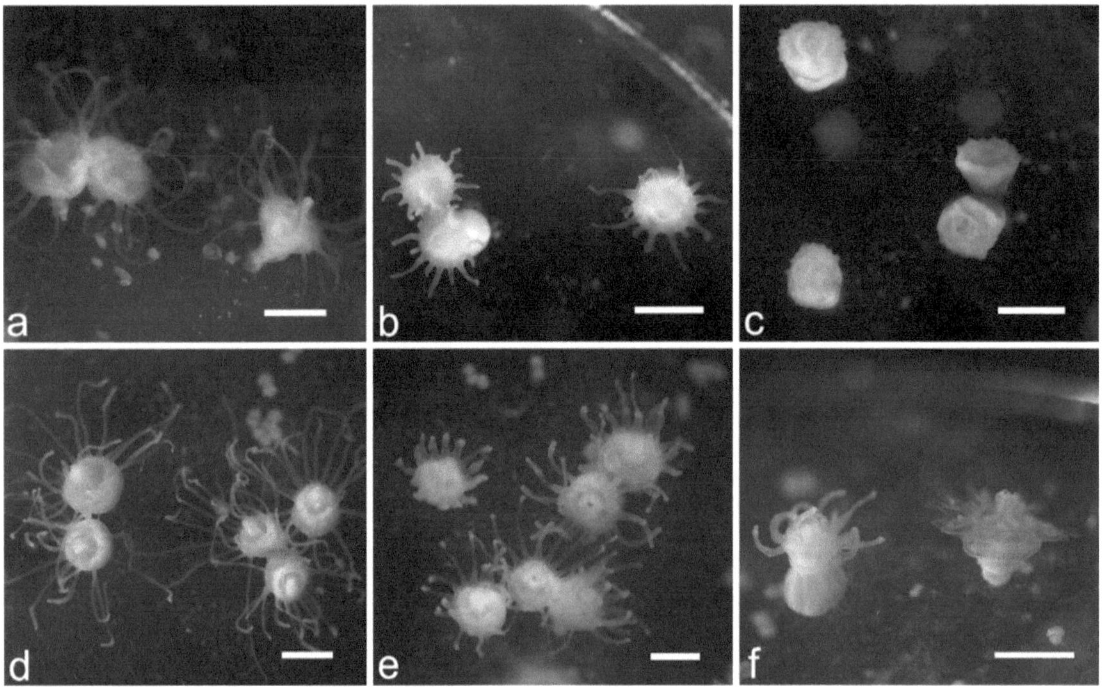

Fig. 5 Polyps cultured in different salinities. **a** *Aurelia aurita* polyps with long tentacles after 19 months in salinity 36; **b** *A. aurita* polyps with short tentacles after 3 months in salinity 12; **c** *A. aurita* polyps with extremely reduced tentacles after 3 months in salinity 8; **d** *Cyanea capillata* polyps with long tentacles after 19 months in salinity 36; **e** *C. capillata* polyps with short tentacles after 3 months in salinity 12; **f** strobilating *C. capillata* polyps after 5 months in salinity 12. Scale bars = 1 mm

76–112% by the end of the experiment in mid-June. Polyps of *C. capillata* increased only slightly (3%) in salinity 12. Polyps of *C. lamarckii* had mortalities of 10–12% in all salinities tested and no polyp reproduction occurred. The highest mortalities occurred during the first month after temperature reduction to 10°C in mid-December, when 5–10% of the polyps disappeared (Fig. 6).

Strobilation

For *A. aurita* and *C. lamarckii*, most ephyrae were produced during the first 2 months of the experiment (15 November to 15 January) in salinities of 36, 28, and 20 (Fig. 7). For *C. capillata*, most strobilations started 3 months later (15 April to 15 May) in salinities of 36, 28, and 20 (Fig. 7). In the lowest salinity (12), most strobilation started later than in the higher salinities for all tested species (Fig. 7).

More than 50% of *A. aurita* polyps strobilated in salinities of 20, 28, and 36 during the experiment (Fig. 8); however, a significantly lower percentage of polyps (11.6%) strobilated in salinity 12 (Table 5). Means of 6.1–8.7 ephyrae were produced by strobilating polyps in salinities of 20–36, respectively (Fig. 8). Significantly fewer ephyrae were produced in salinity 12 than in higher salinities of 28 and 36 (Table 6). H_2 and H_3 were rejected for *A. aurita*. More than 90% of *C. capillata* polyps strobilated in all experiments, with no significant differences among salinities (Fig. 8, Table 5), and H_2 could not be rejected. In the extreme salinities of 12 and 36 (without reduction), significantly fewer ephyrae were released from strobilating polyps than in salinities of 20 and 28 (Fig. 8, Table 6), and H_3 was rejected for *C. capillata*. For *C. lamarckii*, 16.1% of polyps strobilated in salinity 12, which was significantly fewer than in higher salinities where more than 50% strobilated (Fig. 8, Table 5). At the higher salinities tested, more *C. lamarckii* ephyrae were produced per strobila than in salinity 12 (Fig. 8, Table 6); H_2 and H_3 were rejected for *C. lamarckii*.

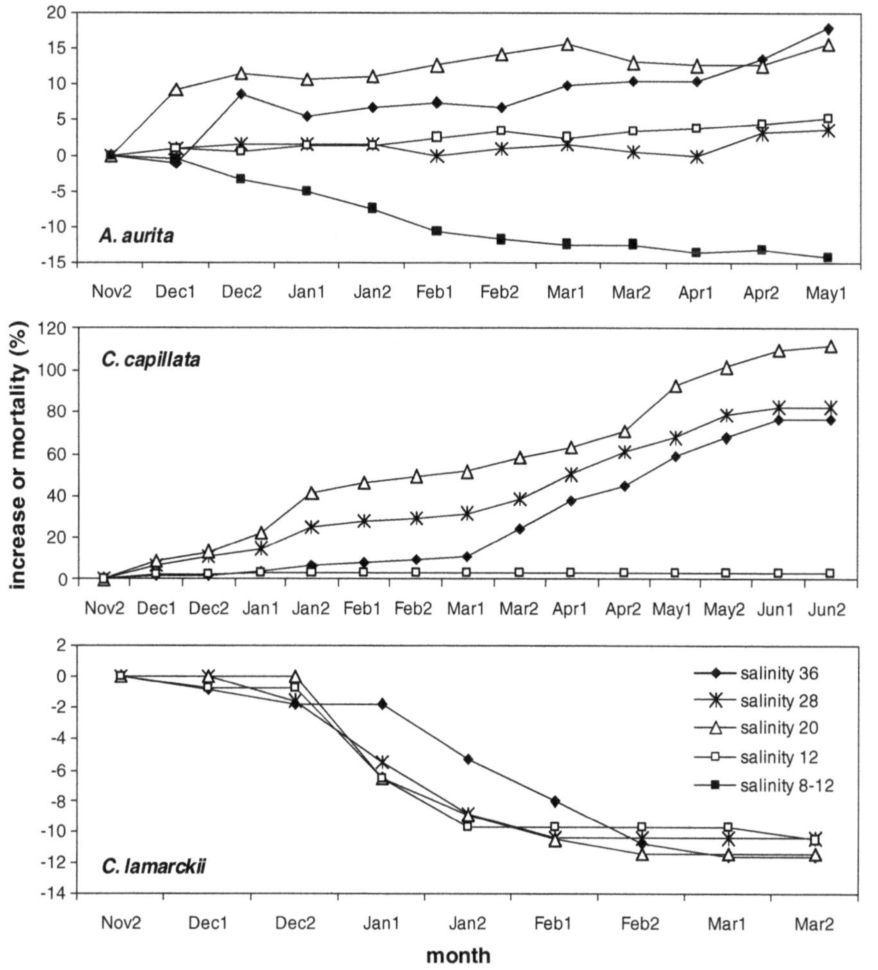

Fig. 6 Percentages of polyp increase and mortality at different salinities for *Aurelia aurita*, *Cyanea capillata*, and *Cyanea lamarckii*. In salinity 8–12, polyps (*A. aurita*) were cultured in salinity 8 from mid-November to mid-February before the salinity was increased to 12. 1 and 2 = beginning and middle of the month, respectively

Investigations on *C. capillata* medusae from Kiel Bight (western Baltic Sea)

Bell diameters of ephyrae and young medusae collected in Strande marina (Kiel Bight) ranged from 11.5 to 85 mm (mean = 31.3 ± 23.0, $n = 13$) in June 2006 and from 9.6 to 14.9 mm (mean = 11.8 ± 1.5, $n = 20$) in May 2009. One of the youngest ephyrae found in the samples had only one tentacle, or tentacle bud, present near each marginal lappet cleft, arising from the subumbrella (Fig. 9).

Planulae released by female medusae collected in late August in salinity 15 in Kiel Bight swam actively after being transferred into sampling beakers. In the laboratory in salinity 15, planulae settled on the bottom and walls of the glass bowl, on the watch glasses, or adhered at the water surface. First settlements occurred after only 1 day. Settled planulae directly developed into young polyps within 1 day, without forming planulocysts.

Discussion

The salinity regime in the Baltic Sea is mainly influenced by river runoff, precipitation, and water exchange with the North Sea. In the central Baltic, the water column is permanently stratified, with the fresher surface water separated from the deeper, more saline water by a halocline. In the shallow south-western area, the water column may be stratified or well-mixed, depending on the conditions. Changes in the salinities of estuarine and coastal waters are expected due to changing precipitation, evaporation, melting ice, and current patterns; general predictions include freshening of mid- and high-latitude waters

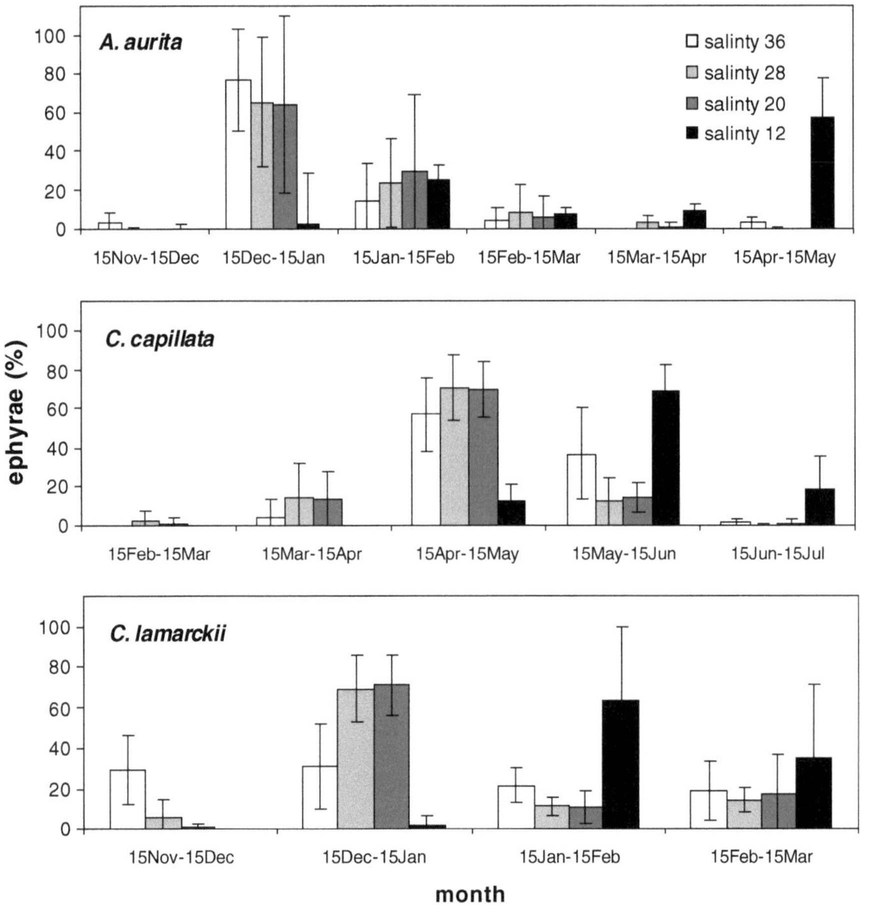

Fig. 7 Percentages of *Aurelia aurita*, *Cyanea capillata*, and *Cyanea lamarckii* ephyrae produced per month in salinities ranging from 36 to 12

but increased salinity in low latitude waters, as well as local-scale changes (IPCC, 2007). Although salinity in the entire Baltic might be significantly lower by the end of the twenty-first century, stability and deep-water ventilation probably would change only slightly (HELCOM, 2007).

The salinity regime of the Baltic Sea has determined its ecological characteristics. The number of animal species of marine origin occurring in the high-salinity waters of the Kattegat area (>850 species) decreases dramatically to the central Baltic (50) and to the innermost Bothnian Bay (<10) (Leppäkoski et al., 2002). The brackish waters do not protect the Baltic from new introductions; it has been called as "sea of invaders" (Leppäkoski et al., 2002). During the last two centuries, about 100 non-indigenous species have been recorded in the Baltic Sea, including some Hydrozoa that were introduced by ballast water or hull fouing (Olenin & Leppäkoski, 1999; Leppäkoski & Olenin, 2000, Leppäkoski et al., 2002). Also the invasive ctenophore, *Mnemiopsis leidyi* Agassiz, 1865, recorded in masses in the Baltic Sea since 2006, presumably was introduced by shipping (Javidpour et al., 2006).

Physiological limitations of cnidarians can restrict their occurrence in brackish waters. Only a few genera of Hydrozoa occur there, perhaps because they lack resistant stages or cannot osmoregulate in a hypotonic environment (Dumont, 1994). Nevertheless, some hydrozoan and ctenophore species tolerate wide ranges of temperature and salinity and colonize brackish habitats (Mills & Vogt, 1984; Purcell et al., 2001; Ma & Purcell, 2005). Scyphozoa are described as osmoconformers, i.e., the proportions of water in the fluids of tissues vary with the salinity of the surrounding water according to osmotic principles; the proportion of water increases or decreases until a lethal level is reached (Arai, 1997). For example, *Chrysaora quinquecirrha* (Desor, 1848) scyphomedusae cannot osmoragulate below salinities of 5

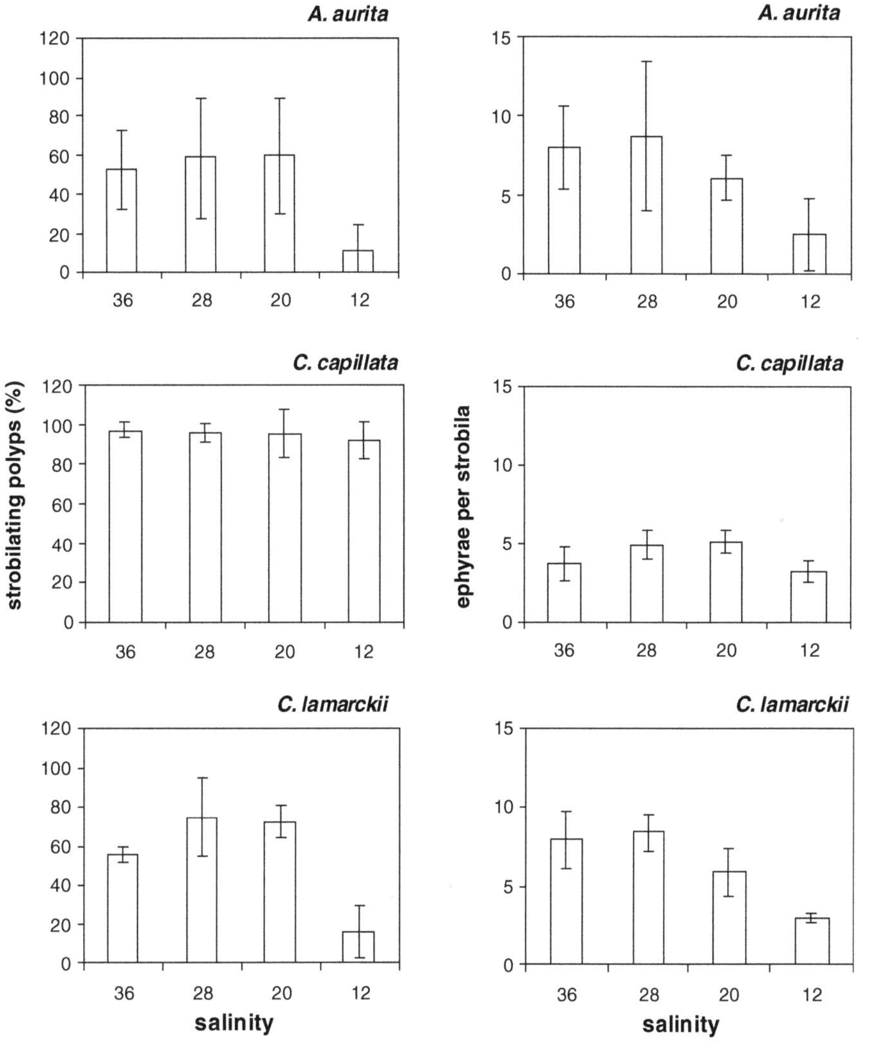

Fig. 8 Percentages of strobilating polyps (*left*) and ephyrae produced per strobila (*right*) in different salinities for *Aurelia aurita*, *Cyanea capillata*, and *Cyanea lamarckii*. Durations of the experiments are listed in Table 4. Data are presented as mean ± SD of six replicates

(Wright & Purcell, 1997).The annual abundance of the scyphomedusae *A. aurita* and *C. capillata* in the Baltic Sea is believed to depend on the inflow of North Sea waters (Möller, 1980; Janas & Witek, 1993; Barz et al., 2006). Increasing salinity is known to allow expansion of *A. aurita* and *C. capillata* medusae northwards in the Baltic Sea, but it is unknown whether the limits are due to direct effects of salinity on the medusae or the sessile stages of the life cycle (Arai, 1997).

A high tolerance of scyphopolyps to low salinities demonstrated in the present study has also been shown for other species (Cargo, 1984; Rippingale & Kelly, 1995; Purcell et al., 1999; Purcell, 2005, 2007). Thus, scyphozoan polyps are able to colonize estuaries and other mesohaline waters, e.g., the Thames estuary (Lambert, 1935), the Chesapeake Bay (Cargo & King, 1990), the Niantic River (Brewer, 1989; Brewer & Feingold, 1991; Colin & Kremer, 2002), and Baltic Sea waters as Kiel Fjord (Thiel, 1962) and Kertinge Nor (Olesen et al., 1994). Scyphozoan polyps can survive and asexually propagate even during prolonged exposure to hypoxic conditions (Condon et al., 2001). The combination of high tolerances of low salinities and low oxygen concentrations may allow scyphozoan polyps to colonize deeper layers of the Baltic Sea basins which provide higher salinities but are inhospitable for longer periods for many marine organisms due to oxygen-poor conditions (Olenin & Leppäkoski, 1999).

In the present study, planulae of all species tested were very tolerant of low salinities. Although salinity

Table 5 Numbers of strobilations (total numbers of six replicates) and percentages of strobilating polyps (mean ± SD of six replicates) at different salinities

Species and temperature	Salinity	Strobilations (tot. no.)	Strobilating polyps (%, mean ± SD)	Test statistic	*P*
Aurelia aurita 10°C	36	83	52.9 ± 20.0	$F_{3,20} = 6.225$	0.004
	28	104	58.4 ± 30.8		
	20	113	59.8 ± 29.3		
	12	25	11.6 ± 13.0		
Cyanea capillata 5°C	36	149	97.5 ± 3.8	$H_3 = 1.342$	0.719
	28	163	96.3 ± 4.4		
	20	153	95.8 ± 12.0		
	12	142	92.1 ± 9,1		
Cyanea lamarckii 10°C	36	62	55.6 ± 3.8	$F_{3,20} = 18.806$	<0.001
	28	83	74.4 ± 20.0		
	20	86	72.2 ± 8.4		
	12	16	16.1 ± 13.5		

Differences among salinities were tested by one-way ANOVA and Kruskal–Wallis one-way ANOVA on ranks

SD standard deviation, tot. no. total number

Table 6 Numbers of ephyrae produced (total numbers of six replicates) and numbers of ephyrae produced per strobila (mean ± SD of six replicates) in different salinities

Species and temperature	Salinity	Ephyrae (tot. no.)	Ephyrae per strobila (mean ± SD)	Test statistic	*P*
Aurelia aurita 10°C	36	687	8.0 ± 2.6	$F_{3,20} = 5.138$	0.009
	28	913	8.7 ± 4.7		
	20	702	6.1 ± 1.5		
	12	106	2.5 ± 2.3		
Cyanea capillata 5°C	36	520	3.7 ± 1.1	$F_{3,20} = 6.645$	0.003
	28	797	4.9 ± 0.9		
	20	756	5.1 ± 0.7		
	12	433	3.2 ± 0.7		
Cyanea lamarckii 10°C	36	459	7.9 ± 1.8	$H_3 = 16.819$	<0.001
	28	690	8.4 ± 1.2		
	20	467	5.9 ± 1.5		
	12	47	2.9 ± 0.3		

Differences among salinities were tested by one-way ANOVA and Kruskal–Wallis one-way ANOVA on ranks

Tot. no. total number

in the treatments was reduced from 32 to 7 within a short acclimation time of only 4 d, planulae of all species still swam at salinities of 10 or less. The low percentages of *C. capillata* planulae that settled within 6 d after release from the medusa are consistent with Russell's (1970) report that they may swim for as long as 4 weeks. Settlement of *C. capillata* planulae was greater at higher salinities and did not occur in salinities below 20; however, planulae sampled from *C. capillata* medusae in salinity 15 settled and developed into polyps. The rapid reduction of salinity in the experiments may explain the inability of planulae to settle at lowest salinities there. In the field, salinity changes may be less severe and planulae able to acclimate over longer periods.

Planula settlement of *Cyanea* sp. was observed in salinities of 21.5–27 in the Niantic River, where <10% of the planulocysts on the settling plates

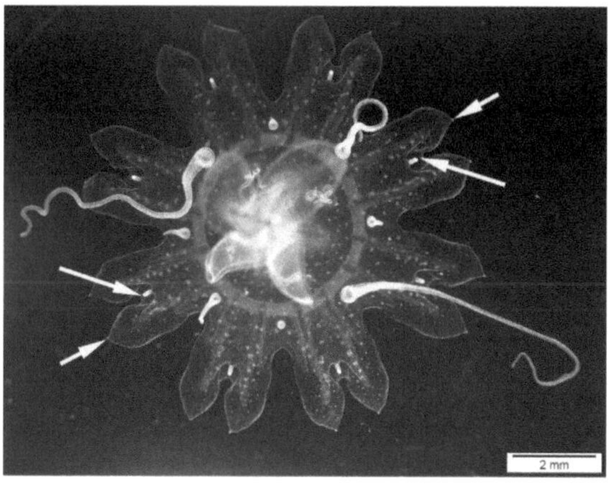

Fig. 9 *Cyanea capillata* ephyra collected in Strande marina (Kiel Bight) in June 2009. Only one tentacle or tentacle bud is developed near the clefts between the marginal lappets, demonstrating the early stage of the ephyra. Diameter between opposite rhopalar lappet tips (*short arrows*) = 9.8 mm; diameter between opposite rhopalia (*long arrows*) = 7.4 mm

excysted immediately to form polyps (Colin & Kremer, 2002), as we saw for *C. lamarckii*. A high number of cysts formed from *C. lamarckii* planulae collected from North Sea medusae at their natural salinity of 32, but no planulocysts formed from *C. capillata* planulae collected from medusae in the same area in this and Holst & Jarms (2007) studies. Similar results were obtained in laboratory and field experiments in the Gullmar Fjord (west Sweden) in salinities from 20 to 33; unlike *C. capillata*, *C. lamarckii* planulae immediately formed cysts after settlement that developed to polyps after 6 months (Gröndahl, 1988). In Scyphozoa, encystment of planulae is known only for the genus *Cyanea* (Brewer, 1976). Planulocysts were suggested to be special protection against extreme temperatures, settlement competitors, or predating species common at settlement (Brewer & Feingold, 1991; Holst & Jarms, 2007). High mortality rates of planulocysts due to overgrowth by algae, barnacles, bryozoans, hydrozoans, and ascidians indicate that the formation of these cysts is not a successful strategy against post-settlement mortality (Gröndahl, 1988; Colin & Kremer, 2002). Cysts produced by polyps (podocysts) are known for most semaeostome and rhizostome polyps with the exception of the suborder Kolpophorae (Holst et al., 2007). Podocyst production contributes to increasing numbers of scyphopolyps and to survival through seasonal periods of reduced food availability or predation (Arai, 2009), but extended observations and field studies are rare for both types of cysts.

All life stages of *A. aurita* show high tolerance of a wide range of salinities. Our results on *A. aurita* polyps, which survived a salinity decrease from 36 to 8 within 1 month, confirm their high tolerance of changing environmental conditions described in previous studies (Halisch, 1933; Miyake et al., 1997). Like the euryhaline medusae, *A. aurita* polyps and planulae are highly tolerant of low salinities, which enables them to colonize the northeastern Baltic Sea up to the Gulf of Finland where the ephyrae occur in salinities of only 5.5 (Wikström, 1932; Palmén, 1954). Planulae from the Baltic Sea even developed into polyps at a salinity of 4 (Thill, 1937). By contrast, *C. lamarckii* and *C. capillata* polyps did not survive salinity reduction below 12 in our study; although they are less tolerant of low salinity than *A. aurita* polyps, their ability to strobilate in salinity 12 after a reduction from 36 demonstrates high tolerance of salinity changes. Numbers of *C. lamarckii* polyps decreased by similar amounts at all tested salinities, which suggests that temperature reduction to 10°C, and not the salinity decrease, may have caused the mortality in the experiments. Restriction in most years of *C. lamarckii* medusae to the west coast of Sweden in the Baltic Sea has led to speculation that young medusae are transported there from the southern part of the North Sea (Gröndahl, 1988). Polyps of *C. lamarckii* probably cannot survive the low winter temperatures in the Baltic Sea. This idea is corroborated by the distributions of scyphomedusae in the North and Irish seas, where *A. aurita* and *C. capillata* medusae are abundant in northern areas but *C. lamarckii*, *Ch. hysoscella*, and *R. octopus* medusae were abundant in southern areas (Verwey, 1942; Russell, 1970; Hay et al., 1990; Doyle et al., 2007; Barz & Hirche, 2007). Rising winter temperatures due to climate changes in the North Sea may shift distributions of the southern species northwards (Franke & Gutow, 2004). If their distributions shift to the northern North Sea, they could reach the Baltic Sea more frequently with currents through the Skagerrak and Kattegat.

Environmental factors, such as temperature, salinity, and food supply, influence asexual polyp and ephyra production rates (Russell, 1970; Purcell et al., 1999; Purcell, 2007; Willcox et al., 2007). Numbers of ephyrae produced per strobila at higher salinities (28–36) in our study were in accordance with previous results. Ephyrae per strobila are usually lower in *C. capillata* than in *A. aurita* and *C. lamarckii* (Lambert, 1935; Berrill, 1949). Strobilation was delayed and fewer ephyrae per strobila were produced in salinity 12 than at higher salinities for all species we tested, which agrees with other studies showing reduced ephyra production by other scyphozoan species at extremes of ambient salinities (Purcell et al., 1999; Purcell, 2007). Fewer ephyrae per strobila at low salinities also was reported by previous authors. In the field, *A. aurita* strobilae had 3–10 ephyrae in salinity 13–16 in Kiel Fjord but only 1–4 ephyrae in salinity 12–13 in nearby Kiel Canal (Thiel, 1962). In the Gulf of Finland, most *A. aurita* strobilae were monodisc, possibly due to the low salinity (Palmén, 1954). The effects of salinity on strobilation seemed to be less important than temperature and food supply on strobilation timing and ephyra production (Purcell et al., 1999; Purcell, 2007).

Constant polyp numbers and high percentages of strobilation in *C. capillata* cultures at the low salinity 12 demonstrated that polyps of this species are much more tolerant of low salinities than expected. The polyps were believed to need salinities of at least 20 for survival and reproduction and that the west coast of Sweden could be the geographical limit for reproduction of *C. capillata* (Cargo, 1984; Båmstedt et al., 1997; Barz et al., 2006). Undoubtedly, large numbers of *C. capillata* medusae drift into the Baltic Sea with North Sea water masses during the summer months (Möller, 1980; Gröndahl & Hernroth, 1987; Janas & Witek, 1993; Barz et al., 2006); however, our collections of *C. capillata* ephyrae from Kiel Bight demonstrate their presence in the western Baltic Sea at low salinity (15). Our data confirm previous collections of even smaller ephyrae (2–3 mm diameter) from the Kiel Bight more than 50 years ago (Mielck & Künne, 1935; Thiel, 1960). These very young ephyrae in this area support the conclusion that the ephyrae probably were released by local polyps. Alternatively, the ephyrae may have drifted into the Baltic Sea with deep water masses from the North Sea or from the west coast of Sweden; however, our results demonstrated that planulae collected from mature medusae from the western Baltic Sea developed into polyps in salinity 15, and that polyps acclimated to salinity 12 could produce ephyrae. Thus, an ephyra-producing polyp population could exist at least in the western Baltic Sea, and occurrence of the medusae in the Baltic Sea would not depend only on water inflow from the North Sea and the west coast of Sweden. If a reproductive population of *C. capillata* polyps exists in the western or southern Baltic Sea, a mass development of the species in this area is possible. Climate change during the twenty-first century will affect species composition, distributions, and interactions in the Baltic Sea (HELCOM, 2007). Environmental changes could initiate jellyfish mass development, since climate changes, eutrophication, overfishing, and increased hard substrates for polyp settlement are possible causes for jellyfish blooms (Purcell, 2005; Holst & Jarms, 2007; Purcell, 2007). *Cyanea capillata* medusae eat the same prey as zooplanktivorous fish (Purcell & Sturdevant, 2001), and nematocysts of *C. capillata* contain a highly effective toxin that causes a painful sting in humans (Heeger et al., 1992; Lassen et al., this volume; Wiebring et al., this volume). Therefore, a mass development of lion's mane medusae in the Baltic Sea, in addition to those drifting in with currents from the North Sea, would have negative consequences for fisheries and tourism.

The present study contributes to our knowledge of the ecology and distribution of scyphozoan polyps. Since one polyp is able to produce several medusae per year, the origin of the conspicuous medusa generation depends on the development of the inconspicuous polyp generation. Moreover, metagenetic scyphozoan species can only invade new habitats permanently if the planulae find suitable conditions for settlement and polyp survival. Therefore, more investigations on the distribution and ecology of polyps are important for understanding, predicting, and mitigating jellyfish blooms and the expansion of metagenetic scyphozoans.

Acknowledgments We thank the Biologische Anstalt Helgoland (Alfred Wegener Institute) for providing research facilities. This work is part of a PhD thesis conducted at the Biocenter Grindel and Zoological Museum in Hamburg (Holst, 2008, http://www.sub.uni-hamburg.de/opus/volltexte/2008/3635) supported by EUROGEL (EUROpean GELatinous Zooplankton, European Commission Contract no. EVK-CT-2002-00074). The assistance of students and technicians is greatly appreciated. We are grateful to Drs. J. D. R. Houghton and J. E. Purcell for English language revisions.

References

Arai, M. N., 1997. A Functional Biology of Scyphozoa. Chapman & Hall, London.

Arai, M. N., 2009. The potential importance of podocysts to formation of scyphozoan blooms: a review. Hydrobiologia 616: 241–246.

Båmstedt, U., H. Ishii & M. B. Martinussen, 1997. Is the scyphomedusa *Cyanea capillata* (L.) dependent on gelatinous prey for its early development? Sarsia 82: 269–273.

Barz, K. & H.-J. Hirche, 2007. Abundance, distribution and prey composition of scyphomedusae in the southern North Sea. Marine Biology 151: 1021–1033.

Barz, K., H. H. Hinrichsen & H.-J. Hirche, 2006. Scyphozoa in the Bornholm Basin (central Baltic Sea) – the role of advection. Journal of Marine Systems 60: 167–176.

Berrill, N. J., 1949. Developmental analysis of scyphomedusae. Bio Review 24: 393–410.

Brewer, R. H., 1976. Larval settling behavior in *Cyanea capillata* (Cnidaria: Scyphozoa). Biological Bulletin 150: 183–199.

Brewer, R. H., 1984. The influence of the orientation, roughness, and wettability of solid surfaces on the behavior and attachment of planulae of *Cyanea* (Cnidaria: Scyphozoa). Biological Bulletin 166: 11–21.

Brewer, R. H., 1989. The annual pattern of feeding, growth, and sexual reproduction in *Cyanea* (Cnidaria: Scyphozoa) in the Niantic River estuary, Connecticut. Biological Bulletin 176: 272–281.

Brewer, R. H. & J. S. Feingold, 1991. The effect of temperature on the benthic stages of *Cyanea* (Cnidaria: Scyphozoa), and their seasonal distribution in the Niantic River estuary, Connecticut. Journal of Experimental Marine Biology and Ecology 152: 49–60.

Cargo, D. G., 1984. Some Laboratory Techniques for the Culture of Scyphozoa. United Nations Environment Programme-Workshop on Jellyfish Blooms in the Mediterranean, Athens: 129–138.

Cargo, D. G. & D. R. King, 1990. Forecasting the abundance of the sea nettle, *Chrysaora quinquecirrha*, in the Chesapeake Bay. Estuaries 13: 486–491.

Colin, S. P. & P. Kremer, 2002. Population maintenance of the scyphozoan *Cyanea* sp. settled planulae and the distribution of medusae in the Niantic River, Connecticut, USA. Estuaries 25: 70–75.

Condon, R. H., M. B. Decker & J. E. Purcell, 2001. Effects of low dissolved oxygen on survival and asexual reproduction of scyphozoan polyps (*Chrysaora quinquecirrha*). Hydrobiologia 451: 89–95.

Danielssen, D. S., E. Svendsen & M. Ostrowski, 1996. Long-term hydrographic variation in the Skagerrak based on the section Torungen-Hirtshals. ICES Journal of Marine Science 53: 917–925.

Doyle, T. K., J. D. R. Houghton, S. M. Buckley, G. C. Hays & J. Davenport, 2007. The broad-scale distribution of five jellyfish species across a temperate costal environment. Hydrobiologia 579: 29–39.

Dumont, H. J., 1994. The distribution and ecology of the fresh- and brackish-water medusae of the world. Hydrobiologia 272: 1–12.

Fonselius, S. & J. Valderrama, 2003. One hundred years of hydrographic measurements in the Baltic Sea. Journal of Sea Research 49: 229–241.

Franke, H. D. & L. Gutow, 2004. Long-term changes in the macrozoobenthos around the rocky island of Helgoland (German Bight, North Sea). Helgoland Marine Research 58: 303–310.

Gröndahl, F., 1988. A comparative ecological study on the scyphozoans *Aurelia aurita*, *Cyanea capillata* and *C. lamarckii* in the Gullmar Fjord, western Sweden, 1982 to 1986. Marine Biology 97: 541–550.

Gröndahl, F. & L. Hernroth, 1987. Release and growth of *Cyanea capillata* (L.) ephyrae in the Gullmar Fjord, western Sweden. Journal of Experimental Marine Biology and Ecology 106: 91–101.

Groth, H. & H. Thede, 1989. Does brackish water exert long-term stress on marine immigrants in the Baltic Sea? Scientia Marina 53: 677–684.

Haahtela, I. & J. Lassig, 1967. Records of *Cyanea capillata* (Scyphozoa) and *Hyperia galba* (Amphipoda) from the Gulf of Finland and the northern Baltic. Annales Zoologici Fennici 4: 469–471.

Halisch, W., 1933. Beobachtungen an Scyphopolypen. Zoologischer Anzeiger 104: 296–304.

Hay, S. J., J. R. G. Hislop & A. M. Shanks, 1990. North Sea scyphomedusae: summer distribution, estimated biomass and significance particularly for 0-group gadoid fish. Netherlands Journal of Sea Research 25: 113–130.

Heeger, T., H. Möller & U. Mrowietz, 1992. Protection of human skin against jellyfish (*Cyanea capillata*) stings. Marine Biology 113: 669–678.

HELCOM, 2007. Climate change in the Baltic Sea area – HELCOM thematic assessment in 2007. Baltic Sea Environment Proceedings 111: 1–49.

Holst, S., 2008. Grundlagen der Populationsentwicklung verschiedener Scyphozoa (Cnidaria) der Deutschen Bucht [online]. Hamburg, Univ., Diss., 2008, URN: urn:nbn:de:gbv:18-36356. http://www.sub.uni-hamburg.de/opus/volltexte/2008/3635/.

Holst, S. & G. Jarms, 2007. Substrate choice and settlement preferences of planula larvae of five Scyphozoa (Cnidaria) from German Bight, North Sea. Marine Biology 151: 863–871.

Holst, S., I. Sötje, H. Tiemann & G. Jarms, 2007. Life cycle of the rhizostome jellyfish *Rhizostoma octopus* (L.) (Scyphozoa, Rhizostomeae), with studies on cnidocysts and statoliths. Marine Biology 151: 1695–1710.

IPCC, 2007. Summary for policymakers. In Solomon, S., D. Qin, M. Manning, Z. Chen, M. Marquis, K. B. Averyt, M. Tignor & H. L. Miller (eds), Climate Change 2007: The Physical Science Basis. Contribution of Working Group I to the Fourth Assessment Report of the Intergovernmental Panel on Climate Change. Cambridge University Press, Cambridge, New York.

Jakobsen, F., 1997. Hydrographic investigations of the Northern Kattegat front. Continental Shelf Research 17: 533–535.

Janas, U. & Z. Witek, 1993. The occurrence of medusae in the southern Baltic and their importance in the ecosystem, with special emphasis on *Aurelia aurita*. Oceanologia 34: 69–84.

Javidpour, J., U. Sommer & T. Shiganova, 2006. First record of *Mnemiopsis leidyi* A. Agassiz 1865 in the Baltic Sea. Aquatic Invasions 1: 299–302.

Khlebovich, V. V., 1969. Aspects of animal evolution related to critical salinity and internal state. Marine Biology 2: 338–345.

Kramp, P. L., 1934. An exceptional occurance of *Rhizostoma octopus* and *Chrysaora hysoscella* in the Danish waters in 1933. Journal du Conseil 9: 211–221.

Lambert, F. J., 1935. Observations on the scyphomedusae of the Thames estuary and their metamorphosis. Travaux de la Station Zoologique de Wimereux 12: 281–307.

Lassen, S., H. Helmholz, C. Ruhnau & A. Prange, Characterisation of neurotoxic polypeptides from *Cyanea capillata* medusae (Scyphozoa). Hydrobiologia. doi:10.1007/s10750-010-0215-x.

Leppäkoski, E. & S. Olenin, 2000. Non-native species and rates of spread: lessons from the brackish Baltic Sea. Biological Invasions 2: 151–163.

Leppäkoski, E., S. Gollasch, P. Gruszka, H. Ojaveer, S. Olenin & V. Panov, 2002. The Baltic – a sea of invaders. Canadian Journal of Fisheries and Aquatic Science 59: 1175–1188.

Ma, X. & J. E. Purcell, 2005. Effects of temperature, salinity, and predators on mortality of and colonization by the invasive hydrozoan *Moerisia lyonsi*. Marine Biology 147: 215–224.

Merck, T., 1989. Untersuchungen zur ökologischen Nische von *Chrysaora hysoscella*. Jahresbericht der Biologischen Anstalt Helgoland: 53–54.

Mielck, W. & C. Künne, 1935. Fischbrut- und Plankton-Untersuchungen auf dem Reichsforschngsdampfer „Poseidon" in der Ostsee, Mai-Juni 1931. Wissenschaftliche Meeresuntersuchungen, Abteilung Helgoland N. F. 19: 1–120.

Mills, C. E. & R. G. Vogt, 1984. Evidence that ion regulation in hydromedusae and ctenophores does not facilitate vertical migration. Biological Bulletin 166: 216–227.

Miyake, H., K. Iwao & Y. Kakinuma, 1997. Life history and environment of *Aurelia aurita*. South Pacific Study 17: 273–285.

Möller, H., 1980. A summer survey of large zooplankton, particularly scyphomedusae, in North Sea and Baltic. Report of Marine Research 28: 61–68.

Olenin, S. & E. Leppäkoski, 1999. Non-native animals in the Baltic Sea: alteration of benthic habitats in coastal inlets and lagoons. Hydrobiologia 292: 233–243.

Olesen, N. J., K. Frandsen & H. U. Riisgård, 1994. Population dynamics, growth and energetics of jellyfish *Aurelia aurita* in a shallow fjord. Marine Ecology Progress Series 105: 9–18.

Palmén, E., 1954. Seasonal occurrence of ephyrae and subsequent instars of *Aurelia aurita* (L.) in the shallow waters of Tvärminne, S. Finnland. Archivum Societatis Zoologicae-Botanicae Fennicae Vanamo 8: 122–131.

Purcell, J. E., 2005. Climate effects on jellyfish and ctenophore blooms: a review. Journal of the Marine Biological Association of the UK 85: 461–476.

Purcell, J. E., 2007. Environmental effects on asexual reproduction rates of the scyphozoan, *Aurelia labiata*. Marine Ecology Progress Series 348: 183–196.

Purcell, J. E. & M. V. Sturdevant, 2001. Prey selection and dietary overlap among zooplanktivorous jellyfish and juvenile fishes in Prince William Sound, Alaska. Marine Ecology Progress Series 210: 67–83.

Purcell, J. E., J. R. White, D. A. Nemazie & D. A. Wright, 1999. Temperature, salinity and food effects on asexual reproduction and abundance of the scyphozoan *Chrysaora quinquecirrha*. Marine Ecology Progress Series 180: 187–196.

Purcell, J. E., T. A. Shiganova, M. B. Decker & E. D. Houde, 2001. The ctenophore *Mnemiopsis* in native and exotic habitats: U.S. estuaries versus Black Sea basin. Hydrobiologia 451: 145–176.

Rasmussen, E., 1973. Systematics and ecology of the Isefjord marine fauna (Denmark). Ophelia 11: 1–507.

Rippingale, R. J. & S. J. Kelly, 1995. Reproduction and survival of *Phyllorhiza punctata* (Cnidaria: Rhizostomeae) in a seasonally fluctuating salinity regime in Western Australia. Marine and Freshwater Research 46: 1145–1151.

Russell, F. S., 1970. The medusae of the British Isles. II. Pelagic Scyphozoa with a supplement to the first volume on hydromedusae. Cambridge University Press, London.

Thiel, M. E., 1960. Beobachtungen über Wachstum, Variationen und Abnormitäten bei *Cyanea capillata* der Ostsee. Abhandlungen und Verhandlungen des Naturwissenschaftlichen Vereins in Hamburg N. F. 4: 89–108.

Thiel, H., 1962. Untersuchungen über die Strobilisation von *Aurelia aurita* LAM. an einer Population der Kieler Förde. Kieler Meeresforschungen 18: 198–230.

Thill, H., 1937. Beiträge zur Kenntnis der *Aurelia aurita* (L.). Zeitschrift für wissenschaftliche Zoologie 150: 51–96.

Tronolone, V. B., A. C. Morandini & A. E. Migotto, 2002. On the occurrence of scyphozoan ephyrae (Cnidaria, Scyphozoa, Semaeostomeae and Rhizostomeae) in the southeastern Brazilian coast. Biota Neotropica 2 http://www.biotaneotropica.org.br/v2n2/en/fullpaper?bn02102022002+en.

Verwey, J., 1942. Die Periodizität im Auftreten und die aktiven und passiven Bewegungen der Quallen. Archives Néerlandaises de Zoologie 6: 363–468.

Wiebring, A., H. Helmholz, S. Lassen, A. Prange & G. Jarms, Separation and analysis of different types of nematocysts from *Cyanea capillata* (L.) medusae. Hydrobiologia. doi: 10.1007/s10750-010-0227-6.

Wikström, D. A., 1932. Beobachtungen über die Ohrenqualle (*Aurelia aurita* L.) in den Schären SW-Finnlands. Memoranda Societatis pro Fauna et Flora Fennica 8: 14–17.

Willcox, S., N. A. Motschaniwskyi & C. Crawford, 2007. Asexual reproduction in scyphistomae of *Aurelia* sp.: effects of temperature and salinity in an experimental study. Journal of Experimental Marine Biology and Ecology 353: 107–114.

Wright, D. A. & J. E. Purcell, 1997. Effect of salinity on ionic shifts in mesohaline scyphomedusae, *Chrysaora quinquecirrha*. Biological Bulletin 192: 332–339.

Xian, W., B. Kang & R. Liu, 2005. Jellyfish blooms in the Yangtze Estuary. Science 307: 41.

Hydrobiologia (2010) 645:69–79
DOI 10.1007/s10750-010-0225-8

Effects of El Niño-driven environmental variability on black turtle migration to Peruvian foraging grounds

Javier Quiñones · Victoria González Carman · Jorge Zeballos · Sara Purca · Hermes Mianzan

Published online: 21 March 2010

Abstract We analyzed sea temperature as an environmental factor, in association with ENSO, affecting the migration of East Pacific black turtle, *Chelonia mydas* (=*Chelonia agassizii* Bocourt), to its foraging areas and its feeding ecology at San Andrés, Peru. A 19-year sea turtle landing database (1970–1988) was constructed to associate landing fluctuations with environmental variability represented by the Peruvian Oscillation Index. A positive correlation between them ($r = 0.75$, $P < 0.05$) indicated that exceptionally large black turtle landings occurred in San Andrés port during El Niño episodes. Warmer waters (SST 22–28°C) approached near the Peruvian coast during El Niño episodes, thus facilitating black turtle access to this area. Furthermore, during El Niño 1987, large juvenile and adult black turtles, known to be primarily herbivorous, fed mainly on the scyphozoan jellyfish *Chrysaora plocamia* Péron & Lesueur, which was very abundant during this event. It is likely that black turtles exploited this resource opportunistically. Inter-annual environmental variability, driven by El Niño Southern Oscillation, has profound consequences for the ecology of the endangered black turtle, which should be considered when evaluating the effects of anthropogenic activities on its population dynamics.

Keywords *Chelonia mydas* (=*Chelonia agassizii*) · ENSO · Trophic migration · Jellyfish · Sea turtle landings · Peru

Guest editors: J.E. Purcell & Dror Angel / Jellyfish Blooms: New Problems and Solutions

J. Quiñones · J. Zeballos
Laboratorio Costero de Pisco, Instituto del Mar del Perú (IMARPE), Av. Los Libertadores A-12, Urb. El Golf, Paracas, Ica, Peru
e-mail: javierantonioquinones@gmail.com

V. González Carman · H. Mianzan (✉)
CONICET-Instituto Nacional de Investigación y Desarrollo Pesquero (INIDEP), Paseo Victoria Ocampo n°1, B7602HSA Mar del Plata, Argentina
e-mail: hermes@inidep.edu.ar

S. Purca
Centro de Investigaciones en Modelaje Oceanográfico y Biológico-Pesquero (CIMOBP), Instituto del Mar del Perú (IMARPE), Esquina Gamarra y General Valle s/n Chucuito, Callao, Peru

Introduction

The coastal ecosystem off Peru and Chile is characterized by strong, persistent wind-driven upwelling of nutrient-rich waters. This upwelling sustains high biological productivity and supports one of the world's largest single-species fisheries (Chavez et al., 1999, 2008; Pennington et al., 2006). Oceanographic and atmospheric conditions fluctuate on seasonal to inter-decadal scales (Wyrtki, 1975; Tourre et al., 2005). Among such variations, inter-annual variability due to the El Niño Southern Oscillation (ENSO) is known to affect physical, chemical, and nutritional conditions of the ocean (Fiedler et al., 1991; Chavez et al., 1999; Ayón et al., 2004). During an ENSO, the equatorial zonal winds weaken, thus driving an upwelling of warmer waters due to a deeper-than-normal main thermocline (Ji & Leetmaa, 1997). Biological responses to this phenomenon are exhibited by a wide variety of organisms. For example, primary production is severely reduced (Barber & Chavez, 1983) and large amounts of benthic macroalgae wash ashore during the strongest ENSO (i.e., 1982–1983). The latter affects the related benthic and grazing fauna (Tarazona et al., 1988; Arntz & Fahrbach, 1996; Fernandez et al., 1999). Furthermore, anchovy (*Engraulis ringens* Jenyns) concentrates in shallower refuge waters (Csirke, 1989; Muck et al., 1989a; Arntz & Fahrbach, 1996) where a large die off occurs due to a drastic habitat reduction (Bertrand et al., 2004). In addition, there is high mortality of top predators like seabirds (i.e., *Phalacrocorax bougainvilli* Lesson and *Sula variegate* Tschudi) and marine mammals (i.e., *Otaria byronia* Shaw) due to limited food availability (Arntz & Fahrbach, 1996; Jahncke et al., 2004). Such examples demonstrate how marine organisms off Peru and Chile respond to ENSO; however, there are few examples of how sea turtles respond.

The black turtle, *Chelonia mydas* (=*Chelonia agassizii* Bocourt), is distributed along the west coasts of North and South America. Main nesting grounds are at Michoacan (Mexico), the Galapagos Archipelago (Ecuador), and the Revillagigedos Islands (Mexico). Coastal primary foraging areas range from San Diego Bay (USA) to Mejillones (Chile) (Marquez, 1990). The feeding ecology of this species, which is widely accepted to be herbivorous, has been studied along most of its distribution. Macroalgae and seagrass were reported in the diet of turtles from Bahía de Los Angeles and Bahía Magdalena (Baja California, Mexico) (Seminoff et al., 2002). A more omnivorous diet, based mainly on algae but with an additional wide variety of animals, such as crustaceans, fish eggs, mollusks, and, to a lesser extent, jellyfish were recorded in turtles from Bahía de Sechura and San Andrés-Pisco (Paredes, 1969; Hays-Brown & Brown, 1982; de Paz et al., 2004; Kelez et al., 2004; Santillán Corrales, 2008). Because the effects of ENSO on marine organisms are seen at several levels of the food web, black turtle diet could change with respect to non-ENSO years; however, the feeding ecology of this species has never been studied in relation to ENSO.

For several sea turtle species, oceanographic conditions at their foraging grounds prior to the nesting season determine their probability of nesting (Broderick et al., 2001; Wallace et al., 2006; Saba et al., 2007, 2008a, b). Within the large black turtle foraging area, it is reasonable to expect that events like ENSO could greatly affect their migration and foraging ecology; however, few reports are available that refer to ENSO-driven conditions on black turtle reproductive ecology (Hurtado, 1984; Zárate, 2003). Most recently, Seminoff et al. (2008) showed that black turtle migrations from the Galapagos to their foraging grounds are restricted by sea surface temperature (SST) $\leq$24–25°C; they suggested that periods of elevated SST (i.e., ENSO) could facilitate access of this species to its foraging habitats along the coast of South America. It is known that during ENSO, warmer waters (>22°C) reach the coast of South America (Wang, 1995), and in this scenario, sea turtle approach to the coast of Peru could be favored. This species might respond to environmental variability driven by ENSO, at least on its nesting grounds and possibly on its migration to foraging areas, but no conclusive evidence has been presented yet.

In Peru, sea turtles have been consumed by men since the pre-Hispanic era (Frazier & Bonavia, 2000). A traditional sea turtle fishery, with a well-developed trade along the southern coast (Hays-Brown & Brown, 1982), existed until 1995 when this fishery was banned (Morales & Vargas, 1996); however, some turtle carapaces are found sporadically on dump sites, suggesting that some captures still occur (Quiñones, pers. obs.). San Andrés was recognized as the main locality of this fishery (Fig. 1) (Aranda & Chandler, 1989; Vargas et al., 1994), while other Peruvian ports did not have fishermen specializing on turtle harvesting (Frazier, 1979). This fishery was artisanal, operating mostly during summer and early fall (Frazier, 1979; Hays-Brown & Brown, 1982). Shore-based boats of 4–8 m with engines of 25–40 HP were employed (Frazier, 1979; Pinedo, 1988). They used tangle-nets of different sizes and mesh, some of which were designed specifically to catch turtles (Frazier, 1979; Hays-Brown & Brown, 1982). In San Andrés port, sea turtle landings showed high inter-annual fluctuations, with some years having

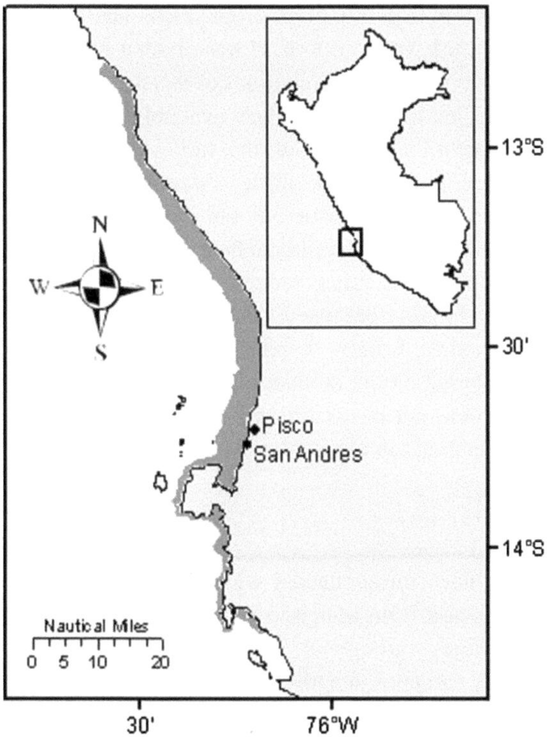

Fig. 1 Study area off the coast off Peru. Fishing grounds of the San Andrés turtle fishery during 1970–1988 are shown in *dark gray*

very large landings and other years having almost none. El Niño (EN), the warm phase of the ENSO, is believed to play an important role in this inter-annual fluctuation.

In this study, we tested the hypothesis that environmental variability, driven by EN, may favor black turtle trophic migration to the San Andrés foraging ground. Because sea turtles are ectotherms, their distribution is restricted by sea temperature (Polovina et al., 2004; Wallace & Jones, 2008). Consequently, it is reasonable to expect that black turtle trophic migration to the San Andrés foraging ground is facilitated during warm events like EN and that their occurrence would increase. We used published data on the EN index and SST isotherms and constructed a sea turtle landing database for the local turtle fishery. Thus, historical records of sea turtle landings allowed us to study the response of this species to EN during a 19-year period. In addition, because black turtles migrate to Peruvian coastal waters to feed mostly on macroalgae (de Paz et al., 2004; de Paz & Alfaro Shigeto, 2008; Kelez et al., 2004), and because the whole system of marine organisms (including macroalgae) responds to EN, we examined the stomach contents of black turtles landed in San Andres during one EN episode to determine if its diet was altered, potentially reflecting changes in food availability. Paradoxically, the information on sea turtle commercial landings served to improve our understanding of the effect of EN-driven environmental variability in the trophic migration of the endangered black turtles and also may provide the foundations for conservation of the species in the East Pacific Ocean.

Materials and methods

In this study, environmental variability driven by EN was represented by the Peruvian Oscillation Index (POI) and SST isotherm location along the Peruvian coast. To correlate these environmental indices with black turtle migration toward southern Peru, we used historical sea turtle fishery landings recorded in San Andrés from 1970 to 1988. Landings were used as a proxy of turtle occurrence and migration to this area. Then, to analyze if food availability also was altered by EN, diet analysis was performed on some of the turtles landed during EN 1987.

Environmental context

ENSO index

The official Oceanic Niño Index (ONI) is the SST of NIÑO 3.4 area (5–5°S, 170–140°W). This oceanic index does not necessarily reflect warming events restricted to the Peruvian coast, so the first principal component of SST of Peru, known as the POI, was considered to be more appropriate for this study. It was calculated from the means of monthly average SST during a 50-year time series (1950–1999) measured at five coastal stations along the Peruvian coast. Therefore, the POI best represents the coastal SST variation in Peru and is positively correlated with other indexes of EN and the Pacific Decadal Oscillation, thus reflecting conditions in the East Tropical Pacific region (Montecinos et al., 2003). In this study, monthly POI data were averaged by year and plotted with phases of EN episodes according to the ONI (onset = −1, peak = 0, and offset = +1) and to warm events for the period 1970–1988. An EN episode is different from a warm event because an EN episode originates in the Central Pacific, while a warm event could originate in the East or West Pacific, and its effect is local.

SST isotherms

To illustrate the approach of warmer waters toward the San Andres foraging ground, the location of SST isotherms during EN episodes (1972, 1982–1983, and 1987) and a non-ENSO episode (1981) were plotted. We chose 1981 as a typical non-ENSO episode because the POI index was almost neutral throughout that year. SST isotherm maps were constructed based on previous works of Anónimo (1972), Zuta et al. (1984), Urquizo et al. (1987), Rivera (1988), and Muck et al. (1989b). For EN episodes of 1972, 1982–1983, and 1987, isotherms from March were represented. In 1981, only an average of the SST isotherms for January, February, and March was available.

Sea turtle landings

A sea turtle landing database for the San Andrés fishery was constructed using published data and interviews conducted for this study. Published data included monthly landings from 1970 to 1988 (Wosnitza-Mendo et al., 1988; Aranda & Chandler, 1989; Flores et al., 1994, 1998). Because landing reports for 1982–1983 were not reliable for the area, interviews with 32 San Andrés turtle fishermen, which represented 50% of all fishermen for 1982–1983, were conducted from April to July 2009. Data recorded during the interviews included the numbers of boats, days, trips, turtles landed, season, area fished, and turtle behavior. Sea turtle landings (L, in tons) for the years 1982–1983 were calculated as follows:

$$\mathrm{L} = a * b * c * d/1000$$
$$a = x/z$$

where a is the mean effective fishing period calculated as the mean fishing period (x, in d) divided by mean trip duration (z, in d). Parameter b is the mean number of fishing boats, c is the mean number of turtles captured per boat, and d is the average turtle weight (in kg) obtained from 1987 landings. To obtain an L value in tons, d was divided by 1000.

There were some limitations of this database. First, no species identification was available; however, it was known that 90% of the turtles landed corresponded to the black turtle (Aranda & Chandler, 1989) and was confirmed by our interviews. Second, monthly resolution could not be assessed in fishermen interviews, so L values plotted for 1982–1983 are not means. Finally, because of the artisanal nature of the San Andrés fishery, a reliable measure of fishing effort based on the number of fishing boats or fishing trips could not be established; therefore, we used the total landings in our analyses.

Sea turtle diet

Dead black turtles landed were recovered from San Andrés port from March to September 1987. Because diet varies throughout the sea turtle life history, curved carapace length (CCL), measured from the middle nucal scute to the posterior middle between supracaudal scutes, and weight were recorded. Adult and juvenile turtles were distinguished according to the minimum breeding size of CCL 60.7 cm for the Galapagos rookery (Zárate, 2003). Necropsies were performed on some of the turtles landed. Esophagus and stomachs were retrieved and immediately transported to the laboratory, where diet samples were analyzed. Prey items were identified to the lowest possible taxon. Plant matter, mollusks, fish, crustaceans, and jellyfish were identified according to Dawson et al. (1964), Alamo & Valdivieso (1987), Chirichigno (1974), Retamal (1981), Bowman & Gruner (1973), Brusca (1981), Stiasny (1937), and Mianzan & Cornelius (1999).

Data analysis

To test for association between EN and sea turtle occurrence and migration toward San Andrés foraging ground, a Spearman correlation was carried out between sea turtle landings and the POI. A type I error (α) of 0.05 was considered significant.

Quantitative assessment of diet was based on the frequency of occurrence (FO) of each diet item and its relative wet weight (WW) in each sample was calculated as follows:

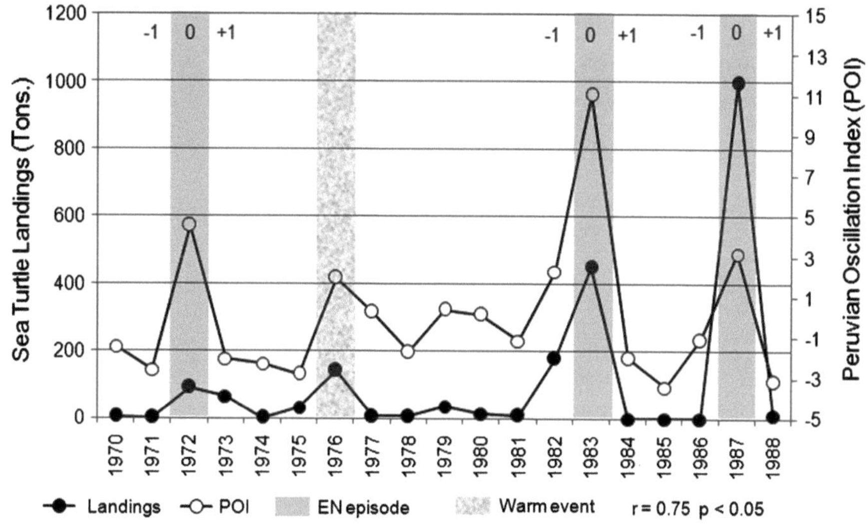

Fig. 2 Historical series of sea turtle landings for the San Andrés turtle fishery during 1970–1988. Correlation between the POI anomalies and sea turtle landings in tons was statistically significant. Warm events and EN phases (−1, 0, +1) are indicated by *gray shading*

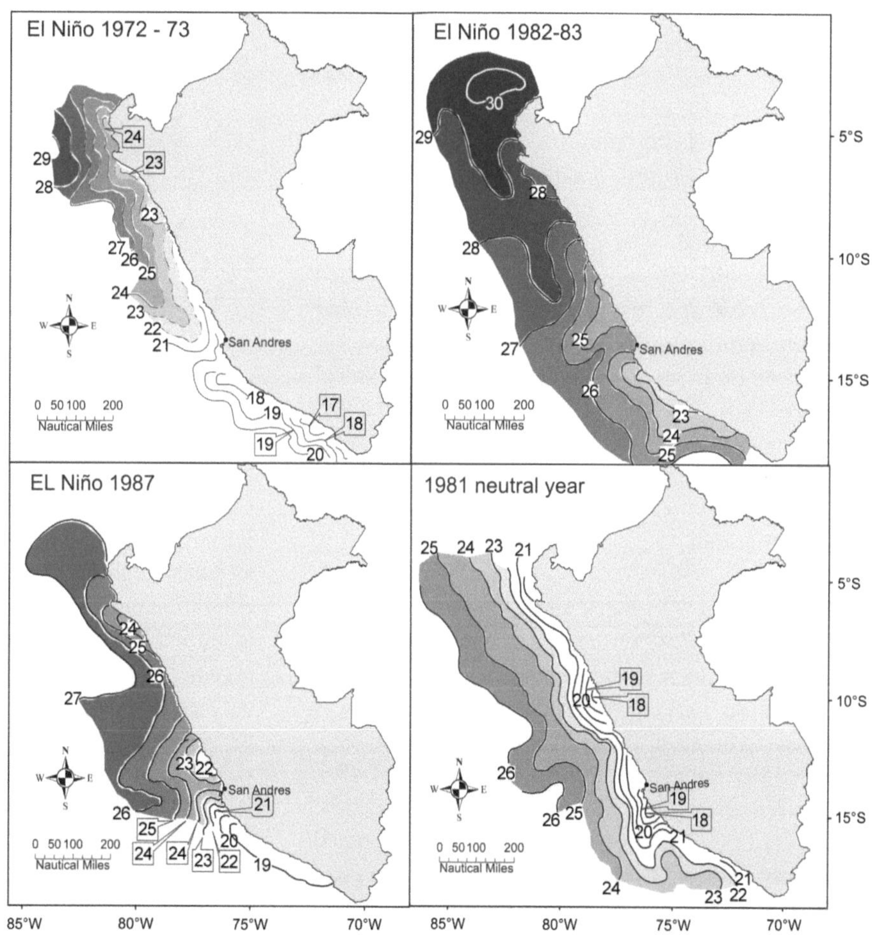

Fig. 3 SST isotherm locations during EN episodes **a** 1972–1973, **b** 1982–1983, **c** 1987, and during a neutral year **d** 1981. Warmer waters can be seen near the coast of Peru during EN episodes

$$\%FO = (\text{number of samples in which diet item observed}/\text{total samples}) * 100$$

$$\%WW = (\text{wet weight of a diet item}/\text{total weight of all items}) * 100.$$

Results

POI, SST, and sea turtle landings

During a 19-year period, the POI varied inter-annually from a maximum of +9.54 to a minimum of −2.99. During this period, three EN episodes were observed; two strong EN occurred in 1972–1973 and 1982–1983 with POI anomalies of +4.43 and +9.36, respectively, and a moderate EN occurred in 1986–1987 with a maximum anomaly of +4.23. In addition, a warm event occurred during 1975–1976 with a maximum anomaly of +2.9 (Fig. 2).

Sea turtle landings in San Andres also increased during EN episodes and the warm event (Fig. 2). Maximum catches were more than 90, 400, and 900 tons in EN 1972–1973, 1982–1983, and 1987, respectively, and 142 tons for the warm event of 1976. POI and sea turtle landings were significantly positively correlated ($r = 0.749$, $P = 0.0002$) during the 19-year period.

Warm-water intrusions (22–28°C) approached near the coast and reached San Andrés and other southern Peruvian areas in every EN during the 19-year period (Fig. 3a–c). In contrast, during the typical summer 1981, warm waters (22–26°C) remained approximately 100 nautical miles off the Peruvian coast (Fig. 3d).

Black turtle diet

From a total of 998 black turtles measured in San Andrés in 1987; 76% of the turtles were of adult size (Fig. 4). Food items in the stomach and esophagus of 192 of those turtles were analyzed: Jellyfish was the prey item most frequently consumed (70.8%), followed by mollusks (62.0%), crustaceans (47.4%), and macroalgae (37.5%). Anthropogenic debris, especially those made of plastic, also was consumed frequently (41.7%; Table 1).

In addition to being the most frequently consumed prey item, jellyfish represented the highest mean WW (49.5%), which was followed by mollusks (17.0%) and macroalgae (13.2%). Jellyfish represented ≥75% of gut content WW for more than 40% of the turtles analyzed (Table 1). The only jellyfish consumed was *Chrysaora plocamia* Péron & Lesueur (Scyphozoa). Jellyfish found in diet samples appeared as pieces and as partly digested liquefied jellyfish. Because of the coloration and consistency, this material could be identified as *C. plocamia*. "Other" items included all organic materials (i.e., pieces of wood, land plants, etc.) and stones that probably were consumed when the turtles ate other prey items. Among mollusks, *Donax marincovichi* Coan was consumed most frequently (67%). *Macrocystis pyrifera* Linnaeus was the most consumed macroalga (97%). Almost all crustaceans (99.9%) were amphipods, *Hyperia* sp. Latreille, which probably were living in, and were consumed with, the jellyfish. Finally, annelids, fishes, sponges, and echinoderms were considered to be

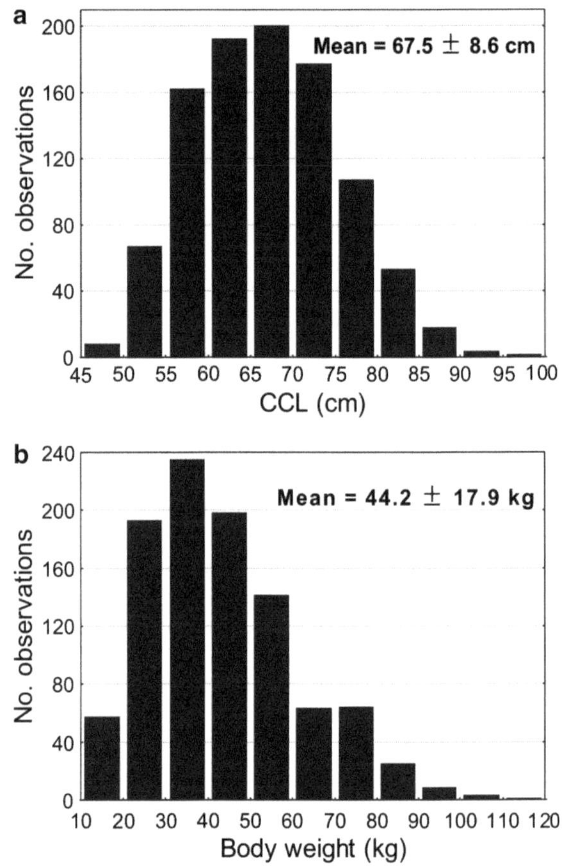

Fig. 4 **a** Curved carapace length and **b** body weight of black turtles, *C. mydas* (=*C. agassizii*), landed in San Andrés during EN 1987

Table 1 Mean percent sample wet weight (%WW) and frequency of occurrence (%FO) of prey groups recovered in esophagus and stomach samples from black turtles *C. mydas* (=*C. agassizii* Bocourt) in San Andrés

Diet item	%W		%FO			
	Mean	SD	Present	≤5%WW	≥50%WW	≥75%WW
Jellyfish *Chrysaora plocamia*	49.5	41.2	70.8	32.3	53.1	41.7
Mollusks[a]	17.0	29.0	62.0	62.0	14.6	8.3
Crustaceans[b]	3.3	10.4	47.4	85.9	1.6	0.5
Anthropogenic debris	10.7	22.9	41.7	71.9	9.9	4.2
Macroalgae[c]	13.2	26.6	37.5	69.3	12.0	6.3
Others	2.9	10.3	36.9	88.5	1.6	1.0
Unknown	0.4	3.4	6.25	100	–	–
Annelids *Diopatra* sp. Audouin & Milne-Edwards	0.6	5.9	3.6	99.0	0.5	0.5
Fishes[d]	0.3	2.8	4.2	98.4	–	–
Sponges	T	–	1.0	100	–	–
Echinoderms *Ophiectis kröyeri* Lütken	T	–	0.5	100	–	–

T denotes presence in trace levels, *SD* standard deviation

[a] *Mulinia edulis* King & Broderip, *Donax marincovichi* Coan, *Nassarius gayi* Kiener, *Discinisca lamellosa* Broderip, *Semimytilus algosus* Gould, *Crepipatella dilatata* Lamarck, *Aulacomya ater* Molina, *Tegula atra* Lesson, *Polinices otis* Broderip & Sowerby, *Fisurella peruviana* Lamarck, *Sinum cymba* Menke

[b] *Hyperia* sp. Latreille, *Ceratothoa gaudichaudii* Milne-Edwards, *Balanus* sp. Da Costa, *Hepatus chilensis* Edwards, *Lepas anatifera* Linnaeus, *Cancer porteri* Rathbun

[c] *Macrocystis pyrifera* Linnaeus, *Rhodymenia flabellifolia* Montagne, *Ulva costata* Howe, *Chondracanthus chamissoi* Kützing, *Gracilariopsis lemaneiformis* E. Y. Dawson, Acleto & Fold

[d] *Engraulis ringens* Jenyns, *Gaelichthys peruvianus* Lütken, *Schroederichthys chilensis* Guichenot

trace items because each group constituted <1%WW and ≤5%FO (Table 1).

Discussion

In this study, we present consistent evidence that warm episodes, including EN, favor black turtle trophic migration to southern Peruvian foraging grounds. High black turtle landings occurred during every EN and warm event observed from 1970 to 1988 (Fig. 2). Since turtles were the target species of this specialized fishery, an increase in sea turtle landings may reflect an increase in sea turtle occurrence during their migration to San Andrés. In fact, abundances of sea turtles changed from being considered a rare visitor most of the time to become common during EN episodes on Peruvian coasts (Arntz & Fahrbach, 1996). Landings alone, however, may not reflect sea turtle abundance because the fishing effort would vary. The scarce literature on the San Andrés fishery revealed that since 1979, the turtle fleet contained 7–11 boats (see Frazier, 1979; Hays-Brown & Brown, 1982; Aranda & Chandler, 1989), indicating that the fishing effort was fairly consistent. Opportunistic fishermen might have fished during EN, probably because the other fishing resources declined. During EN 1987, up to 110 boats were recorded to land turtles in San Andrés (Zeballos, pers. obs.), encompassing the largest sea turtle harvest ever recorded in Peru (Aranda & Chandler, 1989). That year, however, about 14 boats (about the usual fleet size) captured up to 50% of the turtles during 50% of the total fishing trips reported, which suggests that even if the landings by the opportunistic boats are disregarded, very high sea turtle landings occurred during that EN episode (Zeballos, pers. obs.).

Among the various physical and biological factors that might be involved in the high black turtle occurrences in Peruvian coastal waters during EN

Table 2 Summary of diet studies on the black turtle along its distribution in the East Pacific

References	Location	*N*	Sampling year	Frequency of occurrence (%FO)		
				Algae	Jellyfish	Others
Seminoff et al. (2002)	Mexico, Baja California	108	1995–1999	Chlorophyta >24% Phaeophyta >19% Rhodophyta 100%	No	Overall animal matter was 38.5–42.8%
Amorocho & Reina (2007)	Colombia	84	2003–2004	18%	No	Tunicates 74% Mangrove 69% Crustaceans 63%
Paredes (1969)	Peru, San Andrés-Pisco	20	1969	100%	Unident. 60%	Mollusks 50% Fish eggs 60% Crustaceans 25%
Hays-Brown & Brown (1982)	Peru, San Andrés-Pisco	39	1979	51%	Unident. 31%	Mollusks 64% Polychaetes 49% Crustaceans 13%
Kelez et al. (2004)	Peru, San Andrés, and Chimbote	3	2000	100%	Unident. 33%	Mollusks 67%
de Paz et al. (2004)	Peru, San Andrés	13	1999–2000	43%	No	Mollusks 29%
Santillán Corrales (2008)	Peru, Bahía de Sechura	NR	2002–2004	>33%	Hydrom. 13%	Squid eggs 22%

NR not reported, *N* number of turtles analyzed, *unident.* unidentified, *hydrom.* hydromedusae

episodes, we focused on sea temperature. During average years, the west coast of South America is characterized by upwelling of cold subsurface waters (Brink et al., 1983), but during EN episodes, warmer waters propagate poleward along the entire coastlines of Peru and Chile (Hormazabal et al., 2001), as shown by SST fields with cold water extending 100 km offshore of the Peruvian coast in 1981, but warm waters >22°C reaching the San Andrés coast during EN 1982–1983 and 1987. Because the POI is calculated from SST data along the coast of Peru (i.e., Montecinos et al., 2003), it follows that temperature plays a key role determining sea turtle migration. Our results strongly support the idea that warm waters near the shore facilitate black turtle access to the San Andrés foraging grounds (Fig. 3).

Previous work on black turtle diet throughout its distribution showed that macroalgae were consumed most frequently, in combination with several invertebrate species, including jellyfish, to a lesser extent (Table 2). No previous study considered dietary preferences related to ENSO. During EN, primary productivity decreases triggering changes throughout the food web (Barber & Chavez, 1983; Muck et al., 1989a; Arntz & Fahrbach, 1996). Because benthic macroalgae washes ashore during strong EN events (Arntz & Fahrbach, 1996), that food resource probably is limited during EN events, and thus, may affect the turtle's diet. During EN 1987 as reported here, 70.1% of black turtles fed on the large jellyfish *C. plocamia* and only 37.5% fed on macroalgae. Moreover, of the turtles feeding on *C. plocamia*, more than 40% had a stomach full (≥75%WW) of this species (Table 1). During La Niña episodes (i.e., 1999–2001 with negative POI values) and even during neutral or mild EN episodes (1979 and 2002–2003 with a maximum POI of +1.5), turtles probably have an omnivorous diet based mostly on macroalgae with some jellyfish consumed (see Fig. 2 and Table 2). Conversely, during moderate to strong EN episodes (1969, maximum POI of +3.2), jellyfish may predominate in the diet. To test this hypothesis, further research on black turtle diet during EN episodes is needed.

Jellyfish are not trophic dead ends (Mianzan et al., 2001; Arai, 2005), as they are often mistakenly thought to be. Jellyfish have four to six times less gross energy than algae and seagrass (Doyle et al., 2007); however, some of the largest marine vertebrates like the leatherback sea turtle and the oceanic sunfish, *Mola mola* Linnaeus, survive on diets mostly of jellyfishes (Arai, 2005; Houghton et al., 2006; Wallace et al., 2006). For those species, prey quantity probably prevails over quality, as for large terrestrial herbivores (Shrader et al., 2006; Doyle et al., 2007). Interestingly, initial analysis of concurrent jellyfish data suggests that their occurrences may be positively correlated with black turtle landings and EN episodes (Quiñones et al., in prep.). If jellyfishes are most abundant during EN episodes, sea turtles could exploit that resource opportunistically, as in EN 1987.

Most adult turtles foraging in coastal Peruvian waters come from their nesting beaches on the Galapagos Islands and, to a lesser extent, on the mainland of Ecuador (Green, 1984; Aranda & Chandler, 1989). Because both nesting and sea turtle fishing seasons occur from summer to early fall in these latitudes (Aranda & Chandler, 1989), adult turtles feeding in Peru probably do not reproduce that same year in the Galapagos Islands. In fact, during EN 1982–1983 and the warm event of 1975–1976, fewer turtles nested in the Galapagos than in non-EN years (Zárate, 2003). Although nesting records are not complete for the period 1970–1988, the numbers of turtles nesting in the Galapagos (apparently driven by EN) were opposite to the numbers landed in the San Andrés foraging ground. Furthermore, during EN 1987, and possibly during every EN between 1970 and 1988, adult-sized black turtles were captured in southern Peru to the detriment of Galapagos nesting population. Additional research is needed to understand how physical and biological factors governed by ENSO drive the ecology of endangered species like black turtles. The effects of environmental factors need to be considered in combination with the effects of anthropogenic factors, such bycatch or direct capture mortality, on sea turtle population dynamics.

Acknowledgments We thank the San Andrés fishermen for sharing their information, the turtle traders for the samples, and all the technicians of the Pisco laboratory of IMARPE, who helped in the fieldwork in 1987. This project was supported by the Instituto del Mar del Peru at Pisco and partially supported by grants to H. M.: CONICET (PIP 5009), FONCyT 01553, and the Inter-American Institute for Global Change Research (IAI) CRN 2076, which is supported by the US National Science Foundation (grant GEO-0452325). V. G. C. was supported by scholarships from CONICET. This is INIDEP contribution no. 1582.

References

Alamo, V. & V. Valdivieso, 1987. Lista sistemática de moluscos marinos del Perú. Boletín Instituto del Mar del Perú & Deutsche Gesellschaft füf Technische Zusammenarbeit, Volumen Extraordinario, Callao, Perú: 1–205.

Amorocho, D. F. & R. D. Reina, 2007. Feeding ecology of the East Pacific green sea turtle *Chelonia mydas agassizii* at Gorgona National Park, Colombia. Endangered Species Research 3: 43–51.

Anónimo, 1972. Cartas promedio de temperatura superficial del mar en Marzo, Abril, Mayo y Junio de 1972, Vol. 109. Serie de informes especiales del Instituto del Mar del Perú, Callao: 1–15.

Arai, M. N., 2005. Predation on pelagic coelenterates: a review. Hydrobiologia 451: 523–536.

Aranda, C. & M. Chandler, 1989. Las tortugas marinas del Perú y su situación actual. Boletín de Lima 62: 77–86.

Arntz, W. E. & E. Fahrbach, 1996. El Niño experimento climático de la naturaleza, causas físicas y efectos biológicos. Fondo de Cultura Económica, México: 1–311.

Ayón, P., S. Purca & R. Guevara-Carrasco, 2004. Zooplankton volume trends off Peru between 1964 and 2001. ICES Journal of Marine Science 61: 478–484.

Barber, R. T. & F. P. Chavez, 1983. Biological consequences of El Niño. Science 222: 1203–1210.

Bertrand, A., M. Segura, M. Gutierrez & L. Vásquez, 2004. From small-scale habitat loopholes to decadal cycles: a habitat-based hypothesis explaining fluctuation in pelagic fish populations off Peru. Fish and Fisheries 5: 296–316.

Bowman, T. E. & H. E. Gruner, 1973. The families and genera of *Hyperiidea* (Crustacea: Amphipoda). Smithsonian Contributions to Zoology 146: 1–64.

Brink, K. H., D. Halpern, A. Huyer & R. L. Smith, 1983. The physical environment of the Peruvian upwelling system. Progress in Oceanography 12: 185–305.

Broderick, A. C., B. J. Godley & G. C. Hays, 2001. Trophic status drives interannual variability in nesting numbers of marine turtles. Proceedings of the Royal Society 268: 1478–1481.

Brusca, R. C., 1981. A monograph on the isopoda Cymothoidae (Crustacea) of the eastern Pacific. Zoological Journal of the Linnean Society 73: 117–199.

Chavez, F. P., P. G. Strutton, G. E. Friederich, R. A. Feely, G. C. Feldman, D. G. Foley & M. G. McPhaden, 1999. Biological and chemical response of the equatorial Pacific Ocean to the 1997–1998 El Niño. Science 286: 2126–2131.

Chavez, F. P., A. Bertrand, R. Guevara-Carrasco, P. Soler & J. Csirke, 2008. The northern Humboldt current system: brief history, present status and a view towards the future. Progress in Oceanography 79: 95–105.

Chirichigno, N., 1974. Clave para identificar los peces marinos del Perú, Vol. 44. Informe Instituto del Mar del Perú, Callao: 1–388.

Csirke, J., 1989. Changes in the catchability coefficient in the Peruvian anchoveta (*Engraulis ringens*) fishery. In Pauly, D., P. Muck, J. Mendo & I. Tsukayama (eds), The Peruvian Upwelling Ecosystem: Dynamics and Interactions, Vol. 18. ICLARM Studies and Reviews, Manila: 207–219.

Dawson, E. Y., C. Acleto & N. Foldvik, 1964. The seaweeds of Perú. Nova Hedwigia 13: 1–111.

de Paz, N., J. C. Reyes & M. Echegaray, 2004. Capture and trade of marine turtles at San Andres, Southern Peru. In Coyne, M. S. & R. D. Clark (eds), Proceedings of the Twenty-First Annual Symposium on Sea Turtle Biology and Conservation. NOAA Technical Memorandum NMFS–SEFSC-528: 52–54.

de Paz, N. & J. Alfaro Shigeto, 2008. Foraging grounds for sea turtles in inshore Peruvian waters. In Kalb, H., A. Rhode, K. Gayheart & K. Shanker (eds), Proceedings of the Twenty-Fifth Annual Symposium on Sea Turtle Biology and Conservation. NOAA Technical Memorandum NMFS-SEFSC-582: 1–118.

Doyle, T. K., J. D. R. Houghton, R. McDevitt, J. Davenport & G. C. Hays, 2007. The energy density of jellyfish: estimates from bomb-calorimetry and proximate-composition. Journal of Experimental Marine Biology and Ecology 343: 239–252.

Fernandez, E. C., C. Cordova & J. Tarazona, 1999. Condiciones del bosque submareal de *Lessonia trabeculata* en la Isla Independencia durante el evento El Niño 1997–1998. In Tarazona, J. & E. Castillo (eds), El Niño 1997–98 y su impacto sobre los ecosistemas marino y terrestres. Revista Peruana de Biología, Volumen Extraordinario, Lima: 47–59.

Fiedler, P. C., V. Philbrick & F. P. Chavez, 1991. Oceanic upwelling and productivity in the eastern tropical Pacific. Limnology and Oceanography 36: 1834–1850.

Flores, M., S. Vera, R. Marcelo & E. Chirinos, 1994. Estadísticas de los desembarques de la pesquería marina Peruana 1983–1992, Vol. 105. Informe Instituto del Mar del Perú, Callao: 1–202

Flores, M., S. Vera, R. Marcelo & E. Chirinos, 1998. Estadísticas de los desembarques de la pesquería marina Peruana 1970–1982, Vol 134. Informe del Instituto del Mar del Perú, Callao: 1–256.

Frazier, J., 1979. Marine Turtles in Peru and the East Pacific. Office of Zoological Research, National Zoological Park, Smithsonian Institution (unpublished manuscript).

Frazier, J. & D. Bonavia, 2000. Prehispanic marine turtles in Peru: where were they? In Abreu-Grobois, F. A., R. Briseño-Dueñas, R. Marquez & L. Sarti (eds), Proceedings of the Eighteenth International Sea Turtle Symposium. U.S. Department of Commerce, NOAA Technical Memorandum NMFS-SEFSC-436: 243–245.

Green, D., 1984. Long distance movements of Galapagos green turtles. Journal of Herpetology 18: 121–130.

Hays-Brown, C. & W. M. Brown, 1982. Status of sea turtles in the southeastern Pacific: emphasis on Peru. In Bjorndal, K. (ed.), Biology and Conservation of Sea Turtles. Smithsonian Institution Press, Washington, DC: 235–240.

Hormazabal, S., G. Shaffer, J. Letelier & O. Ulloa, 2001. Local and remote forcing of sea surface temperature in the coastal upwelling system off Chile. Journal of Geophysical Research 106: 16657–16672.

Houghton, J. D. R., T. K. Doyle, M. W. Wilson, J. Davenport & G. C. Hays, 2006. Jellyfish aggregation and leatherback turtle foraging patterns in a temperate coastal environment. Ecology 87: 1072–1967.

Hurtado, M., 1984. Registros de anidación de la tortuga negra, *Chelonia mydas*, en las Islas Galápagos, Vol. 6. Boletín Científico y Técnico, Instituto Nacional de Pesca, Guayaquil: 77–104.

Jahncke, J., D. Checkley & G. L. Hunt Jr., 2004. Trends in carbon flux to seabirds in the Peruvian upwelling. Fisheries Oceanography 13: 208–223.

Ji, M. & A. Leetmaa, 1997. Impact of data assimilation on ocean initialization and El Niño prediction. Monthly Weather Review 125: 742–753.

Kelez, S. A., C. Manrique, X. Velez-Suazo & M. Williams, 2004. Green turtle (*Chelonia mydas agassizii*) diet differences in two Peruvian coastal localities. In Coyne, M. S. & R. D. Clark (comps), Proceedings of the Twenty-First Annual Symposium in Sea Turtle Biology and Conservation. NOAA Technical Memorandum NMFS-SEFSC-528: 305–307.

Marquez, R., 1990. Sea turtles of the world. FAO Species Catalogue 125: 1–81.

Mianzan, H. W. & P. F. S. Cornelius, 1999. Cubomedusae and Scyphomedusae. In Boltovskoy, D. (ed.), South Atlantic Zooplankton. Backhuys Publishers, Leiden: 513–559.

Mianzan, H. W., M. Pájaro, G. Alvarez Colombo & A. Madirolas, 2001. Feeding on survival-food: gelatinous plankton as a source of food for anchovies. Hydrobiologia 451: 45–53.

Montecinos, A., S. Purca & O. Pizarro, 2003. Interannual to interdecadal sea surface temperature variability along the western coast of South America. Geophysical Research Letters 30: 1570.

Morales, V. R. & P. Vargas, 1996. Legislation protecting marine turtles in Peru. Marine Turtle Newsletter 75: 22–23.

Muck, P., A. Z. De Moreno & C. Aranda, 1989a. The seasonal dynamics of sea surface temperature and its impact on anchoveta distribution off Peru. In Pauly, D., P. Muck, J. Mendo & I. Tsukayama (eds), The Peruvian Upwelling Ecosystem: Dynamics and Interactions, Vol. 18. ICLARM Studies and Reviews, Manila: 33–44.

Muck, P., I. Tsukayama & E. Ferrandis, 1989b. Temperaturas superficiales del mar a lo largo de la costa Peruana para el período 1950–1988, Vol. 95. Informe Instituto del Mar del Perú, Callao: 1–29.

Paredes, R., 1969. Introducción al estudio biológico de *Chelonia mydas agassizii* en el perfil Pisco. Thesis. Universidad Nacional Federico Villareal, Lima: 1–83.

Pennington, J. T., K. L. Mahoney, V. S. Kuwahara, D. D. Kolber, R. Calienes & F. P. Chavez, 2006. Primary production in the eastern tropical Pacific: a review. Progress in Oceanography 69: 285–317.

Pinedo, M., 1988. La pesqueria artesanal en la caleta de San Andrés - Pisco. In Wosnitza-Mendo, C., M. Espino & M. Veliz (eds), La pesquería artesanal en el Perú durante junio 1986 a junio 1988, Vol 93. Informe Instituto del Mar del Perú, Callao: 127–130.

Polovina, J. J., G. H. Balazs, E. A. Howell, D. M. Parker, M. P. Seki & P. H. Dutton, 2004. Forage and migration habitat of loggerhead (*Caretta caretta*) and olive ridley (*Lepidochelys olivacea*) sea turtles in the central North Pacific Ocean. Fisheries Oceanography 13: 36–51.

Retamal, M. A., 1981. Catalogo ilustrado de los crustáceos decápodos de Chile, Vol. 44. Universidad de Concepción, Gayana: 1–110.

Rivera, T., 1988. Temperaturas del mar frente al Perú durante Enero – Marzo de 1987. In Salzwedel, H. & A. Landa (eds), Recursos y Dinámica del Ecosistema de Afloramiento Peruano. Boletín del Instituto del Mar del Perú, Volumen Extraordinario: 11–15.

Saba, V. S., P. Santidrian-Tomillo, R. D. Reina, J. R. Spotila, J. A. Musik, D. A. Evans & F. V. Paladino, 2007. The effect of the El Niño southern oscillation on the reproductive frequency of eastern Pacific leatherback turtles. Journal of Applied Ecology 44: 395–404.

Saba, V. S., G. L. Shillinger, A. M. Swithenbank, B. A. Block, J. R. Spotila, J. A. Musik & F. V. Paladino, 2008a. An oceanographic context for the foraging ecology of eastern Pacific leatherback turtles: consequences of ENSO. Deep-Sea Research I 55: 646–660.

Saba, V. S., J. R. Spotila, F. P. Chavez & J. A. Musick, 2008b. Bottom-up and climatic forcing on the worldwide population of leatherback turtles. Ecology 89: 1414–1427.

Santillán Corrales, L. A., 2008. Composición del contenido estomacal de *Chelonia mydas* en Bahía de Sechura. In Kelez, S., F. Van Oordt, N. de Paz & K. Forsberg (eds), Libro de Resúmenes II Simposio de tortugas marinas en el Pacífico Sur Oriental, Lima: 16.

Seminoff, J. A., A. Resendiz & W. J. Nichols, 2002. Home range of green turtles *Chelonia mydas* at a coastal foraging area in the Gulf of California, Mexico. Marine Ecology Progress Series 242: 253–265.

Seminoff, J. A., P. Zárate, M. Coyne, D. G. Foley, D. Parker, B. N. Lyon & P. H. Dutton, 2008. Post-nesting migrations of Galápagos green turtles *Chelonia mydas* in relation to oceanographic conditions: integrating satellite telemetry with remotely sensed ocean data. Endangered Species Research 3: 1–16.

Shrader, A. M., N. Owen-Smith & J. O. Ogutu, 2006. How a megagrazer copes with the dry season: food and nutrient intake rates by white rhinoceros in the wild. Functional Ecology 20: 376–384.

Stiasny, G., 1937. Scyphomedusae. John Murray Expedition 1933–1934, Scientific Reports 4: 203–242.

Tarazona, J., C. Paredes, L. Romero & S. Guzmán, 1988. La recolonización de las comunidades de mitílidos en la costa central del Perú después de El Niño 1982–1983. In Salzwedel, H. & A. Landa (eds), Recursos y dinámica del ecosistema de afloramiento Peruano. Boletín Instituto del Mar del Perú. Volumen Extraordinario, Callao: 115–120.

Tourre, Y. M., C. Cibot, L. Terray, W. B. White & B. Dewitte, 2005. Quasi-decadal and inter-decadal climate fluctuations in the Pacific Ocean from a CGCM. Geophysical Research Letters 32: 1–4.

Urquizo, W., C. Wosnitza-Mendo, E. Valdivia & C. Moreno, 1987. Catálogo de datos oceanográficos (T, S y O_2) por estaciones del año en diferentes profundidades frente a la costa Peruana, período 1970–1985. Boletín Instituto del Mar de Perú, Volumen Extraordinario, Callao: 1–445.

Vargas, P., P. Tello & C. Aranda, 1994. Sea turtle conservation in Peru: the present situation and a strategy for immediate action. In Bjorndal, K. A., A. B. Bolten, D. A. Johnston & P. J. Eliazar (eds), Proceedings of the Fourteenth Annual Symposium on Sea Turtle Biology and Conservation. NOAA Technical Memorandum NMFS-SEFSC-351: 159–161.

Wallace, B. P. & T. T. Jones, 2008. What makes marine turtles go: a review of metabolic rates and their consequences. Journal of Experimental Marine Biology and Ecology 356: 8–24.

Wallace, B. P., S. S. Kilham, F. V. Paladino & J. R. Spotila, 2006. Energy budget calculations indicate resource limitation in eastern Pacific leatherback turtles. Marine Ecology Progress Series 318: 263–270.

Wang, B., 1995. Transition from a cold to a warm state of the El Niño-Southern Oscillation cycle. Meteorology and Atmosphere Physics 56: 17–32.

Wosnitza-Mendo, C., M. Espino & M. Veliz, 1988. La pesquería artesanal en el Perú durante Junio 1986 a Junio 1988, Vol 93. Informe Instituto del Mar del Perú, Callao: 127–130.

Wyrtki, K., 1975. El Niño the dynamic response to the equatorial Pacific Ocean to Atmospheric Forcing. Journal of Physical Oceanography 5: 572–584.

Zárate, P., 2003. First results of the East Pacific green turtle, *Chelonia mydas*, nesting population assessment in the Galápagos Islands. In Seminoff, J. A. (ed.), Proceedings of the Twenty-Second Annual Symposium on Sea Turtle Biology and Conservation. NOAA Technical Memorandum NMFS-SEFSC-503: 80–81.

Zuta, S., M. Farfán & O. Morón, 1984. Características principales del mar frente a Perú durante la evolución del evento El Niño 1982–83, Vol. 15. Revista de la Comisión Permanente del Pacífico Sur, Quito: 141–178.

Hydrobiologia (2010) 645:81–96
DOI 10.1007/s10750-010-0217-8

Recurrence of bloom-forming scyphomedusae: wavelet analysis of a 200-year time series

Tjaša Kogovšek · Branko Bogunović · Alenka Malej

Published online: 1 April 2010

Abstract Four meroplanktonic scyphomedusae, *Aurelia aurita*, *Chrysaora hysoscella*, *Cotylorhiza tuberculata* and *Rhizostoma pulmo*, and the holoplanktonic, non-resident *Pelagia noctiluca* have formed blooms in the northern Adriatic over the last 200 years. Published data about the historical occurrences of these five scyphomedusae, in combination with our data, were used to analyse their long-term fluctuations in this northernmost part of the Mediterranean Sea. Analysis of the most recent blooms was complemented with environmental descriptors (temperature, salinity, pH, chlorophyll *a*, zooplankton dry weight and major river discharges). Continuous wavelet transformation analysis of the historical time series of scyphomedusae occurrences and environmental parameters revealed that the five species have been present regularly in the northern Adriatic over the last 200 years, with two major periods of jellyfish proliferations. The first period in the years around 1910 was characterised by significant periodicity of 8–12 years for each species, while the second period from the 1960s onwards was characterised by a shortened significant periodicity of less than 8 years. *Pelagia noctiluca* fluctuations were analysed in greater detail for the last four decades, revealing significant periodicities of ~10 years, 2.5 years, 8–14 months, and 8 months. The significantly marked periodicity of about 10 years in the *P. noctiluca* spectrum indicates a pattern similar to that observed in the western Mediterranean. Wavelet analysis showed that the periodicity of occurrence of five jellyfish species has shortened in recent decades and the recurrence of blooms has increased, particularly for *A. aurita* and *R. pulmo*.

Keywords Time series · Fluctuations · Wavelet analysis · Mediterranean Sea · Jellyfish

Introduction

Ecological responses to climate drivers have been observed in marine communities, including plankton (Hays et al., 2005). In recent years, connections between jellyfish blooms and variations in climate have been emphasised (reviewed in Lynam et al., 2004; Purcell, 2005; Purcell et al., 2007; Molinero et al., 2008). Most of the climate-associated effects on jellyfish have been regime shifts, phenological changes, and distribution expansions over duration of 3–60 years. Very few long-term time series of blooms of gelatinous plankton exist, because these organisms are fragile and difficult to sample with

Guest editors: J. E. Purcell & Dror Angel / Jellyfish Blooms: New Problems and Solutions

T. Kogovšek (✉) · B. Bogunović · A. Malej
Marine Biology Station, National Institute of Biology, Fornace 41, 6330 Piran, Slovenia
e-mail: kogovsek@mbss.org

conventional oceanographic methods. Although large scyphomedusae generally are more robust than other gelatinous taxa, sampling them is difficult due to their size (CIESM, 2001). Therefore, it is unsurprising that long-term data sets for them are rare and were begun mostly in the last few decades. Sampling and observation techniques have improved recently and new ones have been developed [SCUBA diving, submersibles, ROVs (Remotely Operated Vehicles), and aerial and acoustic surveys]. Semi-quantitative data are available from fishery surveys that record jellyfish as by-catch (Brodeur et al., 1999; Graham, 2001) and have proved useful in assessing trends in jellyfish distribution, biomass, and abundance (reviewed in Purcell, 2009). Because most historical data on jellyfish population size are descriptive rather than quantitative, determining a suitable method for time series analysis can be problematic.

Moreover, non-linear, aperiodic, and non-stationary processes are very common in nature (Hsieh et al., 2005). Such properties of time series constitute a limitation when applying most of the analytical techniques to detect periodicity in time series data. Wavelet transformation, a time scale and/or time frequency decomposition of the signal, is a powerful tool for analysing the non-stationary, aperiodic, and noisy signals that often occur in ecological time series (Torrence & Compo, 1998).

In this study, information about the historical occurrences of five scyphomedusae, *Aurelia aurita* (Linnaeus, 1758), *Chrysaora hysoscella* (Linnaeus, 1766), *Cotylorhiza tuberculata* (Macri, 1778), *Pelagia noctiluca* (Forskål, 1775) and *Rhizostoma pulmo* (Macri, 1778), from published sources as well as from our own observations, was availed to analyse their long-term fluctuations in the northernmost part of the Mediterranean Sea. Four of the five species (*A. aurita*, *C. hysoscella*, *C. tuberculata* and *R. pulmo*) have metagenetic life histories and are considered to be native to the northern Adriatic, although polyp locations have been found only for *A. aurita* (Di Camillo et al., 2010). In contrast, the holoplanktonic *P. noctiluca* has not established a viable long-term population there (Malej & Malej, 2004). Our main objective of this study was to assess the periodicity of scyphomedusae occurrence in the northern Adriatic during the past 200 years and in more detail for the recent decades by use of the wavelet time series technique.

Materials and methods

Study area

Long-term fluctuations of five bloom-forming scyphomedusae were studied in the northernmost part of the Adriatic Sea (Fig. 1). The northern Adriatic is a shallow sea with large salinity and temperature variations and strong seasonal vertical stratification. The physical properties and dynamics of the northern Adriatic water masses are characterised mainly by atmospheric forces due to wind stress, heat flux, and fresh water inputs by major rivers along the Italian coast (Cushman-Roisin et al., 2001). The northern Adriatic gains heat at the air–sea surface interface from March to August (Malačič, 1991). The dominant winds are the Bora that blows from the north-east, most frequently in autumn and winter, and the Sirocco from the south-east. Two gyres can be generated by the Bora, a cyclonic gyre in the northern part of the basin and an anticyclonic gyre off the southern Istrian coast (Kuzmić & Orlić, 2006). The cyclonic gyre pushes the Po River freshwater input up to the Istrian coast.

The Po river is the largest river discharging in the northern Adriatic. Its freshwater flux is strongly influenced by the seasonal pattern of precipitation, with two low-water periods in winter and summer

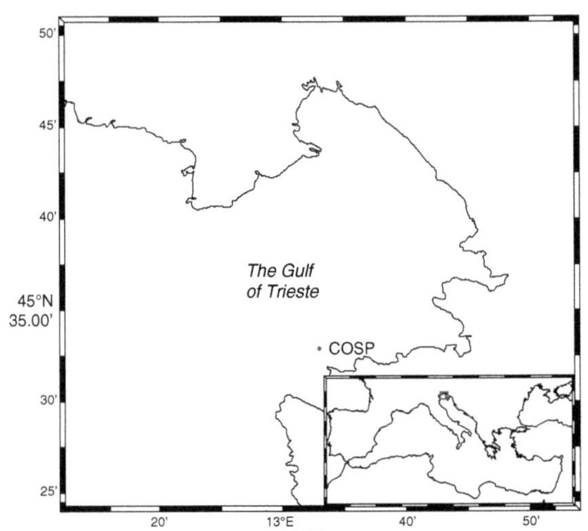

Fig. 1 Location of the study area and a map of the Gulf of Trieste showing the position of the Coastal Oceanographic Station Piran (MBS, NIB), monitoring site for environmental parameters

and two flood periods in late fall and spring (Zanchettin et al., 2008). The Soča River plays a similar role, but in the Gulf of Trieste. After the peak outflow in April and October–November (Comici & Bussani, 2007), and after the southern wind decreases, the Soča plume occupies a large area of the inner part of the Gulf of Trieste (Malačič & Petelin, 2009). Freshwater discharges a significant quantity of nutrients into the northern Adriatic, and the area has been considered as one of the most productive in the Mediterranean Sea (Sournia, 1973). Reduction of the phosphorus load in the Po River in the late 1980s (de Wit & Bendoricchio, 2001), coupled with reduced freshwater discharge (Comici & Bussani, 2007; Zanchettin et al., 2008), has caused oligotrophication of the basin over the last decade (Mozetič et al., 2009).

Data

The reconstructed time series for the historical occurrences of five scyphomedusae species in the northern Adriatic was assembled from published data sources, reports and theses, and supplemented by our own observations (Table 1). Scientific observations are available for about 200 years from marine observatories in the northern Adriatic. The data are predominantly qualitative or semi-quantitative, but quantitative data on jellyfish abundance are rare. We considered every observation reported for each species, and included them in the time series.

In order to avoid the problems associated with descriptive data and to apply robust analyses to this time series, the years of jellyfish occurrence and absence were marked as one and zero, respectively. For *P. noctiluca*, an additional, more detailed analysis was performed from 1976 onwards, because monthly observations were recorded (the jellyfish database of the Marine Biology Station, Piran, Slovenia). Scyphomedusae were monitored by visual observations from the shore and with an underwater camera deployed from boats. Additional information came from fishermen and similar reliable sources (co-workers, biologists and photographs sent by reporters).

Estimation of *P. noctiluca* abundance during the last two bloom periods in the northern Adriatic (1977–1986 and 2004–2007) was attempted, although the sampling methods were not strictly comparable. Estimates were based on plankton net tows from the bottom to surface using an FAO net with 500-μm mesh (Piccinetti & Piccinetti Manfrin, 1991), a WP2 net with 200-μm mesh (Malej, 1989), and a fish trawl with 20-mm stretched net (Miloš, 2009).

Environmental descriptors include Po and Soča river discharges, sea temperature, salinity and pH, phytoplankton biomass estimated as chlorophyll *a* (Chl *a*), and zooplankton biomass expressed in dry weight. Sampling was conducted during daylight monthly from 1989 to 2008 at the Coastal Oceanographic Station Piran (45°32′55.68″N 13°33′1.89″E) (Fig. 1). During 2005–2008, environmental descriptors were sampled twice monthly and the corresponding monthly means calculated. Temperature, salinity, and pH were measured using a CTD probe and were averaged for every 5-m of depth. Chl *a* concentrations, corrected for phaeopigments, were determined fluorometrically (Holm-Hansen et al., 1965) from 90% acetone extracts for every 5 m depth. Zooplankton was sampled with vertical tows from near-bottom to the surface with a WP 2 net (200-μm mesh). Total dry weight was determined on aliquots of the preserved samples (4% buffered formaldehyde). For the detailed analysis of the environmental parameters, we selected the measurements made at 5-m depth, with the exception of zooplankton dry weight for which vertical tows integrated the water column. Monthly values of the Po and Soča river discharges were obtained from Ufficio idrografico del Po, Parma (Ministero dei Lavori Pubblici, Italy) and ARSO, Slovenia.

Data analysis

Nonlinear and non-stationary processes are often characteristic of ecological time series. Therefore, it is essential to understand the limitations of each time series analytical technique and their application in different time series. The Blackman–Tukey method commonly is used in non-stationary time series and is based on Fourier transform; however, it assumes that the data are a composite of sine and cosine waves that are globally uniform in time and have infinite spans. In contrast to the Fourier transform, wavelet transformation uses base functions (wavelets) (Torrence & Compo, 1998). Wavelets are defined as small packets of waves with specific frequencies that approach zero at both ends. The wavelets can be stretched easily and translated to the desired resolution. Thus, changes

Table 1 Historical records of bloom-forming scyphomedusae in the northern Adriatic

Year	Aa	Ch	Ct	Pn	Rp	References
1790				+		Lhotsky (1830) cited in Avian & Rottini Sandrini (1994)
1830				+		Lhotsky (1830) cited in Avian & Rottini Sandrini (1994)
1837	+					Siebold (1839) cited in Avian & Rottini Sandrini (1994)
1838	+					Siebold (1839) cited in Avian & Rottini Sandrini (1994)
1843				+		Will (1944) cited in Avian & Rottini Sandrini (1994)
1874	+	+				Graeffe (1875)
1875	+	+	+		+	Claus (1876) cited in Avian & Rottini Sandrini (1994)
1879				+		Claus (1883) cited in Avian & Rottini Sandrini (1994)
1880	+					Haeckel (1880) cited in Avian & Rottini Sandrini (1994)
1883			+		+	Claus (1883) cited in Avian & Rottini Sandrini (1994)
1884	+					Carus (1885) cited in Avian & Rottini Sandrini (1994)
1895				+		Babić (1913)
1896	+					Babić (1913)
1899	+	+	b		+	Cori & Steuer (1901)
1900	+	+	+		+	Cori & Steuer (1901)
1901	+				+	Steuer (1902)
1902	b	+	+		b	Steuer (1904)
1903					+	Steuer (1904)
1907		b				Babić (1913)
1908	+	+	+		b	Stiasny (1909)
1909	+	+			b	Stiasny (1910)
1910	b	+		+	+	Babić (1913), Stiasny (1911)
1911	+	+	b	+	b	Babić (1913), Stiasny (1912)
1913				+		Stiasny (1914), Stiasny (1919)
1914	+		+	+	b	Szüts (1915)
1921			b		+	Issel (1922)
1922				+		Neppi (1922) cited in Avian & Rottini Sandrini (1994)
1962	b					Vio (unpubl.) cited in Avian & Rottini Sandrini (1994)
1976				+	+	Malej (1982), Avian & Rottini Sandrini (1994)
1977				b	+	Malej (1982), Avian & Rottini Sandrini (1994)
1978	+			b	+	Malej (1982, 1989, unpubl. data), Avian & Rottini Sandrini (1994)
1979	+			b	+	Malej (1982, 1989, unpubl. data), Avian & Rottini Sandrini (1994)
1980	+		b	b	+	Malej (1982, 1989, unpubl. data), Avian & Rottini Sandrini (1994)
1981	+	+	+	b	+	Malej (1982, 1989, unpubl. data), Avian & Rottini Sandrini (1994)
1982	+	+	b	b	+	Malej (2001, unpubl. data), Avian & Rottini Sandrini (1994)
1983	+		+	b	b	Malej (2001, unpubl. data), Giorgi et al. (1985), Piccinetti & Piccinetti Manfrin (1991)
1984	+		b	b	b	Malej (2001, unpubl. data), Giorgi et al. (1985), Piccinetti & Piccinetti Manfrin (1991)
1985	+	+	+	b	b	Malej (2001, unpubl. data), Giorgi et al. (1985), Piccinetti & Piccinetti Manfrin (1991)
1986	+		b	b	+	Malej (2001, unpubl. data), Del Negro et al. (1992), Piccinetti & Piccinetti Manfrin (1991)
1987	b		+		b	Malej (2001), Malej & Malej (2004, unpubl. data), Del Negro et al. (1992)

Table 1 continued

Year	Aa	Ch	Ct	Pn	Rp	References
1988	+		+		+	Malej (2001, unpubl. data), Del Negro et al. (1992)
1989	b	b	+		b	Malej (2001, unpubl. data), Del Negro et al. (1992)
1990	+	+			+	Malej (2001, unpubl. data)
1991	+	+			+	Malej (2001, unpubl. data)
1992	+	+			+	Malej (2001, unpubl. data)
1993						Malej (2001, unpubl. data)
1994		+			+	Malej (2001, unpubl. data)
1995		+				Malej (2001, unpubl. data)
1996	+				b	Malej (2001, unpubl. data)
1997	+				+	Malej (2001, unpubl. data)
1998	+				+	Malej (2001, unpubl. data)
1999	b					Malej (2001, unpubl. data)
2000	+				b	Unpubl. data
2001	b	+	+		+	Unpubl. data
2002	+				+	Unpubl. data
2003	b				b	Miloš (2009, unpubl. data)
2004	+			b	+	Miloš (2009, unpubl. data)
2005	+		+	b	b	Miloš (2009, unpubl. data)
2006	b	+		b	b	Miloš (2009, unpubl. data)
2007	b	+	+	+	+	Unpubl. data
2008	b		+		+	Unpubl. data
2009	b	+	+		+	Unpubl. data

Aa *Aurelia aurita*, Ch *Chrysaora hysoscella*, Ct *Cotylorhiza tuberculata*, Pn *Pelagia noctiluca*, Rp *Rhizostoma pulmo*
+ = present, b = high abundance (bloom)

occurring in wavelets can be mapped into a time-period domain.

Before wavelet analysis, time series of descriptive parameters were checked for gaps in the data. Because one of the requirements for wavelet transformation is the continuity of the time series, linear interpolation was performed on all time series except for data on scyphomedusae and river discharge. The latter data were transformed to natural logarithms to achieve a Gaussian (normal) distribution.

The records for the five scyphomedusae covered about 200 years, but some years had few or almost no observations (world wars). The original data (1 = presence, 0 = absence of particular species in the northern Adriatic) were binary, and interpolation to obtain the missing data was unacceptable. Therefore, for the few years when data were missing, the missing data were treated as 0 (absent) in the specified location.

In order to analyse variations in the occurrence of scyphomedusae and possible oscillations of the environmental parameters in the time series, the continuous wavelet transformation (CWT) was performed using software written in MATLAB® (Grinsted et al., 2004). The CWT, W, was defined as a convolution integral of the time series $x(t)$ with a wavelet function (Percival & Walden, 2000):

$$W_x(a,\tau) = \frac{1}{\sqrt{a}} \int_{-\infty}^{\infty} x(t)\psi^*\left(\frac{t-\tau}{a}\right)\mathrm{d}f$$
$$= \int_{-\infty}^{\infty} x(t)\psi^*_{a,\tau}(t)\mathrm{d}t$$

where * indicates a complex conjugate, parameters a and τ denote dilation (scale parameter related to frequency) and translation (related to time position), respectively. $\psi(t)$ is the *mother wavelet*, a continuous

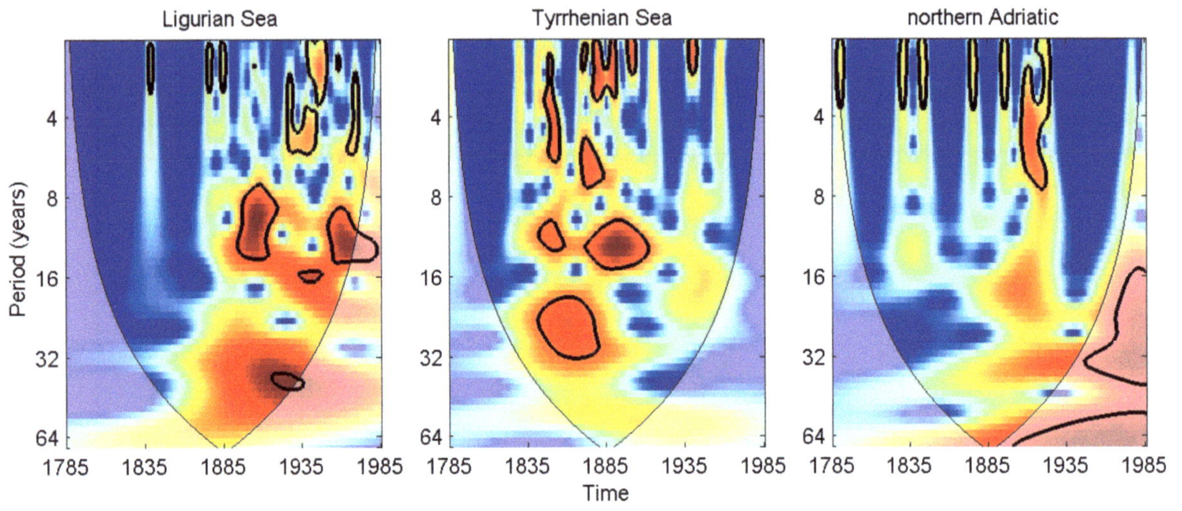

Fig. 2 Wavelet spectra of the long-term time series of *Pelagia noctiluca* occurence in the western Mediterranean Sea from Goy et al. (1989). The time series was divided into two regions: the Ligurian Sea (Villefranche sur mer, Nice and Monaco) and the Tyrrheninan Sea (Naples, Messina and Palermo) data set and represented together with the spectrum from the northern Adriatic covering the same time span. Colours indicate differing intensity of periodicities (warmer colours refer to higher intensity). Significant areas are indicated by solid black lines ($P < 0.05$)

function in both the time and frequency domains, and was expressed as the function of a and τ:

$$\psi_{a,\tau}(t) = \frac{1}{\sqrt{a}}\psi\left(\frac{t-\tau}{a}\right)$$

Several wavelet functions exist for use in the wavelet transform. We used the Morlet function because it best described time series with unknown frequencies. The Morlet function was defined as in Percival & Walden (2000):

$$\psi(t) = \pi^{-\frac{1}{4}}\exp(-i2\pi f_0 t)\exp\left(-t^2/2\right)$$

where f_0 is the central frequency of the Morlet wavelet.

In order to determine the significance level of the wavelet spectra, the spectrum of individual time series was tested against red noise (univariate lag-1 autoregressive process) (Torrence & Compo, 1998). Significant areas (95%) on the wavelet power spectra were indicated by solid black lines. In order to minimise bias due to edge effects, the data were padded with zeros, up to the next highest power of two. Only the area inside the marked cone of influence was considered in the interpretation of wavelet spectra.

In order to test the suitability of the analytical technique, the long-term time series of *P. noctiluca* occurrence in the western Mediterranean Sea (Goy et al., 1989) was re-constructed and re-analysed. The historical data from Villefranche sur mer, Nice and Monaco were assembled for the Ligurian Sea time series and the data from Naples, Messina and Palermo for the Tyrrheninan Sea time series. The data in both sets were converted to 1 = presence, 0 = absence (Fig. 2).

From the CWT of two time series, where a relationship between the two parameters was expected, a cross wavelet transform (XWT) was constructed. The XWT exposed regions with high common power and revealed information about the phase relationship in time–frequency space. The XWT of two time series x_n and y_n is defined as:

$$W^{XY} = W^{W} W^{Y*}$$

where * denotes complex conjugation. For the two time series with theoretical Fourier spectra

P_k^X and P_k^Y, the cross wavelet distribution is as in Torrence & Compo (1998):

$$\frac{\left|W_n^X(s)W_n^{Y*}(s)\right|}{\sigma_X\sigma_Y} \Rightarrow \frac{Z_v(p)}{v}\sqrt{P_k^X P_k^Y}$$

where σ_X and σ_Y are the respective standard deviations, and $Z_v(p)$ is the confidence level associated with the probability p. For $v = 1$ (real wavelets), $Z_1(95\%) = 2.182$, while for $v = 2$ (complex wavelets), $Z_2(95\%) = 3.999$.

For environmental data time series, annual anomalies were calculated after a protocol designed in R language for the EU-funded project SESAME (Berline et al., 2009). The seasonality was subtracted from the linearly interpolated and normally (Gaussian) distributed time series before calculating the annual anomalies. Seasonal anomalies were also calculated for sea temperature and salinity data.

Results

Chronological data of five bloom-forming scyphomedusae in the northern Adriatic

Historical data on the presence and blooms of five scyphomedusae species in the northern Adriatic over the last 200 years showed that three main periods of high scyphomedusae occurred (Table 1). All the five species were present and formed blooms in the periods 1909–1915, 1981–1986 and 2004–2007, while blooms of individual species were also recorded in some other years (Table 1). The chronological data for *Aurelia aurita* proliferation in the northern Adriatic showed that it has been one of the two most common scyphomedusae in the northern Adriatic. *A. aurita* was first recorded during 1837–1838 and has been observed regularly in every decade since the 1960s with blooms increasing in the last two decades. More frequent observations of blooms in recent years may be due partly to improved observation techniques. *Rhizostoma pulmo* also has been very common in the northern Adriatic region. It was first documented in 1875 and has been observed regularly since, except for few decades (1930–1960). *Cotylorhiza tuberculata* and *Chrysaora hysoscella* have formed blooms in the past, although they were not documented very frequently.

Most notable were *P. noctiluca* blooms because they sting painfully when they contact human skin. Its presence in the northern Adriatic was first documented in 1790. The periods around 1915, the late 1970s to the early 1980s, and from 2004 to 2007 were characterised by the most notable bloom events. The two last proliferations of *P. noctiluca* in the northern Adriatic were monitored and the data analysed in detail.

Continuous long-term wavelet spectra analysis of jellyfish in the northern Adriatic

Although the population dynamics and the ability to adapt to different environments differ among the

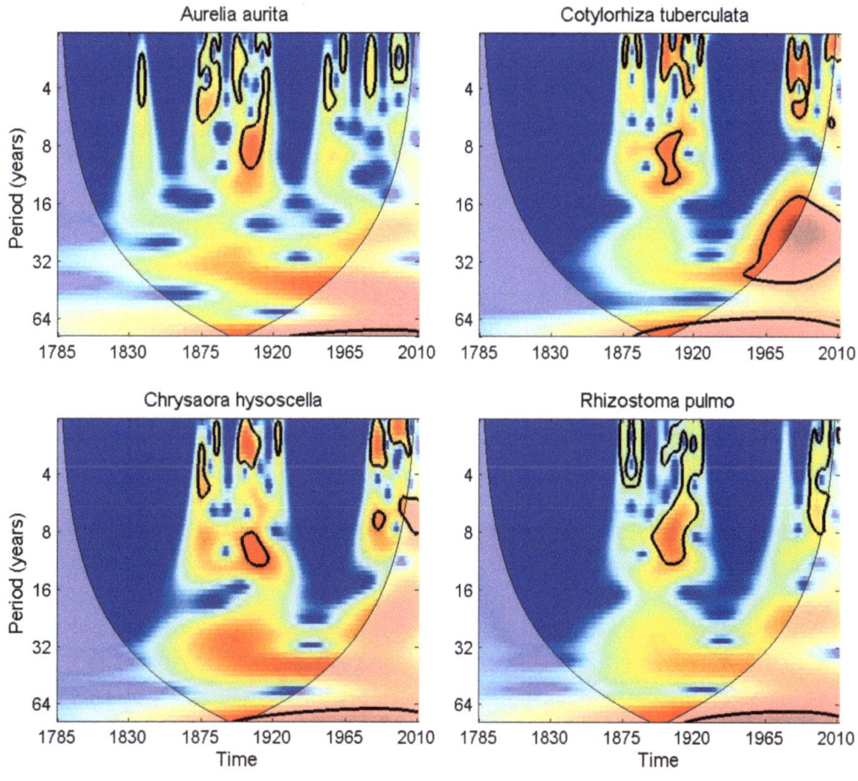

Fig. 3 Wavelet spectra of the four scyphomedusae time series in the northern Adriatic Sea

species, a similar pattern in the wavelet spectra of the four metagenic species (*A. aurita*, *C. hysoscella*, *C. tuberculata* and *R. pulmo*) time series were seen, with two major periods of jellyfish proliferation in the northern Adriatic during the past 200 years (Fig. 3). The first period began around 1875 and lasted until 1922, with increasing occurrences in consecutive years from 1908 to 1915; the second period from the 1970s onward followed nearly 40 years of few jellyfish observations. During those 40 years, the data gaps for 1915–1920 and 1940–1950 included the two world wars.

The XWT (not shown) revealed similar periodicities for all species during the first period of jellyfish proliferation, with periodicities of up to 6 years and a periodicity of 8–12 years around 1910. In contrast, the second period was characterised by a significant periodicity of <8 years and an additional longer term periodicity of 20–40 years; however, the long-term periodicity was significant only for the *C. tuberculata* wavelet spectrum (Fig. 3).

The most frequently reported scyphomedusa in the northern Adriatic throughout the period of study was *A. aurita*, and its recurrence was regular. The periodicity of 6 years, characteristic of its first period of proliferation in the northern Adriatic, appeared to dwindle in recent years. In the 2000s, notably from 2004 onwards, *A. aurita* has been present every year in large numbers from the beginning of February to the end of June.

The most detailed early records of jellyfish in the northern Adriatic probably were of *R. pulmo* blooms, because its high abundance during blooms interfered with fishing. During the first period (1875–1922), blooms of this species showed significant periodicities of 4 and 6–8 years; it has been present every decade after the 1950s, with increasing blooms in the last two decades. It was most abundant during the last decade from 2003 to 2006, after which its numbers decreased markedly.

In contrast, *C. hysoscella* was observed regularly during the 2000s, but rarely in high numbers. It often was present with *A. aurita* blooms from February to May every year. Similarly, *C. tuberculata* rarely formed blooms in the northern Adriatic, although its presence was regularly reported, mostly from August to September. It was present during the first period (1875–1921) with significant periodicities of about 3 and 7–12 years. The second period, from 1978 until

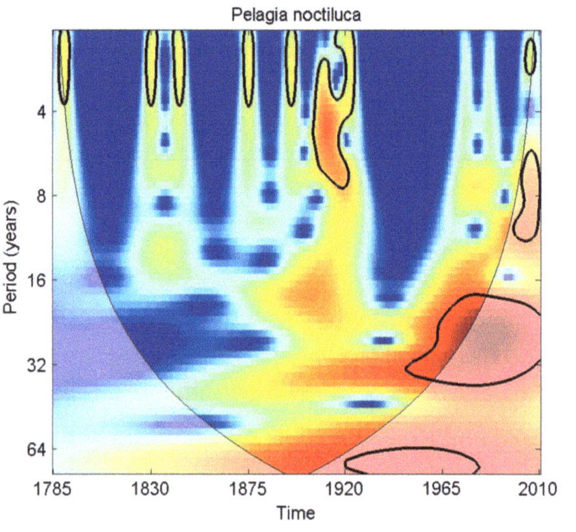

Fig. 4 Wavelet spectrum of the *Pelagia noctiluca* time series in the northern Adriatic Sea

the present, is characterised by periodicities of 6 years and approximately 15–30 years. Blooms of *P. noctiluca* in the late 1970s and 1980s also was marked by a 20–30 years periodicity in the wavelet spectrum (Fig. 4).

The presence of *P. noctiluca* in the northern Adriatic was first recorded in 1790 and noted only four times up to 1909, when a period of 6 consecutive years of proliferation occurred (Fig. 4). A significant periodicity of 8 years appeared in the spectrum in the years around 1910, while the second period, including the blooming period of 10 consecutive years in the late 1970s and early 1980s, was characterised by a relatively strong periodicity of 20–30 years.

Continuous wavelet spectra analysis of *Pelagia noctiluca* recurrence in the western Mediterranean Sea

The long-term time series of *P. noctiluca* occurrence in the western Mediterranean Sea (Goy et al., 1989) was re-analysed with wavelet analysis to test the reliability of the method performed on the binary data and to investigate possible similarities of *P. noctiluca* recurrence in the two areas. The wavelet analysis confirmed the 12-years periodicity previously reported by Goy et al. (1989). Moreover, our analysis enabled definition of the exact time domain of the detected periodicities. In addition to the 12-years periodicity, the wavelet spectra of *P. noctiluca* recurrence in the Ligurian and Tyrrhenian seas showed periodicities of up to 6 years, 20–30 years, and around 45 years (Fig. 2). The spectrum for the Ligurian Sea data set showed two periods of 8–14 years periodicity. The first period corresponded to the years between 1895 and 1925 and the second from 1955 onward. The third period, from 1910 to 1935, occurred as part of a long-term periodicity of approximately 45 years. In contrast, the Tyrrhenian Sea data set had significant periodicities of approximately 12 years between 1845 and 1860 and between 1870 and 1925, but the years between 1840 and 1880 were marked by a long-term periodicity of 20–30 years.

Pelagia noctiluca: decadal analysis

A more detailed wavelet spectrum for *P. noctiluca* was constructed from monthly observations from 1976 to 2009 (Fig. 5). The period 1976–1986 was characterised by three significant periodicities: approximately 2.5 years, 8–14 months, and slightly shorter than 8 months; however, the second period of its proliferation (2004–2007) was distinguished by a short periodicity of up to 6 months. The longest periodicity in Fig. 5 corresponded to 10 consecutive years of jellyfish recurrence.

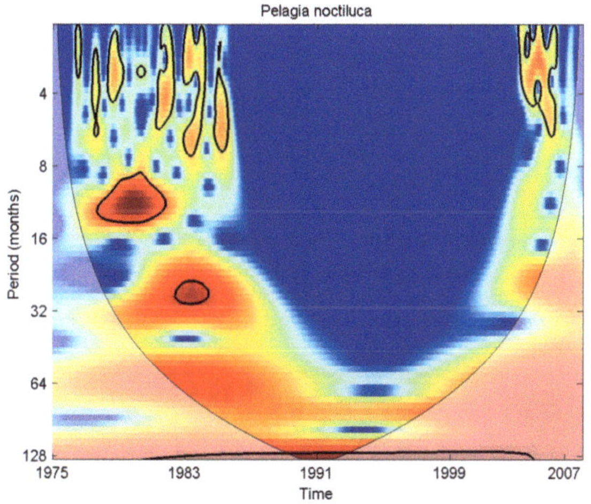

Fig. 5 Wavelet spectrum of monthly observations of *Pelagia noctiluca* covering the period of its last two intense blooming periods (1976–1986 and 2004–2007) in the northern Adriatic Sea

The significantly shorter periodicities during 1976–1986 (Fig. 5) increased with time, indicating prolonged *P. noctiluca* presence over 10 consecutive years. Indeed, it was present only sporadically in August and September in 1976 and 1977, but in the following years, the duration of its presence increased until 1983–1986, when the *P. noctiluca* population had very high densities all year long. After August 1986, the population in the northern Adriatic collapsed and only reappeared in early 2004 after an 18-year absence. During this recent bloom, *P. noctiluca* was most abundant from November 2005 to March 2007, when it suddenly disappeared. The estimated abundance of *P. noctiluca* during this recent bloom was comparable to that in the 1980s, despite differences in sampling techniques (Table 2). The highest densities of *P. noctiluca* populations were observed during March and September in both bloom periods.

In order to understand the changes in the northern Adriatic pelagic ecosystem during the proliferation of the *P. noctiluca* population, the environmental parameters were analysed for the period of the available time series, taking into account the intervals before and after the last bloom (Fig. 6).

River outflows and environmental parameters

Wavelet transformation analysis was performed on Po and Soča river flows and environmental descriptors (temperature, salinity, pH, Chl *a*, and mesozooplankton biomass), even though the time span of the latter did not fully cover the last two episodes of massive *P. noctiluca* abundance. Positive trends in salinity and pH were observed after the decrease of mean annual flows of the Soča and Po rivers from 2002 and 2003 on (Fig. 6). The XWT performed on the CWT of the river discharges and salinity time series (not shown) showed that salinity fluctuations at the sampling location were influenced by both rivers. The CWT of the salinity, which reflected the summed short-term cycles of both inflows, was used to investigate the Chl *a* spectra in relation to river discharge. The XWT (Fig. 7) showed a concomitant yearly cycle until 2003, except during 1996–1998 when a periodicity of 6 months was observed. The cycles in salinity were perfectly accompanied by changes in the Chl *a* concentrations until 2003, when the CWT Chl *a* spectrum shifted from a significant 12-month cycle to significant periodicities of up to

Table 2 Densities of *Pelagia noctiluca* during two bloom periods (1977–1986 and 2004–2007) in the northern Adriatic

Month/year	Density (no. 100 m^{-3})	References
July 1983	1.84	Piccinetti & Piccinetti Manfrin (1991)
July 1984	0.8	Piccinetti & Piccinetti Manfrin (1991)
July 2004	0.001	Miloš (2009)
December 1983	1.02	Piccinetti & Piccinetti Manfrin (1991)
December 1984	0.22	Piccinetti & Piccinetti Manfrin (1991)
December 2004	0.94	Miloš (2009)
December 2005	0.003	Miloš (2009)
March 1984	0.32	Piccinetti & Piccinetti Manfrin (1991)
March 1985	1.15	Piccinetti & Piccinetti Manfrin (1991)
March 1986	19.0	Piccinetti & Piccinetti Manfrin (1991)
February 2005	1.16	Miloš (2009)
Sept.–Nov. 1984	1.9	Malej (1989)
September 1985	2.13	Piccinetti & Piccinetti Manfrin (1991)
October 2005	0.01	Miloš (2009)

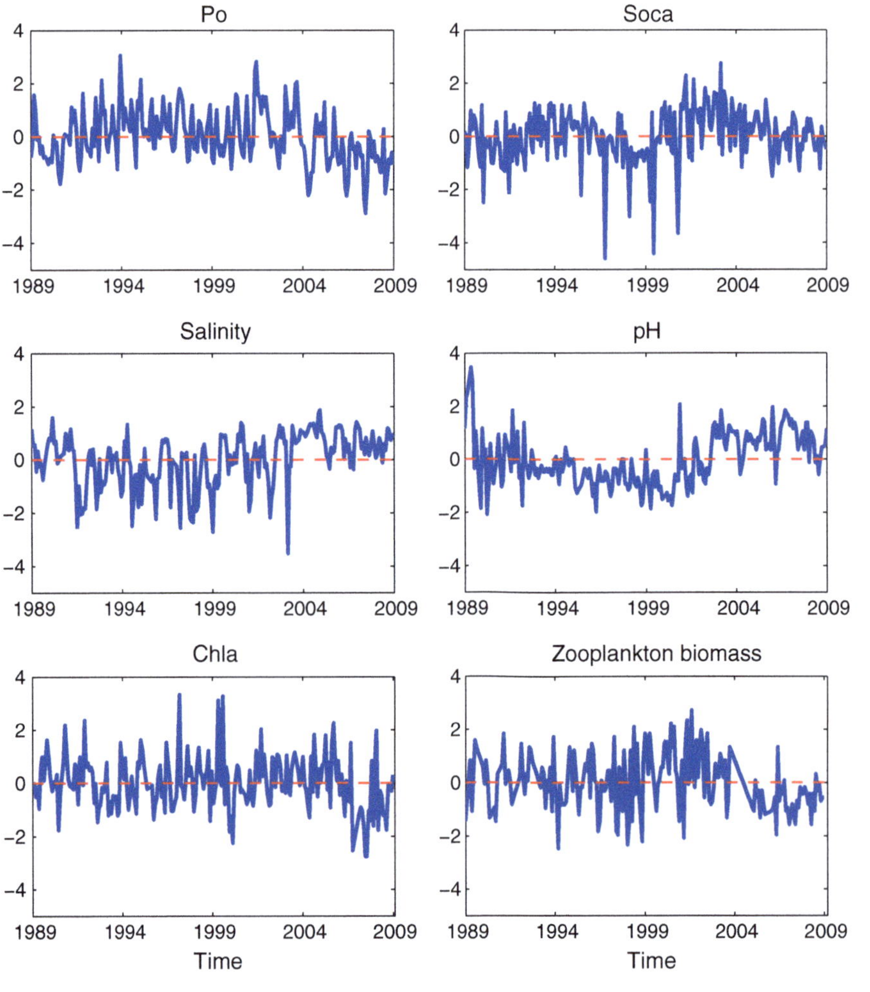

Fig. 6 Normalized data by standard deviation of environmental descriptors (Po and Soča river inflows, salinity, pH, Chl *a* and mesozooplankton dry weight) for the period 1989–2007 in the northern Adriatic Sea

Fig. 7 Cross wavelet spectra of salinity—Chl *a*, Chl *a*—mesozooplankton biomass, Chl *a*—*Pelagia noctiluca*, and mesozooplankton biomass—*Pelagia noctiluca* time series in the northern Adriatic Sea

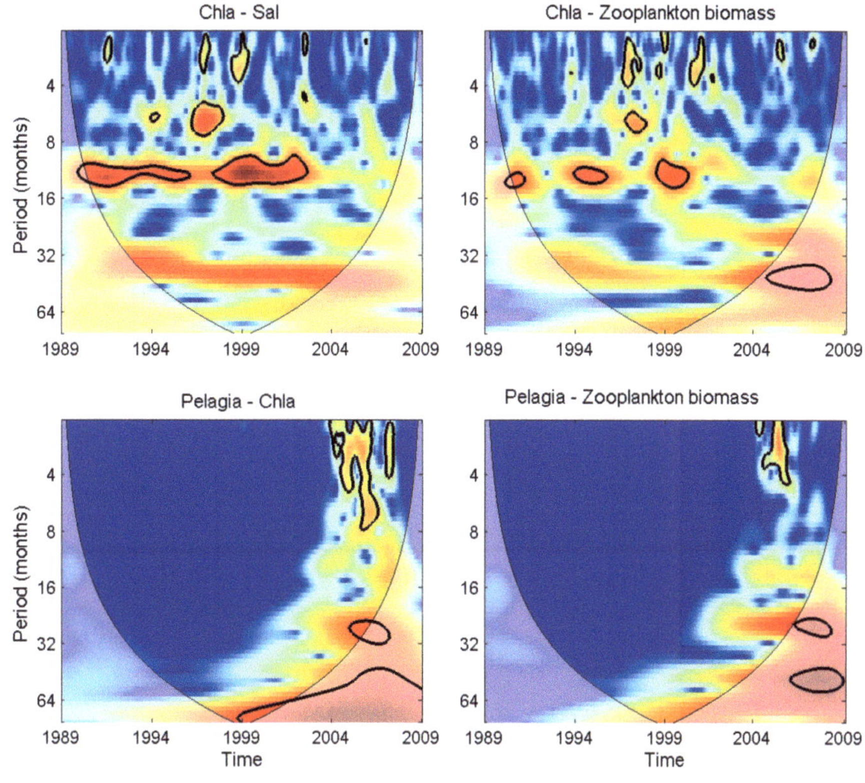

4 months. A change in the next-higher trophic level from 2003 onwards was deduced from the XWT of Chl *a* and mesozooplankton dry weight time series (Fig. 7). Periodicities of 12 months for both parameters were significant until 2003, but thereafter, no significant periodicity in mesozooplankton dry weight spectra was observed. Moreover, the annual anomalies in zooplankton dry weight data showed a negative trend from 2001 to 2008, with a clear shift to negative anomalies in 2004.

The periodicities in CWT spectra <12 months reflected the annual dynamics of the parameters observed. We paid additional attention to periods longer than a year, although the length of the time series analysed did not allow recognition of periodicities longer than 3 years. In all environmental descriptor spectra, except for river discharges and pH, non-significant periodicities of about 1 and 3 years were observed. Three-year cycles of increasing salinity, observed from 1989 to the present, concurred with observations in Malačič et al. (2006). After 2002, the annual anomalies were above the long-term means but the 3-year cycles persisted. No obvious cycles were seen in pH, although a marked rise above the long-term mean was detected.

In addition to yearly trends, seasonal anomalies for temperature and salinity were calculated to delineate seasonality and the eventual stronger ingression of water masses from the more southern part into the northern Adriatic (Fig. 8). Indeed, positive salinity and negative temperature anomalies were seen from autumn 2003 until autumn 2005, which indicated waters of southern origin. The most pronounced anomalies occurred in autumn 2003, which was a year with unusually warm, dry weather.

Discussion

The time series analysis of the occurrences of five scyphomedusae, *A. aurita*, *C. hysoscella*, *C. tuberculata*, *P. noctiluca* and *R. pulmo*, showed two major periods of jellyfish proliferation in the northern Adriatic Sea over the last 200 years, one in the years around 1910, and the second from 1970 onward. The wavelet spectra analysis revealed several significant

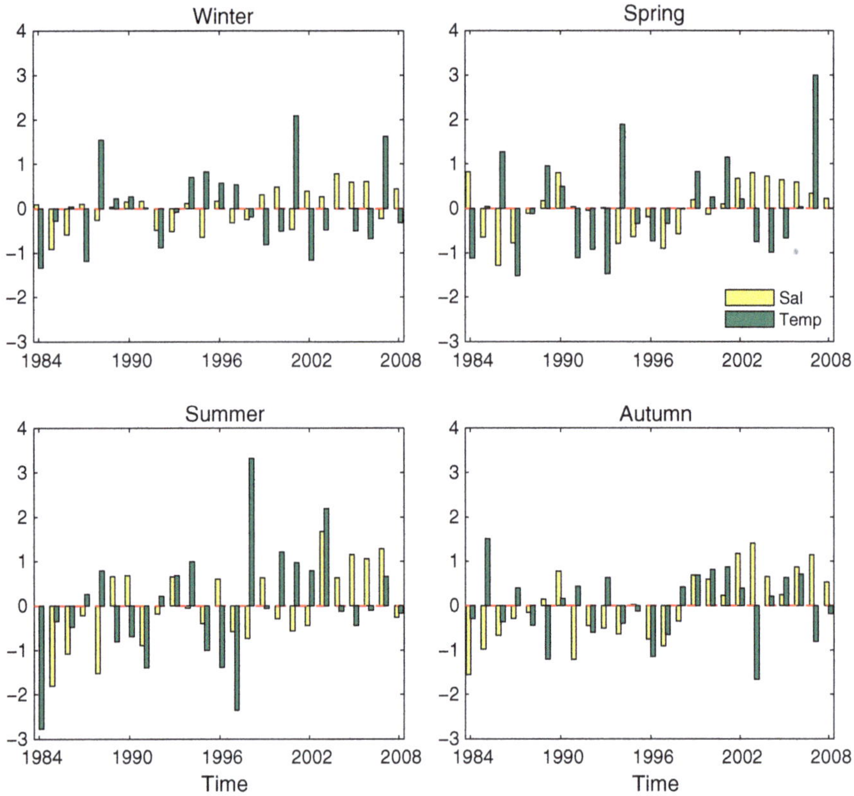

Fig. 8 Seasonal anomalies of temperature (Temp) and salinity (Sal) during 1984–2008 in the northern Adriatic Sea

periodicities common to all species: (1) a periodicity of around 3 years, corresponding approximately to the 3-year cycle of salinity fluctuations; (2) an 8–12-years periodicity that corresponded with the 12-years periodicity of *P. noctiluca* occurrences in the western Mediterranean Sea (Goy et al., 1989); (3) a periodicity of 20–40 years, which was significant only for the *P. noctiluca* and *C. tuberculata* spectra, although present in the other species' spectra.

Aurelia aurita and *R. pulmo* were the most common species in the northern Adriatic. Their intense blooming over the last decade coincided with a phytoplankton community shift (Mozetič et al., 2009). In the last decade, a negative trend in Chl *a* concentrations was observed in the northern Adriatic (Mozetič et al., 2009) in conjunction with reduced river discharges (Comici & Bussani, 2007; Zanchettin et al., 2008). The trend reflected changes in the phytoplankton community structure, with a delayed and reduced early spring diatom bloom that was replaced by a heterogenic April–May bloom composed mostly of nanoflagellates (Mozetič et al., 2009). The change in timing and composition of primary producers would affect the higher trophic levels, as seen by a change in the zooplankton community towards smaller copepod species that prefer to graze on small phytoplankton (Kamburska & Fonda Umani, 2006). Stable isotope analysis showed *A. aurita* to be at a slightly higher trophic level than copepods, and gut content analyses, along with enclosure experiments, showed small copepods and microzooplankton as its main food (Kohama et al., 2006; Malej et al., 2007).

The highest abundances in *R. pulmo* in the last decade occurred in 2003 and 2004. After this period *R. pulmo* numbers decreased, as also reported for *Rhizostoma* spp. occurrences in European seas (Lilley et al., 2009). Their blooms were confined to semi-enclosed bays that receive appreciable freshwater and nutrient input from rivers. Indeed, *R. pulmo* in the northern Adriatic is more abundant and more frequently observed along the western coast where fresh water influence is more pronounced.

Pelagia noctiluca has been the most studied scyphomedusa in the northern Adriatic, due to its painful sting. It was present and formed blooms during the years 1909–1915, 1976–1986 and 2004–2007. Wavelet analysis on monthly *P. noctiluca* occurrence showed several significant periodicities. The periodicity of 8–14 months can be explained by the two main reproductive seasons in spring and autumn (Rottini Sandrini & Avian, 1991) and by high mortality over a 10-month period (Piccinetti Manfrin & Piccinetti, 1986). Malej & Malej (1992) reported the highest mortality from December to January and estimated a lifespan of about 9 months for *P. noctiluca* in the northern Adriatic. Modified Leslie matrix and genetic algorithms applied to data from 1983 and 1984 showed that the population of *P. noctiluca* was halved in approximately 12 months and became locally extinct after 8 years (Malej et al., 2007). Field evidence and modelling indicate that the northern Adriatic population cannot sustain itself without re-introductions from southern water masses. Therefore, *P. noctiluca* in the northern Adriatic, as in some other Mediterranean coastal areas (e.g. Villfranche Bay, Naples Bay, Tunis Bay), may be considered as sink populations (Gaggiotti & Hanski, 2004). On the other hand, the significantly marked periodicity of 8–10 years in the *P. noctiluca* spectrum of the last four decades indicates a pattern similar to that observed in the western Mediterranean by Goy et al. (1989) and suggests that large-scale factors are responsible for successful introduction and proliferation of *P. noctiluca* in the northern Adriatic. The long-term time series of *P. noctiluca* occurrences of Goy et al. (1989) was re-analysed by wavelet analysis (Fig. 2); in addition to the 12-year periodicity, longer periodicities of 20–30 years and about 45 years were found for the Tyrrhenian and Ligurian Seas, respectively. These durations correspond approximately to the 24- and 47-years harmonics reported by Goy et al. (1989). From the wavelet spectra (Fig. 2), we conclude that the *P. noctiluca* bloom patterns differed between the western Mediterranean basins and the northern Adriatic, as also observed by Goy et al. (1989) who excluded the Adriatic Sea data from their analysis.

The major source of *P. noctiluca* bloom-forming populations in the northern Adriatic appears to be outside the Adriatic Sea since genetic analysis (Stopar et al., submitted) showed that the northern Adriatic population is more closely related to that of the western Mediterranean Sea than that of the southern Adriatic. The *P. noctiluca* population in the Mediterranean Sea appears to maintain itself in several source regions, specified by characteristic hydrographic conditions that facilitate their concentration. When occasional hydro-climatic events and/or when population size increases several-fold, advection drives *P. noctiluca* to the coastal regions, where high productivity promotes growth and reproduction and causes local *P. noctiluca* blooms. The frequency of these coastal blooms depends on the distance from the source regions and circulation patterns.

The year 2003 experienced unusual weather conditions in southern Europe and the Mediterranean region, which affected the oceanographic properties in the Adriatic Sea (Celio et al., 2006; Lyons et al., 2006; Malačič et al., 2006; Grbec et al., 2007). The high air temperature in June and extremely low precipitation and river runoff between February and August caused the surface salinity during spring and summer to be at least one standard deviation above average in the central Adriatic. The bottom salinity also was above average, due to the inflow of the water masses from the southern Adriatic and/or Mediterranean into the Adriatic (Grbec et al., 2007). These extreme conditions in 2003 supported three-times more warm-water species in the Adriatic during 2003 than in previous years (Dulčić et al., 2004). The enhanced intrusions of more southern water masses into the northern Adriatic could have introduced large numbers of *P. noctiluca*, leading to the local bloom during 2005–2007.

High *P. noctiluca* abundance during the 1980s blooms was shown to have significant influences on the pelagic ecosystem. Changes in zooplankton were described as a shift in the community towards some gelatinous zooplankton and also an increased abundance of cladocerans relative to copepods. As a consequence, the >500-μm size fraction increased and the <500-μm size fraction decreased significantly in years of *P. noctiluca* blooms. The same pattern was detected in the change of zooplankton biomass as dry weight (Malej, 1989). The zooplankton community structure survey for recent years has not been completed; however, a significant effect of *P. noctiluca* on the mesozooplankton biomass was detected (unpublished data).

Conclusions

Wavelet analysis showed that the periodicity of occurrence of five jellyfish species has dwindled in recent decades and the recurrence of blooms has increased, particularly for *A. aurita* and *R. pulmo*. Spectra for the meroplanktonic scyphomedusae, excluding *C. tuberculata*, differed from that of the holoplanktonic *P. noctiluca*. The similarity in spectra of those two species indicates some congruence in their blooms in the northern Adriatic. At least two conditions must be fulfilled for *P. noctiluca* blooms to occur in the northern Adriatic: (1) that large numbers of medusae are driven northward by favourable hydro-climatological conditions. Specifically, the *P. noctiluca* population in the northern Adriatic depends on re-introduction from a large-scale persistent source population. This conclusion is further supported by the genetic structure of *P. noctiluca* from different Mediterranean and NE Atlantic locations. (2) *P. noctiluca* is introduced into the area, where favourable conditions stimulate growth and reproduction and promote longer retention. Our wavelet transformation analyses indicate that these conditions have been fulfilled more often in the last 50 years than in the previous 150 years.

Acknowledgements This research was supported by the Slovenian Research Agency Program P1-0237, and European Commission 6th Framework Programme: SESAME project (Contract No GOCE-036949).

References

Avian, M. & L. Rottini Sandrini, 1994. History of Scyphomedusae in the Adriatic Sea. Bolletino della Societa Adriatica di Scienze 75: 5–12.

Babić, K., 1913. Planktonički celenterati iz Jadranskog mora. Rad Jugoslavenske Akademije Znanosti i Umjetnosti 200: 186–202.

Berline, L., F. Ibanez & P. Grosgean, 2009. Roadmap for Zooplankton Time Series Analysis. SESAME WP 1. Unpublished.

Brodeur, R. D., C. E. Mills, J. E. Overland, G. E. Walters & J. D. Schumacher, 1999. Evidence for a substantial increase in gelatinous zooplankton in the Bering Sea, with possible links to climate change. Fisheries Oceanography 8: 296–306.

Celio, M., V. Malačič, A. Bussani, B. Čermelj, C. Comici & B. Petelin, 2006. The coastal scale observing system component of ADRICOSM: Gulf of Trieste network. Acta Adriatica 47(Suppl): 65–79.

CIESM, 2001. Gelatinous Zooplankton Outbreaks: Theory and Practice. CIESM Workshop Series 14, Monaco: 112 pp.

Comici, C. & A. Bussani, 2007. Analysis of the River Isonzo discharge (1998–2005). Bolletino di Oceanologia Teorica ed Applicata 8: 435–454.

Cori, D. J. & A. Steuer, 1901. Beobachtungen über das Plankton des Triester Golfes in den Jahren 1899 und 1900. Zoologischer Anzeiger 24: 111–116.

Cushman-Roisin, B., M. Gačić, P. M. Poulain & A. Artegiani, 2001. Physical Oceanography of the Adriatic Sea: Past, Present and Future. Kluwer Academic Publisher, Dordrecht.

de Wit, M. & G. Bendoricchio, 2001. Nutrient fluxes in the Po basin. Science of the Total Environment 273: 147–161.

Del Negro, P., F. Kokelj, A. Tubaro & R. Della Loggia, 1992. *Chrysaora hysoscella* in the Gulf of Trieste: presence, evolution and cutaneous toxicity in man. Science of the Total Environment (Suppl): 427–430.

Di Camillo, C. G., F. Betti, M. Bo, M. Martinelli, S. Puce & G. Bavestrello, 2010. Contribution to the understanding of seasonal cycle of *Aurelia aurita* (Cnidaria, Scyphozoa) scyphopolyps in the northern Adriatic Sea. Journal of the Marine Biological Association of the United Kingdom. doi:10.1017/S0025315409000848.

Dulčić, J., B. Grbec, L. Lipej, G. Beg Paklar, N. Supić & T. Smirčić, 2004. The effect of the hemispheric climatic oscillations on the Adriatic ichthyofauna. Environmental Bulletin 13: 293–298.

Gaggiotti, O. E. & I. Hanski, 2004. Mechanisms of population extinction. In Hanski, I. & O. E. Gaggiotti (eds), Ecology, Genetics, and Evolution of Metapopulations. Elsevier Academic Press, Amsterdam: 337–366.

Giorgi, R., M. Avian, L. Rottini Sandrini & A. Troian, 1985. Monitoraggio meduse e danni alla pesca: analisi a breve termine del Golfo di Trieste. Nova Thalassia 7: 55–62.

Goy, J., P. Morand & M. Etienne, 1989. Long term fluctuations of *Pelagia noctiluca* (Cnidaria, Scyphomedusae) in the western Mediterranean Sea, Prediction by climatic variables. Deep-Sea Research 36: 269–279.

Graeffe, E., 1875. Seethierfauna des Golfes von Triest. III. Coelenteraten, Cnidaria. Arbeiten aus dem Zoologischen Institut der Universität Wien und der Zoologischen Station in Triest 5: 333–362.

Graham, W. M., 2001. Numerical increases and distribution shifts of *Chrysaora quinquecirrha* (Desor) and *Aurelia aurita* (Linné) (Cnidaria: Scyphozoa) in the northern Gulf of Mexico. Hydrobiologia 451: 97–111.

Grbec, B., I. Vilibić, A. Bajić, M. Morović, G. Bec Paklar & V. Dadić, 2007. Response of the Adriatic Sea to the atmospheric anomaly in 2003. Annales Geophysicae 25: 835–846.

Grinsted, A., J. C. Moore & S. Jevrejeva, 2004. Application of the cross wavelet transform and wavelet coherence to geophysical time series. Nonlinear Processes in Geophysics 11: 561–566.

Hays, G. C., A. J. Richardson & C. Robinson, 2005. Climate change and marine plankton. Trends in Ecology and Evolution 20: 337–344.

Holm-Hansen, O., C. J. Lorenzen, R. W. Holmes & J. D. H. Strickland, 1965. Fluorometric determination of chlorophyll. Journal du Conseil permanent International pour l' Exploration de la Mer 30: 3–15.

Hsieh, C. H., S. M. Glaser, A. J. Lucas & G. Sugihara, 2005. Distinguishing random environmental fluctuations from ecological catastrophes for the North Pacific Ocean. Nature 435: 336–340.

Issel, R., 1922. Nuove indagini sul plankton nelle acque di Rovigno. Rapporti Comitato Talassografico Italiano, Memoria 102: 1–36.

Kamburska, L. & S. Fonda Umani, 2006. Long-term copepod dynamics in the Gulf of Trieste (northern Adriatic Sea): recent changes and trends. Marine Ecology Progress Series 31: 195–203.

Kohama, T., N. Shinya, N. Okuda, H. Miyasaka & H. Takeoka, 2006. Estimation of trophic level of *Aurelia aurita* using stable isotope ratios in Uwa Sea, Japan. In Proceedings of COE International Symposium 2006 (Ehime University, Matsuyama, Japan). Pioneering Studies of Young Scientists on Chemical Pollution and Environmental changes.

Kuzmić, M. & M. Orlić, 2006. Modeling the northern Adriatic double-gyre response to intense bora wind: a revisit. Journal of Geophysical Research. doi:10.1029/2005JC 003377.

Lilley, M. K. S., J. D. R. Houghton & G. C. Hays, 2009. Distribution, extent of inter-annual variability and diet of the bloom-forming jellyfish *Rhizostoma* in European waters. Journal of the Marine Biological Association of the United Kingdom 89: 39–48.

Lynam, C. P., S. J. Hay & A. S. Brierley, 2004. Interannual variability in abundances of North Sea jellyfish and links to the North Atlantic Oscillation. Limnology and Oceanography 49: 637–643.

Lyons, D. M., I. Janeković, R. Precali & N. Supić, 2006. Northern Adriatic Sea hydrographic conditions from October 2002–September 2003 including the climatic heating anomaly of summer 2003. Acta Adriatica 47(Suppl): 81–96.

Malačič, V., 1991. Estimation of the vertical eddy diffusion coefficient of heat in the Gulf of Trieste (Northern Adriatic). Oceanologica Acta 14: 23–32.

Malačič, V. & B. Petelin, 2009. Climatic circulation in the Gulf of Trieste (northern Adriatic). Journal of Geophysical Research 114. doi:10.1029/2008JC004904.

Malačič, V., M. Celio, B. Čermelj, A. Bussani & C. Comici, 2006. Interannual evolution of seasonal thermohaline properties in the Gulf of Trieste (northern Adriatic) 1991–2003. Journal of Geophysical Research 111. doi:10.1029/ 2005JC003267.

Malej, A., 1982. Unusual occurrence of *Pelagia noctiluca* in the Adriatic. Acta Adriatica 23(1/2): 97–102 (in Slovenian).

Malej, A., 1989. Shifts in size-classes and structure of net zooplankton in the presence of the jellyfish *Pelagia noctiluca* (Scyphozoa). Biološki Vestnik 37: 35–46.

Malej, A., 2001. Are irregular plankton phenomena getting more frequent in the northern Adriatic Sea? In Gelatinous Zooplankton Outbreaks: Theory and Practice. CIESM Workshop Series 14, Monaco: 16–18.

Malej, A. & M. Malej, 1992. Population dynamics of the jellyfish *Pelagia noctiluca* (Forsskål, 1775). In Colombo, G., et al. (eds), Marine Eutrophication and Population Dynamics. Proceedings of the 25th EMBS. Olsen and Olsen, Fredensborg: 215–219.

Malej, A. & A. Malej, 2004. Invasion of the jellyfish *Pelagia noctiluca* in the northern Adriatic: a non-success story. In Dumont, H., et al. (eds), Aquatic Invasions in the Black, Caspian, and Mediterranean Seas. NATO ASI Series 5, Vol. 35. Kluwer Academic Publisher, Dordrecht: 273–285.

Malej, A., V. Turk, D. Lučić & A. Benović, 2007. Direct and indirect trophic interactions of *Aurelia* sp. (Scyphozoa) in a stratified marine environment (Mljet Lakes, Adriatic Sea). Marine Biology 151: 827–841.

Malej M., A. Malej & A. Malej, 2007. The use of modified Leslie matrix and genetic algorithms to model jellyfish population (*Pelagia noctiluca*, Scyphozoa). In Proceedings of 6th European Conference on Ecological Modelling, ECEM '07: 326–327.

Miloš, Č., 2009. Biometry and elemental analysis of scyphomedusae in the Gulf of Trieste with emphasis on *Rhizostoma pulmo*. M.Sc. thesis, University of Zagreb: 81 pp.

Molinero, J. C., M. Casini & E. Buecher, 2008. The influence of the Atlantic and regional climate variability on the long-term changes in gelatinous carnivore populations in the northwestern Mediterranean. Limnology and Oceanography 53: 1456–1467.

Mozetič, P., C. Solidoro, G. Cossarini, G. Socal, R. Precali, J. France, F. Bianchi, C. De Vittor, N. Smodlaka & S. Fonda Umani, 2009. Recent trends towards oligotrophication of the northern Adriatic: evidence from chlorophyll *a* time series. Estuaries and Coast. doi:10.1007/s1223-009-9191-7.

Percival, D. B. & A. T. Walden, 2000. Wavelet Methods for Time Series Analysis. Cambridge University Press, Cambridge, UK.

Piccinetti, C. & G. Piccinetti Manfrin, 1991. Considerations sur la distribution de *Pelagia noctiluca* (Forskål) dans l'Adriatique de 1976 a 1987. 2nd Workshop on Jellyfish in the Mediterranean, Trieste, 1987. MAP Technical Reports Series 47: 133–140 (in French).

Piccinetti Manfrin, G. & C. Piccinetti, 1986. Distribuzione di *Pelagia noctiluca* in Adriatico dal 1983 e 1984. Nova Thalassia 8(Suppl): 103–106.

Purcell, J. E., 2005. Climate effects on formation of jellyfish and ctenophore blooms: a review. Journal of Marine Biology Association of United Kingdom 85: 461–476.

Purcell, J. E., 2009. Extension of methods for jellyfish and ctenophore trophic ecology to large-scale research. Hydrobiologia 616: 23–50.

Purcell, J. E., S.-I. Uye & W.-T. Lo, 2007. Anthropogenic causes of jellyfish blooms and direct consequences for humans: a review. Marine Ecology Progress Series 350: 153–174.

Rottini Sandrini, L. & M. Avian, 1991. Reproduction of *Pelagia noctiluca* in the central and northern Adriatic. Hydrobiologia 216(217): 197–202.

Sournia, A., 1973. La production primaire planctonique en Méditerranée. Essai de mise a jour. Bulletin Etude en commun de la Méditerranée 5: 1–128.

Steuer, A., 1902. Beobachtungen über das Plankton des Triester Golfes im Jahre 1901. Zoologischer Anzeiger 25: 371–372.

Steuer, A., 1904. Beobachtungen über das Plankton des Triester Golfes im Jahre 1902. Zoologischer Anzeiger 27: 145–148.

Stiasny, G., 1909. Beobachtungen über die marine Fauna des Triester Golfes im Jahre 1908. Zoologischer Anzeiger 34: 289–294.

Stiasny, G., 1910. Beobachtungen über die marine Fauna des Triester Golfes im Jahre 1909. Zoologischer Anzeiger 35: 583–587.

Stiasny, G., 1911. Beobachtungen über die marine Fauna des Triester Golfes während des Jahres 1910. Zoologischer Anzeiger 37: 517–522.

Stiasny, G., 1912. Beobachtungen über die marine Fauna des Triester Golfes im Jahre 1911. Zoologischer Anzeiger 39: 604–608.

Stiasny, G., 1914. Zwei neue Pelagien aus der Adria. Zoologischer Anzeiger 44: 529–533.

Stiasny, G., 1919. Die Scyphomedusen-Sammlung des Naturhistoricshen Reichmuseum in Leiden. Stauromedusae, Coronatae, Semaeostomae. Zoologische Mededelingen 5: 66–99.

Szüts, A., 1915. Biologische Beobachtungen wahrend der ersten und zweiten terminfahrt des Ungarischen Adriavereins an S.M.S. "Najade" im Oktober 1913 und im April–Mai 1914. Zoologischer Anzeiger 45: 432–442.

Stopar, K., A. Ramšak, P. Trontelj & A. Malej, submitted. Molecular evidence for panmixia in mauve stinger, *Pelagia noctiluca*, in European seas.

Torrence, C. & G. P. Compo, 1998. A practical guide to wavelet analysis. Bulletin of the American Meteorological Society 79: 61–78.

Zanchettin, D., P. Traverso & M. Tomasino, 2008. Po River discharges: a preliminary analysis of a 200-year time series. Climatic Change 89: 411–433.

Hydrobiologia (2010) 645:97–111
DOI 10.1007/s10750-010-0213-z

JELLYFISH BLOOMS

Behavior of *Nemopsis bachei* L. Agassiz, 1849 medusae in the presence of physical gradients and biological thin layers

Jessica R. Frost · Charles A. Jacoby · Marsh J. Youngbluth

Published online: 21 March 2010

Abstract In pelagic systems, thin layers (discontinuities with narrow vertical extents and high concentrations of organisms) create patches of food, and aggregations of gelatinous zooplankton can exploit such resources. The establishment, maintenance, and trophic effects of these functional relationships depend on behavioral responses to thin layers by individuals, which remain largely unexplored. In this study, we used laboratory experiments to test the hypothesis that a common and abundant hydromedusa predator, *Nemopsis bachei* L. Agassiz, 1849, would respond similarly to salinity gradients with and without thin layers of algae and copepods. Approximately 75% of the hydromedusae remained in both types of discontinuities. These distributions were not created solely by passive responses related to osmoconformation or an inability to swim through salinity gradients because approximately 25% of hydromedusae swam through or away from salinity gradients or biological thin layers. Biological thin layers stimulated feeding. Feeding success was related directly to encounter rates and it was independent of swimming, as expected for an ambush predator. Feeding increased at higher prey densities, and capture, handling time, and ingestion were not saturated even at 150–200 copepods l^{-1}. The proportion of *N. bachei* that ceased feeding and began swimming increased when encounters with prey decreased to approximately 2 encounters hydromedusa^{-1} 10 min^{-1}. Thus, hydromedusae may seek new patches of prey once encounter rates and subsequent feeding success fall below a threshold. Exposing *N. bachei* to salinity gradients with and without biological thin layers indicated that these hydromedusae will remain in discontinuities and exert predation pressure that should be considered when assessing trophic webs and estimating carbon flux.

Keywords Jellyfish · Hydromedusae · Zooplankton · Discontinuities · Feeding · Aggregation

---

Guest editors: J. E. Purcell & Dror Angel / Jellyfish Blooms: New Problems and Solutions

---

J. R. Frost (✉)
Institute for Hydrobiology and Fisheries Science, University of Hamburg, Olbersweg 24, 22767 Hamburg, Germany
e-mail: frost.jessica.r@gmail.com

C. A. Jacoby
Program in Fisheries and Aquatic Sciences, School of Forest Resources and Conservation, University of Florida, 7922 NW 71st Street, Gainesville, FL 32653, USA
e-mail: cajacoby@ufl.edu

M. J. Youngbluth
Harbor Branch Oceanographic Institute at Florida Atlantic University, 5600 U.S. Highway 1 North, Fort Pierce, FL 34946, USA
e-mail: youngbluth@yahoo.com

Introduction

Spatial and temporal heterogeneity affect movement and resource use by animals (Levin et al., 2000). In particular, behaviors that allow organisms to locate and exploit patches of food represent key influences on trophodynamics. For plankton, external forcing (e.g., currents, shear, and turbulence) combines with behavioral responses to determine the structure and function of assemblages, access to food, and the flow of carbon across a range of spatial and temporal scales (e.g., Bochdansky & Bollens, 2004).

At a micro-scale, thin layers represent persistent, spatially coherent patches of plankton (Donaghay et al., 1992; Cowles et al., 1998). Thin layers that are decimeters to meters thick, extend several kilometers horizontally, and persist for minutes, hours or days have been identified with the aid of technologically advanced instruments (e.g., acoustic Doppler current profilers; Flagg & Smith, 1985; Cowles et al., 1990; Cowles & Desiderio, 1993; Holliday et al., 1998). The formation of thin layers requires the creation of a stable discontinuity, typically associated with a pycnocline at the boundary of two water masses, and threefold increases in densities of phytoplankton and zooplankton that arise passively, primarily due to neutral buoyancy, or actively, as an outcome of behavior (Osborn, 1998; Dekshenieks et al., 2001; Alldredge et al., 2002; McManus et al., 2003; Gallagher et al., 2004; Rawlinson et al., 2004). In thin layers, competition and predation can be intensified, rates of nutrient uptake can be enhanced, chemical wastes and toxins can accumulate, and microbial degradation and remineralization can be higher (Mason et al., 1993; Sieburth & Donaghay, 1993; Johnson et al., 1995; Donaghay & Osborn, 1997; Cowles et al., 1998). The chemistry, physics, and biological compositions of thin layers have been documented (Donaghay et al., 1992; Hanson & Donaghay, 1998; Holliday et al., 1998; Widder et al., 1999; McManus et al., 2005). As yet, relatively few studies have examined the behavioral responses that allow individual organisms to remain within and exploit thin layers, in spite of the recognition that such coupling can alter the abundance and species composition of plankton assemblages (Denman & Powell, 1984; Mackas et al., 1985; Yamazaki et al., 2002; Clay et al., 2004; Ignoffo et al., 2005; Woodson et al. 2005, 2007a, b). In addition, we are unaware of any study that has examined the influence of thin layers on the behavior of individual gelatinous zooplankton.

Gelatinous zooplankton, including ctenophores, medusae, and siphonophores, tend to aggregate at density, salinity, and temperature discontinuities that develop in coastal transition zones, estuaries, and fjords (Owen, 1989; Graham et al., 2001). Aggregations of gelatinous species can arise if discontinuities act as a physical or physiological barrier or if detectable boundary conditions stimulate appropriate behavioral responses. For example, field surveys showed that *Sarsia tubulosa* (M. Sars, 1835) aggregated at a discontinuity (Hansen, 1951), and in the laboratory, this species' ability to cross salinity and temperature discontinuities was limited by the magnitude and rate of its physiological adaptation (Arai, 1973, 1976). In contrast, *Aglantha digitale* (O. F. Müller, 1776) and *Proboscidactyla ornata* (McCrady, 1857) ceased their typical vertical migrations in response to discontinuities (Hansen, 1951; Smedstad, 1972; Moreira, 1973; Southward & Barrett, 1983). Although laboratory and field studies have provided information about the behavioral responses of medusae to discontinuities, they have not focused at the fine scales typical of thin layers (Grindley, 1964; Hamner et al., 1975; Williams, 1985; Tiselius et al., 1994; Lougee et al., 2002).

Hydromedusae represent a widespread and diverse group of jellyfish (Boero et al., 2008). These medusae have been documented to be abundant predators in some systems (Huntley & Hobson, 1978; Matsakis & Conover, 1991; Costello, 1992; Purcell & Nemazie, 1992). One hydromedusa of particular interest is *Nemopsis bachei*, which seems to have extended its range. Populations of *N. bachei* have existed in Chesapeake Bay, the Atlantic coastal waters of North America, and the Gulf of Mexico (Purcell & Nemazie, 1992); and populations recently have become established in the North Sea, in the coastal waters of the Netherlands, and off the coast of Belgium (Dumoulin, 1997; ICES, 2006). Furthermore, *N. bachei* has been reported to be the numerically dominant gelatinous predator in North Inlet, South Carolina, USA during January 2006 (Marshalonis & Pinckney, 2008) and the most abundant and important gelatinous predator in Chesapeake Bay, USA during spring (Purcell & Nemazie, 1992). Results from investigations in Chesapeake Bay showed that *N. bachei* generated greater predation pressure on copepods than has been reported for other predatory jellyfish (Fulton & Wear, 1985; Daan, 1986, 1989).

As an "ambush" predator, *Nemopsis bachei* does not swim as it feeds; rather it remains stationary and relies on prey colliding with its extended tentacles (Gerritsen & Strickler, 1977; Costello & Colin, 2002; Costello et al., 2008; Marshalonis & Pinckney, 2008). In contrast to "cruising" predatory jellyfish, like *Aurelia aurita* (Linnaeus, 1758), that increase swimming when prey are detected (Bailey & Batty, 1983; Costello, 1992), and copepods that employ area-restricted search behavior (Woodson et al., 2005, 2007a, b), *N. bachei* medusae may display a different set of behaviors in order to balance feeding with swimming that allows them to remain in or near a thin layer.

In combination, the abundance, range extension and foraging mode of *Nemopsis bachei* make this species a useful example for studies elucidating functional relationships between jellyfish and thin layers. To examine such relationships, laboratory experiments tested the hypothesis that *N. bachei* respond similarly to 5-cm salinity gradients with or without biological thin layers of adult copepods, *Acartia tonsa* Dana, 1849, and green algae, *Nannochloropsis* sp.

Materials and methods

Collection and maintenance of hydromedusae

Individual *Nemopsis bachei* were collected from the surface in approximately 1 m of water along the coastline of a bay located near Meldorf, Germany (54°05′32″N, 8°57′01″E) on 25 June; 2, 14, and 29 July; and 21 August 2008. During summer 2008, this hydromedusa was abundant at the site with collections being most successful in calm weather during 2-h period before high tides. Individual hydromedusae were collected by hand using wide-mouthed, glass jars. The hydromedusae were transferred from the jars into a 20-l bucket containing ambient seawater with a temperature of 21.8–24.8°C and salinity of 24.0–25.8, as measured on the practical salinity scale.

The bucket containing hydromedusae was nested in a container of ambient seawater for transport to a laboratory at the University of Hamburg Institute for Hydrobiology and Fisheries Science (UHH–IHF). Transport was completed within 2.5 h, and water temperature in the bucket increased by no more than 1°C. The bucket with hydromedusae in seawater from the collection site was placed in a climate-controlled room at 18 ± 0.5°C and subjected to a 13:11 light:dark cycle until the hydromedusae were used in experiments. Initial attempts to modify the salinity of the ambient seawater by adding fresh water or filtered North Sea water of salinity 35.0 induced mortality. The water was aerated with minimal bubbling to ensure adequate oxygenation without causing physical damage to the hydromedusae.

Hydromedusae used for experiments were not fed. Initial experiments were conducted 24 h after each sampling. Subsequent experiments were conducted daily for a maximum holding time of 1 week. This maximum holding time was well below the 3–8 week periods of starvation tolerated by other hydromedusae (Arai, 1992), and the hydromedusae were deemed sufficiently healthy if they were swimming actively. Each hydromedusa was used only once; each experiment involved 10 individuals.

Sources and maintenance of water and organisms for thin layers

All water used in cultures and experiments was collected from the North Sea and held with constant aeration in storage tanks at UHH–IHF. Filtered North Sea water was prepared using a two-step process: (1) water was filtered sequentially through three, in-line, wound cotton-fiber filters with nominal pore sizes of 100, 10, and 1 μm and (2) the filtrate was filtered sequentially through Whatman glass fiber (GF/C) and membrane filters with nominal pore sizes of 1.0 and 0.1 μm, respectively. Experimental salinities of 18.0, 22.0, and 28.0 ± 0.1 were made by adding fresh water to filtered North Sea water with a salinity of 35.0. Salinities were measured with a WTW TetraCon® 325 conductivity meter. The water for experiments was stored in 113 l, plastic buckets inside the climate-controlled room. The salinities used in experiments were chosen to bracket the salinity recorded at the collection site and to create narrow and distinct discontinuities.

Acartia tonsa was chosen as a biological component of experimental thin layers because this calanoid copepod is cultivated at UHH–IHF and *Nemopsis bachei* is known to feed on both copepodites and adults of *A. tonsa* in situ (Purcell & Nemazie, 1992). Resting eggs produced by copepods collected in the Kiel Bight (54°N, 10°E) during August 2003 were stored at 4°C until they were hatched in gently aerated, filtered North Sea water lowered to a salinity of 18.0. The newly hatched copepods were subjected to a 12:12 light:dark cycle and fed *Rhodomonas* sp. daily until they were adults and at least 14-day-old. At this stage, a sub-sample from the culture of *A. tonsa* then was transferred to a bucket containing filtered North Sea water of salinity 22.0, which was stored with continuous aeration in the same climate-controlled room as the hydromedusae.

Thin layers typically involve several trophic levels (e.g., Cowles et al., 1998); therefore, small (2-μm spherical diameter), unicellular green algae, *Nannochloropsis* sp., were added to the thin layers as food for *Acartia tonsa*. In preliminary trials, visual observations indicated that *A. tonsa* aggregated near and fed readily on *Nannochloropsis* sp., which remained in the thin layers, whereas *Rhodomonas* sp. sank (Frost, unpublished data). *Nannochloropsis* sp. used in the experiments were drawn from cultures maintained at UHH–IHF. The inoculum was obtained in 2007 from Dansk Skaldyrcenter, a shellfish hatchery located near Limfjord, Denmark. The culture medium comprised filtered and autoclaved North Sea water of salinity 22.0, nutrient and vitamin solutions (Walne, 1970), and 50 mg l^{-1} biotin (Støttrup & Jensen, 1990). The culture was continuously aerated and stored in the same climate-controlled room as the hydromedusae and copepods.

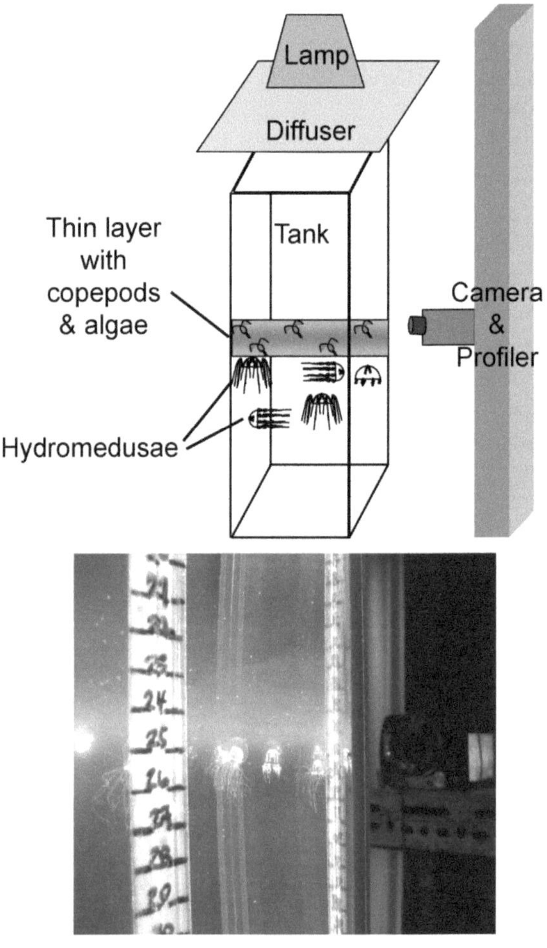

Fig. 1 Experimental system for testing the behavior of the hydromedusa, *Nemopsis bachei*. A salinity gradient with a biological thin layer of algae and copepods is shown. Salinities of the water were 18 above, 20–27 within, and 28 below the gradient. Algae and copepods were absent in experiments involving a salinity gradient without a biological thin layer. Inset is a photograph of an experiment using a biological thin layer. Figure is not drawn to scale

Experimental system

Each experiment occurred in a 1-cm thick, columnar, glass tank (15 × 15 × 50 cm; Fig. 1). A scale with 1-cm markings was affixed to the outside of the tank along one edge. In each experiment, a consistent light regime was supplied by an 11-W lamp transmitting light through a diffuser comprising a pane of glass fitted with a sheet of vellum paper. This diffuser functioned like milk glass commonly used in photography. A black, fabric backdrop enhanced contrast in video images captured with a SONY Handycam® HDR-SR12E high definition camcorder operating at 25 frames s^{-1} and a resolution of 1,920 × 1,080 pixels. The camcorder was mounted on a LINOS™ vertical motion, rail profiler in a position that yielded a 13 × 13 × 7.5 cm or 1.2–1.4 l field of view. The camcorder was panned manually every 10–15 min to scan the vertical extent of the tank. When not panning, the camcorder was focused on the biological thin layer or salinity gradient that had been established in the water column as described below. Video records and direct observations were conducted for 5 h so that delayed responses to salinity gradients and food, such as osmoconformation and satiation, could affect more immediate responses, such as swimming and feeding (Tiselius, 1992).

All experiments involved a stratified water column with a salinity gradient representing the physical discontinuity. Gradients were created by carefully and slowly pouring water of salinity 28.0 along the walls of the tank while it was held at an angle, to avoid generating air bubbles, and then layering water of salinity 18.0 into the upper portion of the tank. The less saline water was dispensed from a carboy fitted with a tap and tubing that ended in a t-valve. At the end of this process, the lower and upper portions of the water column each were approximately 22-cm high, and these layers were separated by a 5-cm thick salinity gradient. Each salinity gradient was profiled using a WTW TetraCon® 325 conductivity meter at the end of the observations to confirm its persistence.

Half of the 16 experiments with salinity gradients included biological thin layers containing green algae, *Nannochloropsis* sp., and adult copepods, *Acartia tonsa*. One 100-ml aliquot of *Nannochloropsis* sp. was prepared for each of these experiments. The aliquots contained an average of 3.04×10^6 cells ml^{-1} (particle diameters 2.38–5.00 μm) as measured with a Beckman Multisizer™ 3 Coulter Counter® 20 min or less before the experiment. In addition, a 100-ml sample of *A. tonsa* was concentrated on a sieve (280-μm mesh), transferred to a Bogorov chamber, and enumerated using magnification from a dissecting microscope to determine the aliquot needed to obtain 400–500 individuals. Thin layers of algae and copepods were formed by slowly introducing the algae and then the copepods into the salinity gradient with a 60-ml, plastic syringe barrel connected to tubing that ended in a t-valve. Flow was controlled with a tubing clamp. On average, the resulting biological thin layers were 3–4 cm thick with volumes of 0.51–0.68 l.

After preparation of the stratified water column with or without a biological thin layer, 10 hydromedusae were introduced individually into the bottom portion of each water column using a polyvinylchloride (PVC) tube. For all 16 experiments, the lower boundaries of salinity gradients or biological thin layers were located near the center of the tank (i.e., 24–27 cm from the top). In order to minimize disruption to the stratified water columns and biological thin layers, the PVC tube was inserted and removed very slowly against one corner of the tank.

The two types of experiments compared and contrasted responses of hydromedusae to sharply defined physical discontinuities and to biological thin layers comprising similar micro-scale discontinuities with high concentrations of organisms (Dekshenieks et al., 2001; Cowles, 2004).

Video analysis

Video streams were split and analyzed using the Sony Picture Utility software Picture Motion Browser Ver.3.0.00. Metrics recorded included the vertical distribution of hydromedusae and any reliably discernible activities, especially movement and feeding. In addition, the vertical distribution of copepods was recorded to track the availability of potential copepod prey.

Activities of hydromedusae were recorded from 1-min video segments beginning 10, 30, 50, 70, 90, 110, 180, and 300 min after each experiment began, and the vertical distributions of hydromedusae were determined from manual pans beginning 30 min after the start of each experiment and recurring every 10–15 min for 5 h. Tentacular bulbs at the intersections between the tentacles and umbrellar margins served as reference points for measuring vertical locations of hydromedusae.

The vertical distributions of copepod prey were quantified in video frames taken at 1, 3, and 5 h to show each of the three sections of the water column (6–14 cm or above, 19–27 cm or in, and 36–44 cm or below the biological thin layer). The positions of copepods were marked on a plastic transparency laid on top of each video frame, and the marks were counted and converted to numbers of copepods l^{-1} using the volume of the field of view. In addition, the rates of changes in prey densities were not predictable within or between experiments. Therefore, the prey densities used in analyses of *Nemopsis bachei* behaviors were derived from counts of copepods in biological thin layers made from individual video frames captured in each of the eight time intervals allocated to observing the hydromedusae.

Data analyses

Analyses of variance (ANOVAs) were used to assess changes in the vertical distributions of copepods and hydromedusae, as well as differences in the activities exhibited by hydromedusae over time and versus varying copepod densities. Prior to each ANOVA, normality was assessed with Ryan–Joiner tests, and homoscedasticity was assessed with Cochran's tests. Data were transformed as necessary to satisfy these assumptions ($P > 0.05$). In cases where the assumptions were not met, the potential for increased Type I error was considered during interpretation of the ANOVAs. All statistical analyses were performed using MiniTab 15.1.00 Solutions (MiniTab Inc.) and functions within Microsoft Excel 2003 (Microsoft Inc.). In general, only significant F values and relevant means ± standard errors (SEs) were reported.

In experiments with a biological thin layer, the vertical distributions of *Acartia tonsa* were compared among time intervals in order to track the availability of prey for the hydromedusae. Data comprised the proportions of copepods in the biological thin layer, and these data were arcsin transformed prior to analysis. The ANOVA treated time intervals as a random factor with three levels, 1, 3, and 5 h.

The vertical distributions of *Nemopsis bachei* were compared between experiments with and without a biological thin layer, among five time intervals (30–60, 61–90, 91–120, 121–180, and ≥181 min), and among copepod densities in the biological thin layers. Data comprised the proportions of hydromedusae aggregated at the biological thin layers, located within the depth range 19–27 cm. These data were calculated from the total number of hydromedusae observed during vertical pans, and they were arcsin transformed before analyses. In a two-way ANOVA, the presence and absence of a biological thin layer were treated as levels of a fixed factor, and time intervals were treated as levels of a random factor. In a one-way ANOVA, prey densities were treated as levels of a random factor.

Wherever appropriate, variations in the activities displayed by hydromedusae were compared between experiments with and without a biological thin layer, among time intervals, and among prey densities. In ANOVAs, the presence or absence of biological thin layers were treated as levels of a fixed factor and time intervals and prey densities were treated as levels of random factors, with the number and span of the levels chosen to yield balanced replication.

Results

Vertical distributions of copepods and hydromedusae

Profiles of salinity indicated that gradients were located between 19 and 27 cm from the top of the tank (Fig. 2). Within the gradients, which were approximately 5-cm deep, salinities increased from 20.0 to 27.0. In experiments with biological thin layers, visual observations confirmed that *Nannochloropsis* sp. remained concentrated at the salinity gradients throughout the 5-h period of observation, but *Acartia tonsa*, the prey provided for hydromedusae, dispersed during the same time periods (Fig. 2). A total of 72 observations indicated that on average 22.6 ± 1.8, 73.8 ± 2.5, and 3.6 ± 1.0% of the copepods were above, in, and below the layer of concentrated algae during the first hour, and these proportions shifted to 53.2 ± 4.4, 33.8 ± 4.8, and 13.0 ± 3.7% during the fifth hour. In fact, the proportion of copepods in the biological thin layer decreased significantly through time ($F_{2,21} = 24.60$, $P < 0.01$). The changes in copepod distributions occurred gradually over a relatively long time interval, which indicated that neither negative nor positive phototaxis was a dominant influence. These changes provided a basis for examining variation in the behavior of *Nemopsis bachei* from another viewpoint, i.e., relative to the changes in the density of prey remaining in the biological thin layers.

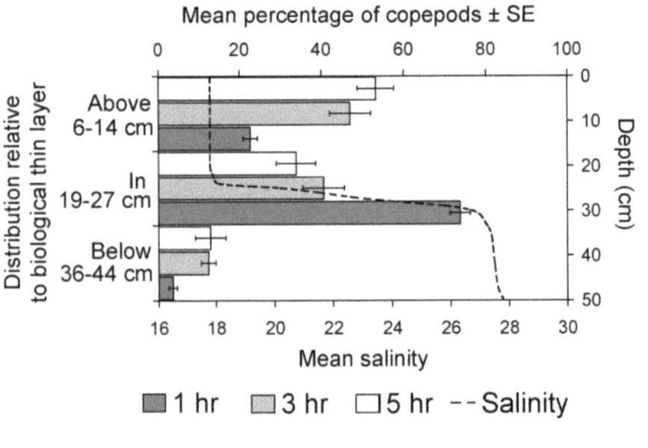

Fig. 2 Vertical distribution of available copepod prey (*Acartia tonsa*) relative to a salinity gradient with and without thin layers of algae during tests of the behavior of the hydromedusa, *Nemopsis bachei*. The percentages of copepods in different depth intervals and the salinities represent means from 16 experiments. *SE* standard error

The vertical distributions of *Nemopsis bachei* medusae were determined from a total of 332 video scans equally divided between experiments with and without biological thin layers. The hydromedusae were strongly concentrated in the salinity gradients and in the thin layers (Fig. 3). Nevertheless, they were capable of traversing the entire water column, passing through the salinity gradient or biological thin layer into more or less saline seawater as indicated by the eventual presence of hydromedusae in all depth intervals. The vertical distribution of hydromedusae differed significantly among time intervals ($F_{4,69} = 6.03$, $P < 0.05$) but not between experiments with or without biological thin layers or among the combinations of these two factors. Thus, the overall trend was similar in experiments with and without biological thin layers, and the mean proportions of hydromedusae in the depth intervals could be averaged across all experiments. These mean proportions decreased from 95.5 ± 1.9% at the start of experiments to 75.6 ± 4.4% after 3 h. Approximately half of the dispersal occurred after 2 h of observation, which suggested that dispersal of *N. bachei* was not caused primarily by a negative or positive phototaxis.

Activities of hydromedusae

During the 16 experiments, a total of 128, 1-min video segments were collected over a total of 80 h. *Nemopsis bachei* were classified as active (swimming or feeding) or non-active. Velocities (cm s^{-1}) were measured for swimming hydromedusae, with swimming defined as a minimum of five pulses yielding movement in any direction. Without these guidelines, swimming could be confused with pulses that regulated position in the water column. Furthermore, observations confirmed that swimming was not an integral part of prey capture, which was consistent with reports for other ambush predators (Costello, 1992). The numbers of hydromedusae feeding were counted in fields of view that included biological thin layers in those eight experiments. Steps in the feeding process were defined as contact and retention of a

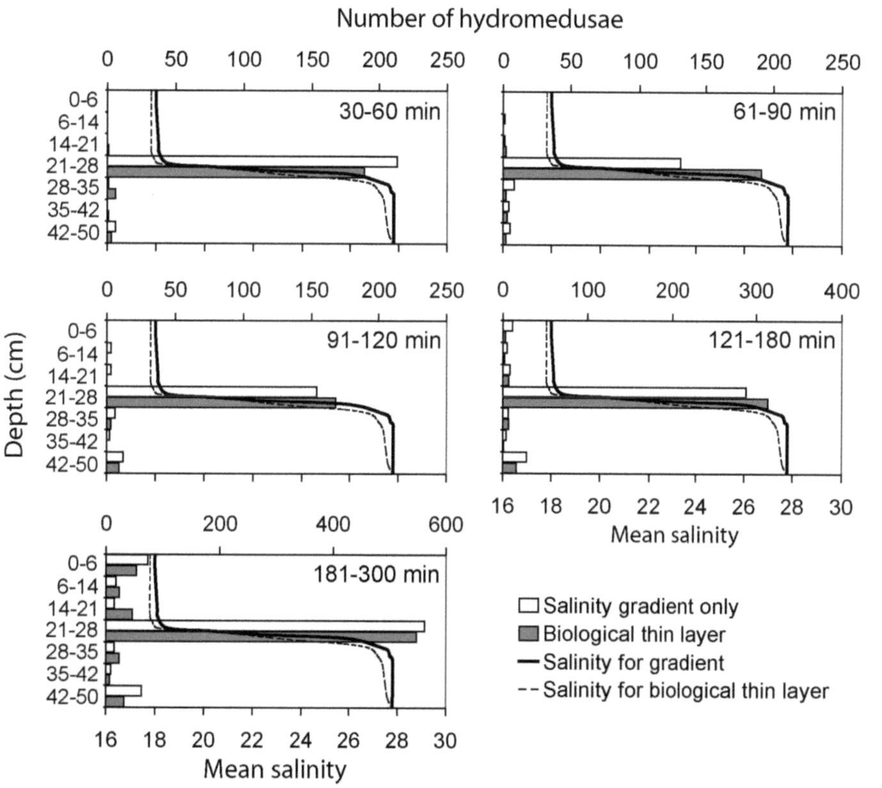

Fig. 3 Vertical distributions of the hydromedusa, *Nemopsis bachei*, in experiments testing its behavior relative to a salinity gradient versus relative to a salinity gradient with a biological thin layer of algae and copepods. The numbers of hydromedusae represent the sums of all individuals in each depth interval across the appropriate experiments, and the salinities represent means across those same experiments

copepod on a tentacle, transfer of the copepod into the mouth, and return of the empty tentacle to an outstretched, sit-and-wait position. Every copepod captured was transferred successfully to the mouth of the hydromedusa; therefore, efficiencies of all steps in the feeding process were 100%. Additional information about feeding behavior was derived from encounter rates, defined as the elapsed time between successive captures by individual hydromedusae, and handling times, defined as the elapsed time from capture to transfer into the mouth.

The proportions of active hydromedusae (swimming or feeding) were examined with two ANOVAs. In one analysis, these proportions were compared among experiments with and without biological thin layers, among time intervals and among combinations of these two factors. The interaction term in this ANOVA was significant ($F_{7,112} = 5.99$, $P < 0.01$), indicating that the proportions of active hydromedusae differed among time intervals and that the pattern differed between experiments with and without biological thin layers. At 10–11 min, an average of 67.1 ± 7.3% and 42.1 ± 7.1% of hydromedusae were active in experiments with and without biological thin layers, respectively (Fig. 4a). At 300–301 min, an average of 92.6 ± 3.9% of hydromedusae were active in experiments with biological thin layers, but only 1.6 ± 1.6% of hydromedusae were active in experiments without biological thin layers (Fig. 4a). In experiments with biological thin layers, the proportions of hydromedusae recorded as active did not differ significantly among prey densities ($F_{6,57} = 2.21$, $P > 0.05$). On average, 92.4 ± 2.6% of hydromedusae were active at prey densities of 0–10 copepods l^{-1} and 62.5 ± 13.8% were active at prey densities of ≥151 copepods l^{-1} (Fig. 4b).

Swimming velocities were examined with two ANOVAs that paralleled those above. Swimming velocities differed significantly between experiments with and without biological thin layers ($F_{1,7} = 5.45$, $P < 0.05$), but not among time intervals or the interaction of these two factors. Although statistically different, the mean swimming velocities were very similar, i.e., 1.24 ± 0.28 and 0.95 ± 0.22 cm s^{-1} for hydromedusae in experiments with and without biological thin layers, respectively. In addition, swimming velocities did not differ significantly across the range of prey densities ($F_{6,38} = 0.94$, $P > 0.05$). Mean swimming velocities varied from

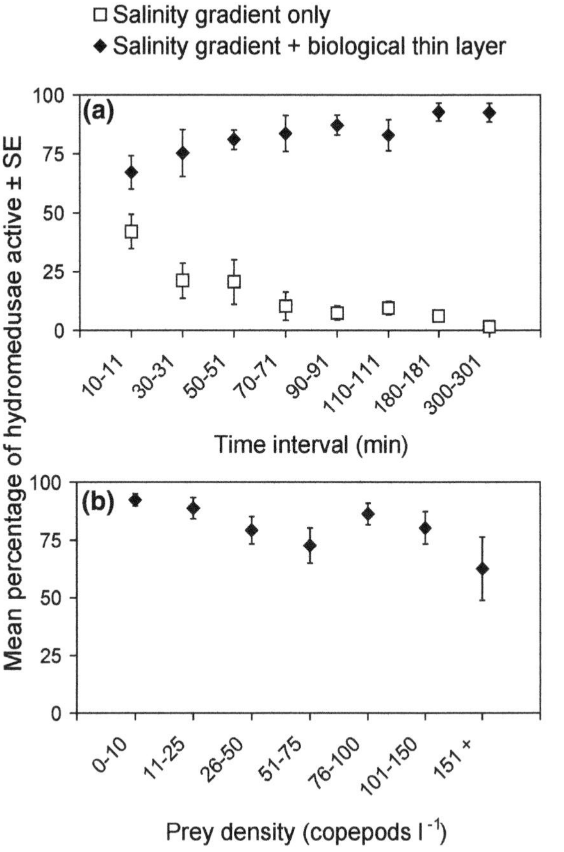

Fig. 4 Mean percentages of the hydromedusa, *Nemopsis bachei*, recorded as active (swimming or feeding) during experiments testing its behavior relative to a salinity gradient versus relative to a salinity gradient with a biological thin layer of algae and copepods. Means were calculated for (**a**) time intervals and (**b**) prey densities. Densities of copepod prey varied independently of elapsed time. *SE* standard error

0.72 ± 0.30 cm s^{-1} at prey densities of 26–50 copepods l^{-1} to 1.32 ± 0.30 cm s^{-1} at 51–75 copepods l^{-1}. The average bell height of *Nemopsis bachei* used in the experiments was 1 cm; therefore, the mean swimming velocities across all experiments ranged from 0.72 to 1.32 bell heights s^{-1}.

Numbers of hydromedusae feeding were compared with one-way ANOVAs. The numbers feeding differed significantly among time intervals ($F_{7,56} = 9.57$, $P < 0.01$) and across the range of prey densities ($F_{6,86} = 6.52$, $P < 0.01$). The mean number of hydromedusae feeding decreased from 3.9 ± 1.4 at 10–11 min to 0.4 ± 0.1 at 70–71 min (Fig. 5a). After that time, the mean number of hydromedusae feeding remained at or below 0.2 ± 0.1 for the duration of the experiments. In contrast, the mean number of hydromedusae feeding increased significantly from 0.2 ± 0.1 to 3.0 ± 0.1 across prey densities ranging from 0–10 copepods l^{-1} to ≥151 copepods l^{-1} (Fig. 5b).

Mean encounter rates (encounters hydromedusa^{-1} 10 min^{-1}) also were analyzed with one-way ANOVAs, which were interpreted cautiously because the transformed data were not normally distributed. The total number and distribution of encounters, which could not be controlled, meant that pooling was required to yield more balanced replication among time intervals and prey densities. The mean encounter rate at 10–11 min was 4.7 ± 1.1 encounters hydromedusa^{-1} 10 min^{-1}, and the mean encounter rates at 30–31 min and beyond were ≤1.8 ± 0.5 encounters hydromedusa^{-1} 10 min^{-1}, which represented a significant difference ($F_{4,19} = 3.96$, $P = 0.02$; Fig. 6a). In addition, encounter rates were significantly higher when hydromedusae were exposed to ≥101 copepods l^{-1} ($F_{3,29} = 3.21$, $P = 0.04$). Encounter rates rose to 3.9 ± 0.6 encounters hydromedusa^{-1} 10 min^{-1} at ≥101 copepods l^{-1} from 1.5 ± 0.16, 1.7 ± 0.26, and 1.8 ± 0.7 encounters hydromedusa^{-1} 10 min^{-1} at 0–25, 26–75, and 76–100 copepods l^{-1}, respectively (Fig. 6b).

Mean handling times (s) were compared among time intervals and prey densities, with the non-significant results considered reliable even though transformed data were not normally distributed. Handling times did not differ significantly among time intervals or across the range of prey densities ($F_{4,41} = 0.37$, $P > 0.05$; $F_{5,40} = 1.64$, $P > 0.05$; respectively). Across all time intervals and prey

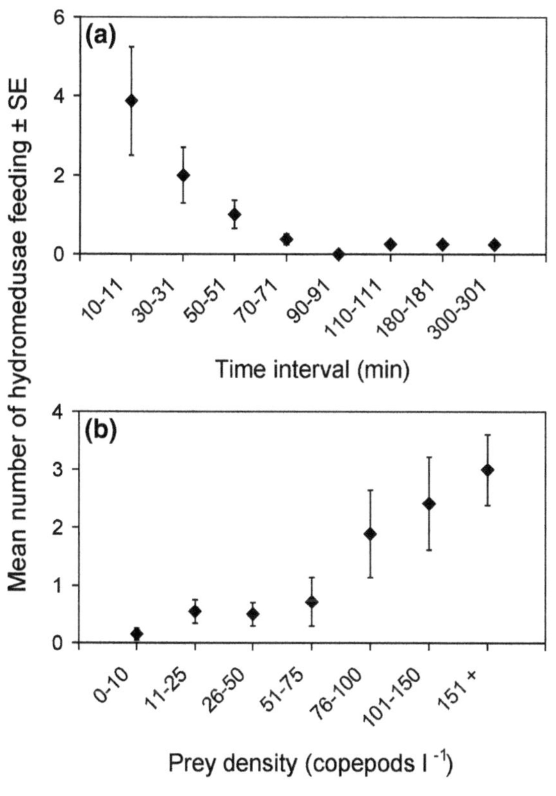

Fig. 5 Mean numbers of the hydromedusa, *Nemopsis bachei*, feeding during experiments testing its behavior relative to a salinity gradient versus relative to a salinity gradient with a biological thin layer of algae and copepods. Means were calculated for (**a**) time intervals and (**b**) prey densities. Densities of copepod prey varied independently of elapsed time. *SE* standard error

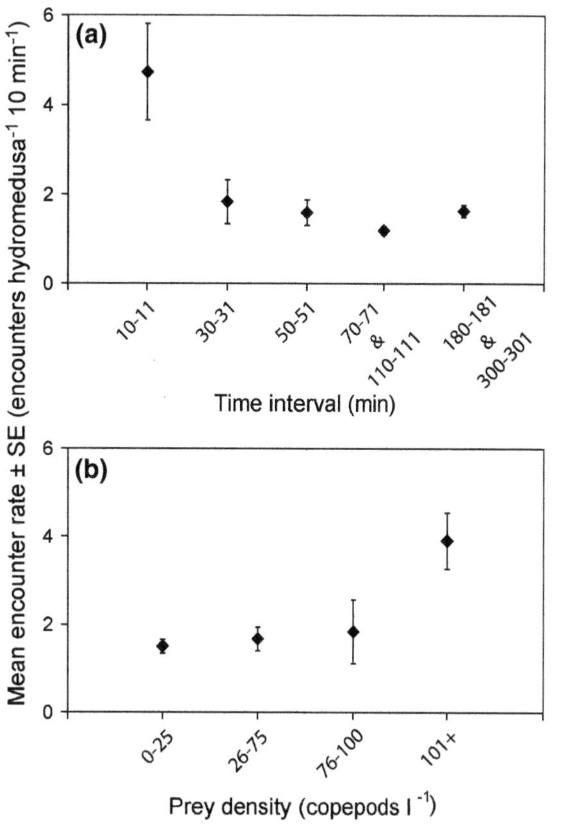

Fig. 6 Mean encounter rates between the hydromedusa, *Nemopsis bachei*, and its copepod prey, *Acartia tonsa*, during experiments testing the behavior of hydromedusae relative to a salinity gradient versus relative to a salinity gradient with a biological thin layer of algae and copepods. Means were calculated for (**a**) time intervals and (**b**) prey densities. Densities of copepod prey varied independently of elapsed time. *SE* standard error

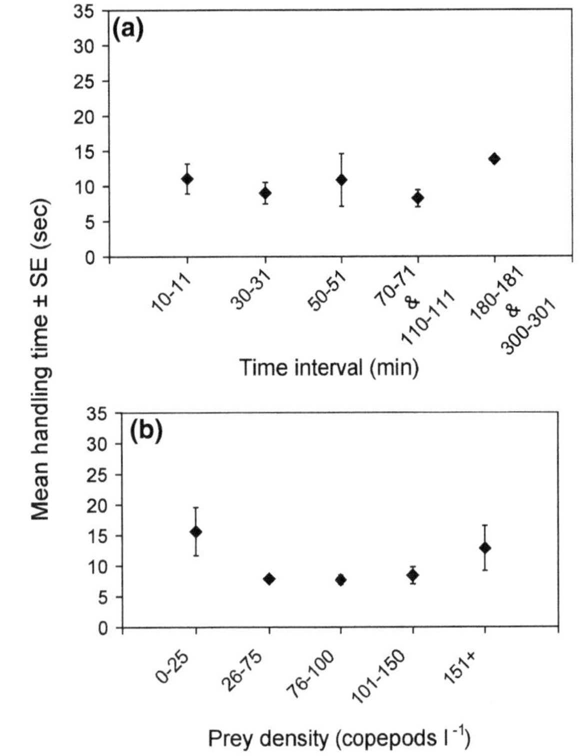

Fig. 7 Mean handling times for the hydromedusa, *Nemopsis bachei*, to transfer its prey, *Acartia tonsa*, into its mouth during experiments testing the behavior of hydromedusae relative to a salinity gradient versus relative to a salinity gradient with a biological thin layer of algae and copepods. Means were calculated for (**a**) time intervals and (**b**) prey densities. Densities of copepod prey varied independently of elapsed time. *SE* standard error

densities, mean handling times varied from 7.7 ± 0.8 to 15.6 ± 4.0 s (Fig. 7a, b).

Discussion

Dense concentrations of living material in thin layers have the potential to influence behaviors of organisms, feeding success, growth rates, reproductive rates, and predation pressure in coastal systems (McManus et al., 2003). The factors that drive the formation and maintenance of these layers are varied and largely unquantified, but nevertheless, the presence and persistence of thin layers are likely to alter biogeochemical cycles, define food webs, enhance fisheries productivity, and influence responses to climate change (Dekshenieks et al., 2001; Stacey et al., 2007; Birch et al., 2008).

Similarly, gelatinous zooplankton, a diverse and sometimes invasive faunal group, also are known to shape marine ecosystems (e.g., Boero et al., 2008). Unfortunately, the distribution and abundance of gelatinous zooplankton are generally underestimated, especially when these voracious consumers bloom or aggregate at discontinuities (Arai, 1992; Graham et al., 2001). In their review of the ecological value of gelatinous zooplankton, Boero et al. (2008) state that the life cycles and life histories of gelatinous zooplankton allow them to exploit temporarily abundant resources with irrefutable, but often overlooked, consequences for marine food webs that include direct effects from predation and indirect effects from disruption of trophic links. Thin layers represent patches of abundant resources that could be exploited by gelatinous zooplankton, such as jellyfish, but existing literature on thin layers is limited and largely focused on phytoplankton, microzooplankton, and ichthyoplankton, rather than predatory jellyfish.

This study addressed the nexus between these two issues by determining how a common and increasingly widespread hydromedusa predator, *Nemopsis bachei,* responded to salinity discontinuities with and without other plankton. Salinity gradients and biological thin layers were established successfully, and they persisted throughout all experiments. Careful handling, suitable experimental conditions, a 50-cm-tall experimental column, and multi-hour experiments provided ample space and time for hydromedusae to respond without a significant decrease in their health. In fact, there were no moribund hydromedusae, rather *N. bachei* medusae increased their activity, i.e., swimming or feeding, over time when biological thin layers were present. There also was no statistically significant change in swimming velocities over time, which further indicated that hydromedusae remained healthy. Finally, there also was evidence of feeding by hydromedusae throughout the duration of experiments with biological thin layers.

Distributions of hydromedusae

Over 75% of the hydromedusae aggregated at the salinity gradients with and without biological thin layers, with no statistically significant difference in their distribution between those conditions. These results indicate that the salinity gradient, not the presence of copepod prey, was the primary influence on the vertical distribution of the hydromedusae.

The results also indicate that hydromedusae did not remain aggregated solely due to osmoregulatory limitations or their ability to swim. *Nemopsis bachei* medusae would have been expected to be able to conform to the salinity differential of 7 that they confronted during the 5-h experiments, because nine other species of hydromedusae conformed to salinity differentials of 3 within an hour (Mills, 1984). In fact, the salinity gradient did not prevent vertical excursions by the hydromedusae, which traversed the entire water column and gradually dispersed to yield minor concentrations at 0–6 cm (lower salinity) and 42–50 cm (higher salinity). In addition, the mean swimming velocities recorded during all experiments did not change significantly over 5 h; therefore, the hydromedusae remained capable of moving away from the salinity gradient.

The sustained aggregation of hydromedusae at salinity gradients was not solely passive or strongly influenced by the presence of organisms, including potential copepod prey. Regardless of the mechanism generating the aggregation, the results of experiments in the present study support previous reports that hydromedusae and other jellyfish will aggregate at discontinuities (Arai, 1992; Graham et al., 2001), which may enhance reproductive success, advection or retention, and feeding.

Changes in activity in response to biological thin layers

Although the vertical distributions of hydromedusae did not differ significantly between experiments with and without biological thin layers, results did indicate significant differences in behavior. The mean percentages of hydromedusae that were active, defined as feeding or swimming, differed significantly between experiments with and without biological thin layers. More hydromedusae were active at the beginning of experiments with biological thin layers (67% vs. 42%), and activity increased over time until 92% of hydromedusae were active after 5 h. In experiments without prey in biological thin layers, activity decreased to approximately 2% over 5 h. The decreased activity through time equates to decreased swimming in experiments without biological thin layers, because feeding could not occur. Nevertheless, most hydromedusae remained in or near the salinity gradient, which involved limited pulsing of their bells to maintain their positions. In experiments with biological thin layers, the numbers of hydromedusae feeding decreased through time, which means that swimming must have increased in order for the levels of activity to increase. Thus, the biological thin layer did affect the behavior of hydromedusae beyond feeding, but the hydromedusae tended to aggregate in or near a physical gradient regardless of whether or not potential prey were present.

The relative effects of the physical and biological components of thin layers vary among planktonic fauna. For example, *Acartia hudsonica* Pinhey, 1926 aggregated consistently at salinity discontinuities, with exposure to thin layers of diatoms modifying this behavior, probably by inducing area-restricted searching and feeding (Bochdansky & Bollens, 2004). In contrast, other adult copepods displayed various responses to salinity gradients (three species crossed gradients, but four species did not; two species increased residence time in gradients, but seven species did not), with variations hypothesized to relate to species-specific sensory capabilities and adaptive responses to evolutionary pressures deriving from use of habitats with differing salinity regimes (Woodson et al., 2005, 2007a, b). All species remained longer in patches of food or exudates from food, with data suggesting area-restricted searching or slower swimming while feeding as the causes of aggregation, although not all statistical tests were significant. Thus, further research is required to elucidate how proximal cues and evolutionary pressures interact with the sensory capabilities, physiological limitations, and behavioral responses of organisms to yield aggregations at discontinuities like thin layers.

Feeding behavior and foraging strategy

As an ambush predator, *Nemopsis bachei* remains stationary when it feeds; therefore, the encounter rates and encounter radii that determine feeding rates depend heavily on the density or movement of prey (Gerritsen & Strickler, 1977; Costello, 1992; Costello et al., 2008). In experiments with biological thin layers, encounter rates for *N. bachei* reflected changes in the densities of copepod prey. These changes in encounter rates equate to changes in the mean number of hydromedusae feeding because *N. bachei* displayed 100% capture, transfer, and ingestion efficiencies (as defined by Regula et al., 2009). Increased feeding with increased prey densities also has been reported for other hydromedusae (Fulton & Wear, 1985; Daan, 1986; Matsakis & Nival, 1989), ctenophores (Reeve et al., 1978), scyphomedusae (Båmstedt et al., 1994), and siphonophores (Purcell, 1982; Purcell & Kremer, 1983).

Results of the present study indicate that prey densities were not high enough to saturate the feeding process (Purcell & Nemazie, 1992; Regula et al., 2009). Ingestion rates, as estimated from encounter rates and 100% ingestion efficiency, increased at higher prey densities, with no evidence of a decrease that would indicate saturation. In addition, handling times did not vary significantly through time or across the full range of prey densities. In combination, these results indicate that hydromedusae were capable of consuming every copepod captured by the tentacles.

Given the observed efficiency of feeding and the lack of saturation, encounter rates can be extrapolated to estimate daily ingestion rates for *Nemopsis bachei*. The range of mean encounter rates, 1.5–4.7 encounters hydromedusa^{-1} 10 min^{-1}, translates to ingestion rates of approximately 9–28 adult copepods h^{-1}, respectively. The high concentration of prey in thin layers may be why these ingestion rates were considerably higher than estimates of 2–15 *Acartia tonsa* copepodites d^{-1} derived from laboratory experiments with *N. bachei* and 1–41 *A. tonsa* copepodites d^{-1} derived from field sampling for *N. bachei* exposed to <25 prey l^{-1} (Purcell & Nemazie, 1992). Ultimately, field surveys or experiments that account for variations in body size, prey densities, temperatures, salinities, and other potential influences will be needed to determine if ingestion rates vary consistently when gelatinous zooplankton encounter high concentrations of prey at narrow discontinuities.

Encounter rates also influenced the allocation of energy to feeding or swimming, which involves tradeoffs because swimming repositions hydromedusae while limiting feeding. Hydromedusae increased their swimming activity as encounter rates decreased to approximately 2 encounters hydromedusa^{-1} 10 min^{-1} and feeding success waned. In contrast, swimming did not vary significantly across the range of prey densities created in experiments, which spanned an order of magnitude. Thus, hydromedusae increased swimming in response to decreased encounters with prey rather than variations in absolute densities of prey. This behavioral response may be the basis of a successful foraging strategy for a predator that remains stationary while ambushing prey, i.e., hydromedusae move from areas yielding low encounter rates and remain in areas yielding higher encounter rates, regardless of whether encounter rates are determined by the density or movement of prey.

Conclusions

This study examined the responses of an ecologically important hydromedusa, *Nemopsis bachei*, to salinity gradients and thin layers containing algae and copepods. The vertical distribution of hydromedusae was most strongly influenced by salinity gradients. The aggregation of hydromedusae at salinity gradients was not solely a passive response resulting from their inability to osmoconform or swim through the gradients. The presence of copepods stimulated feeding, which varied with encounter rates as expected for these ambush predators. Capture, transfer, and ingestion efficiencies were 100%, and feeding increased at higher prey densities without saturation. These results suggest that *N. bachei* can exert significant predation pressure on patches of prey found in thin layers. The frequency of swimming increased when encounters with prey decreased, which suggests that these hydromedusae may swim to new positions when encounter rates and feeding success decrease below some threshold. Such a strategy should increase their chances of finding and exploiting patches of prey, such as those found in thin layers, by balancing feeding with swimming to search for prey, which precludes feeding. In summary, hydromedusae can alter their distribution and feeding in response to salinity gradients and biological thin layers resulting in patches of enhanced predation that would affect the locations and magnitudes of carbon flux in the oceans.

Acknowledgments Special thanks go to Dr. Ferdinando "Nando" Boero for confirming the species identification of the hydromedusae and Dr. Thomas Kiørboe for his immeasurable guidance and support in the preparation of this manuscript. We thank Marco Gauger for his invaluable assistance in the field and laboratory throughout this study. We also would like to thank Drs. Rabea Diekmann and Janna

Peters for helpful discussions and comments as we began preparing this manuscript, as well as Dr. Cinzia Gravili for providing elusive and valuable literature on *Nemopsis bachei*. Dr. J. E. Purcell and two anonymous reviewers provided helpful comments on this manuscript. This article is a contribution to EUR-OCEANS Network of Excellence funded by the European Commission (WP4-SYSMS-1101). J. R. Frost was supported by a EUR-OCEANS fellowship.

References

Alldredge, A. L., T. J. Cowles, S. MacIntyre, J. E. B. Rines, P. L. Donaghay, C. F. Greenlaw, D. V. Holliday, M. A. Dekshenieks, J. M. Sullivan & J. R. Zaneveld, 2002. Occurrence and mechanisms of formation of a dramatic thin layer of marine snow in a shallow Pacific fjord. Marine Ecology Progress Series 233: 1–12.

Arai, M. N., 1973. Behavior of the planktonic coelenterates, *Sarsia tubulosa*, *Phialidium gregarium* and *Pleurobrachia pileus* in salinity discontinuity layers. Journal of the Fisheries Research Board of Canada 30: 1105–1110.

Arai, M. N., 1976. Behavior of the planktonic coelenterates in temperature and salinity discontinuity layers. In Mackie, G. O. (ed.), Coelenterate Ecology and Behavior. Plenum Press, New York: 211–218.

Arai, M. N., 1992. Active and passive factors affecting aggregations of hydromedusae: a review. Scientia Marina 56: 99–108.

Bailey, K. M. & R. S. Batty, 1983. A laboratory study of predation by *Aurelia aurita* on larval herring (*Clupea harengus*): experimental observations compared with model predictions. Marine Biology 72: 295–301.

Båmstedt, U., M. B. Martinussen & S. Matsakis, 1994. Trophodynamics of the two scyphozoan jellyfishes, *Aurelia aurita* and *Cyanea capillata*, in western Norway. ICES Journal of Marine Science 51: 369–382.

Birch, D. A., W. R. Young & P. J. S. Franks, 2008. Thin layers of plankton: formation by shear and death by diffusion. Deep-Sea Research I 55: 277–295.

Bochdansky, A. B. & S. M. Bollens, 2004. Relevant scales in zooplankton ecology: distribution, feeding and reproduction of the copepod *Acartia hudsonica* in response to thin layers of the diatom *Skeletonema costatum*. Limnology and Oceanography 49: 625–639.

Boero, F., J. Bouillon, C. Gravili, M. P. Miglietta, T. Parsons & S. Piraino, 2008. Gelatinous plankton: irregularities rule the world (sometimes). Marine Ecology Progress Series 356: 299–310.

Clay, T. W., S. M. Bollens, A. B. Bochdansky & T. R. Ignoffo, 2004. The effects of thin layers on the vertical distribution of larval Pacific herring, *Clupea pallasi*. Journal of Experimental Marine Biology and Ecology 305: 171–189.

Costello, J., 1992. Foraging mode and energetics of hydrozoan medusae. Scientia Marina 56: 185–191.

Costello, J. H. & S. P. Colin, 2002. Prey resource use by coexistent hydromedusae from Friday Harbor, Washington. Limnology and Oceanography 47: 934–942.

Costello, J. H., S. P. Colin & J. O. Dabiri, 2008. Medusan morphospace: phylogenetic constraints, biomechanical solutions, and ecological consequences. Invertebrate Biology 127: 265–290.

Cowles, T. J., 2004. Planktonic layers: physical and biological interactions on the small scale. In: Handbook of Scaling Methods in Aquatic Ecology: Measurements, Analysis, Simulation. CRC Press, Boca Raton, FL: 31–49.

Cowles, T. J. & R. A. Desiderio, 1993. Resolution of biological microstructure through in situ fluorescence emission spectra. Oceanography 6: 105–111.

Cowles, T. J., R. A. Desiderio, J. N. Mourn, M. L. Myrick, D. Garvis & S. M. Angel, 1990. Fluorescence microstructure using a laser/fiber optic profiler. Ocean Optics X, Proceedings of SPIE 1302: 336–345.

Cowles, T. J., R. A. Desiderio & M. E. Carr, 1998. Small-scale planktonic structure: persistence and trophic consequences. Oceanography 11: 4–9.

Daan, R., 1986. Food intake and growth of *Sarsia tubulosa* (Sars, 1835), with quantitative estimates of predation on copepod populations. Netherlands Journal of Sea Research 20: 67–74.

Daan, R., 1989. Factors controlling the summer development of copepod populations in the Southern Bight of the North Sea. Netherlands Journal of Sea Research 23: 305–322.

Dekshenieks, M. A., P. L. Donaghay, J. M. Sullivan, J. E. B. Rines, T. R. Osborn & M. S. Twardowski, 2001. Temporal and spatial occurrence of thin phytoplankton layers in relation to physical processes. Marine Ecology Progress Series 223: 61–71.

Denman, K. L. & T. M. Powell, 1984. Effects of physical processes on planktonic ecosystems in the coastal ocean. Oceanography and Marine Biology. An Annual Review 22: 125–168.

Donaghay, P. L. & T. R. Osborn, 1997. Toward a theory of biological–physical control of harmful algal bloom dynamics and impacts. Limnology and Oceanography 42: 1283–1296.

Donaghay, P. L., H. M. Rines & J. M. Sieburth, 1992. Simultaneous sampling of fine scale biological, chemical and physical structure in stratified waters. Archiv für Hydrobiologie 36: 97–108.

Dumoulin, E., 1997. Het invasieachtig voorkomen in de zuidelijke Noordzee van de hydromedusen *Nemopsis bachei* L. Agassiz, 1849 en *Eucheilota maculata* Hartlaub, 1894 in augustus-september 1996 (met aanvullende data voor 1997) (Hydrozoa: Athecata, Thecata). De Strandvlo 17: 102–126.

Flagg, C. N. & S. L. Smith, 1985. On the use of the acoustic Doppler current profiler to measure zooplankton abundance. Deep-Sea Research 36: 455–474.

Fulton III, R. S. & R. G. Wear, 1985. Predatory feeding of the hydromedusae *Obelia geniculata* and *Phialella quadrata*. Marine Biology 87: 47–54.

Gallagher, S. M., H. Yamazaki & C. S. Davis, 2004. Contribution of fine-scale vertical structure and swimming behavior to formation of plankton layers on Georges Bank. Marine Ecology Progress Series 267: 27–43.

Gerritsen, J. & J. R. Strickler, 1977. Encounter probabilities and community structure in zooplankton: a mathematical model. Journal of the Fisheries Research Board of Canada 34: 73–82.

Graham, W. M., F. Pagès & W. M. Hamner, 2001. A physical context for gelatinous zooplankton aggregations: a review. Hydrobiologia 451: 199–212.

Grindley, J. R., 1964. Effect of low-salinity water on the vertical migration of zooplankton. Nature 203: 781–782.

Hamner, W. M., L. P. Madin, A. L. Alldredge, R. W. Gilmer & P. P. Hamner, 1975. Underwater observations of gelatinous zooplankton: sampling problems, feeding biology and behavior. Limnology and Oceanography 20: 907–917.

Hansen, K. V., 1951. On the diurnal migration of zooplankton in relation to the discontinuity layer. Journal du Conseil International pour l'Exploration de le Mer 17: 231–241.

Hanson, A. K. Jr., & P. L. Donaghay, 1998. Micro- to fine-scale chemical gradients and layers in stratified coastal waters. Oceanography 11: 10–17.

Holliday, D. V., R. E. Pieper, C. F. Greenlaw & J. K. Dawson, 1998. Acoustical sensing of small-scale vertical structures in zooplankton assemblages. Oceanography 11: 18–23.

Huntley, M. E. & L. A. Hobson, 1978. Medusa predation and plankton dynamics in a temperate fjord, British Columbia. Journal of the Fisheries Research Board of Canada 35: 257–261.

ICES, 2006. Working Group on Introductions and Transfers of Marine Organisms (WGITMO), 16–17 March 2006, Oostende, Belgium. ICES CM 2006/ACME:05: 334 pp.

Ignoffo, T. R., S. M. Bollens & A. B. Bochdansky, 2005. The effects of thin layers on the vertical distribution of the rotifer, *Brachionus plicatilis*. Journal of Experimental Marine Biology and Ecology 316: 167–181.

Johnson, P. W., P. L. Donaghay, E. B. Small & J. M. Sieburth, 1995. Ultrastructure and ecology of *Perispira ovum* (Ciliophora: Litostomatea): an aerobic, planktonic ciliate that sequesters the chloroplasts, mitochondria and paramylon of *Euglena proxima* in a micro-oxic habitat. Journal of Eukaryotic Microbiology 42: 323–335.

Levin, P. S., N. Tolimieri, M. Nicklin & P. F. Sale, 2000. Intergrating individual behavior and population ecology: the potential for habitat-dependent population regulation in a reef fish. Behavioral Ecology 11: 565–571.

Lougee, L. A., S. M. Bollens & S. R. Avent, 2002. The effects of haloclines on the vertical distribution and migration of zooplankton. Journal of Experimental Marine Biology and Ecology 278: 111–134.

Mackas, D. L., K. L. Denman & R. A. Abbott, 1985. Plankton patchiness: biology in the physical vernacular. Bulletin of Marine Science 37: 652–674.

Marshalonis, D. & J. L. Pinckney, 2008. Grazing and assimilation rate estimates of hydromedusae from a temperate tidal creek system. Hydrobiologia 606: 203–211.

Mason, R. P., W. F. Fitzgerald, P. L. Donaghay & J. M. Sieburth, 1993. Mercury speciation and cycling in the Pettaquamscutt Estuary. Limnology and Oceanography 36: 1227–1241.

Matsakis, S. & R. J. Conover, 1991. Abundance and feeding of medusae and their potential impact as predators on other zooplankton in Bedford Basin (Nova Scotia, Canada) during spring. Canadian Journal of Fisheries and Aquatic Sciences 48: 1419–1430.

Matsakis, S. & P. Nival, 1989. Elemental composition and food intake of *Phialidium* hydromedusae in the laboratory. Journal of Experimental Marine Biology and Ecology 130: 277–290.

McManus, M. A., A. L. Alldredge, A. H. Barnard, E. Boss, J. F. Case, T. J. Cowles, P. L. Donaghay, L. B. Eisner, D. J. Gifford, C. F. Greenlaw, C. M. Herren, D. V. Holliday, D. Johnson, S. MacIntyre, D. M. McGehee, T. R. Osborn, M. J. Perry, R. E. Pieper, J. E. B. Rines, D. C. Smith, J. M. Sullivan, M. K. Talbot, M. S. Twardowski, A. Weidemann & J. R. Zaneveld, 2003. Characteristics, distribution and persistence of thin layers over a 48 hour period. Marine Ecology Progress Series 261: 1–19.

McManus, M. A., O. M. Cheriton, P. J. Drake, D. V. Holliday, C. D. Storlazzi, P. L. Donaghay & C. F. Greenlaw, 2005. Effects of physical processes on structure and transport of thin zooplankton layers in the coastal ocean. Marine Ecology Progress Series 301: 199–215.

Mills, C. E., 1984. Density is altered in hydromedusae and ctenophores in response to changes in salinity. Biological Bulletin Marine Biological Laboratory, Woods Hole 166: 206–215.

Moreira, G. S., 1973. On the diurnal vertical migration of hydromedusae off Santos, Brazil. Publications from the Seto Marine Biological Laboratory 20: 537–566.

Osborn, T. R., 1998. Fine structure, microstructure and thin layers. Oceanography 11: 36–43.

Owen, R. W., 1989. Microscale and finescale variations of small plankton in coastal and pelagic environments. Journal of Marine Research 47: 197–240.

Purcell, J. E., 1982. Feeding and growth of the siphonophore *Muggiaea atlantica* (Cunningham 1893). Journal of Experimental Marine Biology and Ecology 62: 39–54.

Purcell, J. E. & P. Kremer, 1983. Feeding and metabolism of the siphonophore *Sphaeronectes gracilis*. Journal of Plankton Research 5: 95–106.

Purcell, J. E. & D. A. Nemazie, 1992. Quantitative feeding ecology of the hydromedusan *Nemopsis bachei* in Chesapeake Bay. Marine Biology 113: 305–311.

Rawlinson, K. A., J. Davenport & D. K. A. Barnes, 2004. Vertical migration strategies with respect to advection and stratification in a semi-enclosed lough: a comparison of mero- and holozooplankton. Marine Biology 144: 935–946.

Reeve, M. R., M. A. Walter & T. Ikeda, 1978. Laboratory studies of ingestion and food utilization in lobate and tentaculate ctenophores. Limnology and Oceanography 23: 740–751.

Regula, C., S. P. Colin, J. H. Costello & H. Kordula, 2009. Prey selection mechanism of ambush-foraging hydromedusae. Marine Ecology Progress Series 374: 135–144.

Sieburth, J. M. & P. L. Donaghay, 1993. Planktonic methane production and oxidation within the algal maximum of the pycnocline: seasonal fine scale observations in an anoxic estuarine basin. Marine Ecology Progress Series 100: 3–15.

Smedstad, O. M., 1972. On the biology of *Aglantha digitale rosea* (Forbes) (Coelenterata: Trachymedusae) in the inner Oslofjord. Norwegian Journal of Zoology 20: 111–135.

Southward, A. J. & R. L. Barrett, 1983. Observations on the vertical distribution of zooplankton, including post-larval-teleosts, off Plymouth in the presence of a thermocline and a chlorophyll-dense layer. Journal of Plankton Research 5: 599–618.

Stacey, M. T., M. A. McManus & J. V. Steinbuck, 2007. Convergences and divergences and thin layer formation and maintenance. Limnology and Oceanography 52: 1523–1532.

Støttrup, J. G. & J. Jensen, 1990. Influence of algal diet on feeding and egg-production of the calanoid copepod *Acartia tonsa* Dana. Journal of Experimental Marine Biology and Ecology 141: 87–105.

Tiselius, P., 1992. Behavior of *Acartia tonsa* in patchy food environments. Limnology and Oceanography 37: 1640–1651.

Tiselius, P., G. Nielsen & T. G. Nielsen, 1994. Microscale patchiness of plankton within a sharp pycnocline. Journal of Plankton Research 16: 543–554.

Walne, P. R., 1970. Studies on the food value of nineteen genera of algae to juvenile bivalves of the genera *Ostrea*, *Crassostrea*, *Mercenaria*, and *Mytilis*. Fishery Investigations, London, Series 26: 1–62.

Widder, E. A., S. Johnsen, S. A. Bernstein, J. F. Case & D. J. Nielson, 1999. Thin layers of bioluminescent copepods found at density discontinuities in the water column. Marine Biology 134: 429–437.

Williams, R., 1985. Vertical distribution of *Calanus finmarchicus* and *C. helgolandicus* in relation to the development of the seasonal thermocline in the Celtic Sea. Marine Biology 86: 145–149.

Woodson, C. B., D. R. Webster, M. J. Weissburg & J. Yen, 2005. Response of copepods to physical gradients associated with structure in the ocean. Limnology and Oceanography 50: 1552–1564.

Woodson, C. B., D. R. Webster, M. J. Weissburg & J. Yen, 2007a. Cue hierarchy and foraging in calanoid copepods: ecological implications of oceanographic structure. Marine Ecology Progress Series 330: 163–177.

Woodson, C. B., D. R. Webster, M. J. Weissburg & J. Yen, 2007b. The prevalence and implications of copepod behavioral responses to oceanographic gradients and biological patchiness. Integrative and Comparative Biology 47: 831–846.

Yamazaki, H., D. L. Mackas & K. L. Denman, 2002. Coupling small scale physical processes with biology: towards a Lagrangian approach. In Robinson, A. R., J. J. McCarthy & B. J. Rothschild (eds), The Sea: Biological–Physical Interaction in the Ocean, Vol. 12. John Wiley & Sons, New York: 51–112.

Hydrobiologia (2010) 645:113–124
DOI 10.1007/s10750-010-0218-7

Avoidance of hydrodynamically mixed environments by *Mnemiopsis leidyi* (Ctenophora: Lobata) in open-sea populations from Patagonia, Argentina

Hermes W. Mianzan · Patricia Martos · John H. Costello · Raúl A. Guerrero

Published online: 21 March 2010

Abstract Biomass distributions of the ctenophore, *Mnemiopsis leidyi*, were collected during a week-long survey program in a tidal front along the Patagonian shelf in December 1989. Average ctenophore biomass concentrations varied significantly along a north–south gradient and in stratified compared to unstratified waters. The relative vertical distribution of *M. leidyi* biomass appeared to be constrained by surface levels of vertical shear. Vertical distributions of ctenophore biomass were highly variable at low levels of vertical shear (<4 s^{-1}) at the surface, but at higher levels of surface vertical shear ctenophores occurred deeper in the water column where shear levels were lower. These results indicate that physical conditions are important factors influencing the distribution of *M. leidyi* along the Patagonian shelf during summer months.

Keywords Tidal front · Native habitats · Vertical distribution · Horizontal distribution · Tidal currents · Surface winds · Turbulence · Atlantic Ocean

Guest editors: J. E. Purcell & Dror Angel / Jellyfish Blooms: New Problems and Solutions

H. W. Mianzan (✉)
CONICET-INIDEP, P. Victoria Ocampo n° 1,
7600 Mar del Plata, Argentina
e-mail: hermes@inidep.edu.ar

P. Martos · R. A. Guerrero
Department of Marine Sciences, Universidad Nacional de Mar del Plata – INIDEP, Mar del Plata, Argentina
e-mail: pmartos@inidep.edu.ar

R. A. Guerrero
e-mail: raul.guerrero@inidep.edu.ar

J. H. Costello
Biology Department, Providence College, Providence, RI 02918-0001, USA
e-mail: costello@providence.edu

Introduction

The ctenophore, *Mnemiopsis leidyi* Agassiz 1865, has a broad native distribution along the Atlantic coasts of North and South America (from 40°N to 46°S) that encompasses a range of coastal estuaries from Buzzards Bay, USA to Blanca Bay, Argentina (Harbison & Volovik, 1994; Mianzan, 1999). *M. leidyi* has earned notoriety for its invasion of the Black (Shiganova et al., 2003), Azov (Studenikina et al., 1991) and Caspian Seas (Ivanov et al., 2000), as well as its potential for invasion of other regions (GESAMP, 1997). This potential has recently been realized in the Mediterranean (Shiganova & Malej, 2009; Fuentes et al., 2009), the Baltic (Hansson, 2006; Javidpour et al., 2006; Kube et al., 2007) and North seas (Faasse & Bayha, 2006; Boersma et al., 2007; Oliveira, 2007). The wide range of potential prey (Waggett & Costello, 1999; Rapoza et al., 2005) and high ingestion rates (Kremer, 1975; Reeve & Walter, 1978) contribute to the influential community role of *M. leidyi* in both endemic (Purcell & Decker, 2005) and introduced (Shiganova et al., 1998) environments.

Despite its considerable ecological importance, the factors determining spatial distribution patterns of *M. leidyi* are only recently becoming clear. Kremer (1994) described temperature and prey abundance patterns as major factors affecting its seasonality patterns in of its native temperate regions; Purcell & Decker (2005) quantified the role of predation as a factor limiting copepod populations; Costello et al. (2006) described factors influencing seasonal population growth in a coastal region. However, the vertical distribution patterns of established *M. leidyi* populations are seldom described, and the factors that determine their spatial, and particularly, vertical distribution are not well defined. The available information describing vertical distribution of *M. leidyi* frames a somewhat confusing picture. Water depth and degree of vertical stratification not only appear to be influential (Purcell et al., 2001b), but may also reflect a range of behavioral responses within local populations. In some exotic habitats like the Black Sea, the bulk of *M. leidyi* seem to be located above the pycnocline during day and night (Mutlu, 1999; Purcell et al., 2001b), where evidence of vertical migration that is restricted to the upper layers is perceptible (Zaika & Sergeeva, 1991). In shallow vertically stratified habitats like Chesapeake Bay (USA), *M. leidyi* also was found above the pycnocline during day and night (Purcell et al., 1994). In vertically homogeneous water columns, such as that of Pamlico Sound (USA), *M. leidyi* dispersed throughout the water column during the night but aggregated near-surface during the day (Miller, 1974). In Puerto Madryn, Argentina, *M. leidyi* was observed aggregating in a dense layer close to the bottom and close to the surface during daylight hours on the same day (Costello & Mianzan, 2003). Miller (1974) observed that *M. leidyi* retreated from surface layers during periods with high surface winds and choppy sea conditions. These observations suggest that certain conditions, such as water column stratification and diurnal migrations, may influence vertical distribution; however, other factors such as surface disturbance may modify the distributions.

The northern Patagonian coastal region of Argentina provides a unique area to evaluate some of the important variables influencing horizontal and vertical distribution patterns of *M. leidyi*. The region possesses a strong, tidally dominated frontal region characterized by high micro- and mesozooplankon standing stocks (Viñas & Ramírez, 1996; Sabatini & Martos, 2002; Acha et al., 2004), anchovy, hake and squid spawning grounds (see Acha et al., 2004, and references therein) and abundant *M. leidyi* (Mianzan & Guerrero, 2000). The seasonal frontal system is located near Peninsula Valdés and extends southward along the Patagonian coast from ca. 42° to 45°S (Carreto et al., 1986; Glorioso, 1987; Bakun & Parrish, 1991; Sánchez et al., 1998; Sabatini & Martos, 2002). This region is characterized by high tidal dissipation rates (Miller, 1966) which generate strong vertical mixing that may homogenize the whole water column in near-shore waters. This feature, combined with the stratification of shelf waters induced by seasonal surface warming, leads to the spring and summer occurrence of shelf sea fronts separating highly mixed inshore waters from stratified offshore waters. Salinity, ranging from 33.5 to 33.65, does not generally cause vertical or horizontal stratification in the region. The position of the front is predictable (Glorioso & Simpson, 1994; Sánchez et al., 1998) and generally aligned in a NE–SW orientation that closely follows bathymetric contours. The front is located offshore in the northern area, on average 80 km from the coastline, and approaches the shore in the south to within 50 km from the coast (Sabatini & Martos, 2002). Turbulent dissipation rates in the water column generally decrease from north to south (Glorioso & Simpson, 1994; Glorioso & Flather, 1995; Glorioso, 2000), which may be influenced by the position of the 75–80 m isobath over the middle shelf in the north to near the coastline (Sabatini & Martos, 2002) in the south. Water column energy dissipation is highly concentrated near the northeast of Península Valdés (Glorioso & Simpson, 1994). Topographic shoals located southeast and northeast of Península Valdés also intensify tidal mixing (Glorioso, 1987). A cruise series through the region provided the opportunity to relate physical features of the water column with distribution patterns of *M. leidyi* inhabiting the open sea.

Methods

The field sampling was carried out on board the German R/V 'Meteor' during 6–12 December 1989. The comprehensive sampling in a Patagonian tidal front allowed the study of *M. leidyi* vertical distribution patterns for 6 days. Environmental data and plankton samples were taken across the tidal front off Peninsula Valdés on four transects each of which was surveyed three times. The first transect series collected physical data used to determine stations for biological sampling. The remaining two series collected physical data and plankton samples at these stations during a continuous sampling design during day and night. Transect orientation used real-time satellite imagery to direct the transect perpendicular to the thermal front (Fig. 1).

Conductivity–temperature–depth (CTD, Neil Brown Mark III, with rosette sampler General Oceanics) profiles ($n = 72$) were collected during each transect. The stations were established at a distance of every five nautical miles along the transects. A low-resolution ship-mounted Acoustic Doppler Current Profiler (ADCP, RD Instruments, 150 kHz) was used to define the tidal front structure at 55 stations. Profiles with 5-m vertical resolution of horizontal velocity components $\boldsymbol{u}$ and $\boldsymbol{v}$ were obtained. Vertical Velocity Shear (VVS) was calculated from the velocity profiles at each station and correlated with *M. leidyi* biomass. VVS was used as an indicator of Turbulent Kinetic Energy (TKE) in the water column due to vertical friction between layers (Lozovatsky et al., 2006). Residual currents for the area were almost an order of magnitude lower than tidal currents (Piola & Rivas, 1997), so instantaneous velocity measurements from the ADCP (total barotropic currents) were used directly and residual currents assumed to be of negligible relative importance. More details on the current regime are in Nellen (1990). Raw CTD data for temperature were extracted every meter but averaged over 2 m intervals and the meter data of this study were filtered over 4 m intervals (see Nellen, 1990; Alheit et al., 1991). The stratification of the water column, estimated from the stability parameter Φ (Simpson, 1981), allowed the estimation of a critical value of 40 J m^{-3} (Martos & Sánchez, 1997), which separate mixed from stratified waters. This value represented the inner, coastal border of the fronts, beyond which the water column was clearly stratified. Wind data, measured in Beaufort scale at the location of each plankton sample, was converted to m s^{-1}.

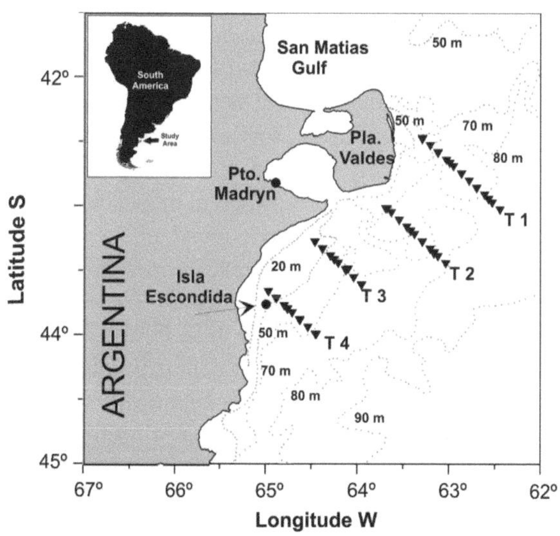

Fig. 1 Sample locations and transects in the tidal front region off Península Valdés, Argentina during 6–12 December 1989. *Triangles* represent locations of physical measurements made prior to biological sampling for each of the four transects (labelled T1–T4)

Plankton was collected using a multiple opening–closing BIOMOC net (1 m^2 mouth area, 1 mm-mesh cod end, nine nets). The samples were taken at five stations along each transect (six stations for the northern-most transect 1), and each transect was surveyed twice in a continuous sampling design (42 stations in total). A total of 255 net hauls were made at different depths with this gear (Fig. 1). The average filtered volume was 248.83 m^3 (range 82–2,278 m^3). The ctenophores collected in the net hauls were immediately sorted and lengths and wet weights (ww) measured. Comparisons of ctenophore biomass were based on the *M. leidyi* ww m^{-3} determined for each depth of a station along a transect.

Relationships between physical and biological variables were analysed statistically using nonparametric methods within the software package Statistica (Statsoft Inc.). Nonparametric methods, primarily the Kruskal–Wallis (K–W) nonparametric analogue to ANOVA and Spearman correlation (Spearman) based on ranks, were chosen due to significant deviations of important variables from normal distributions and non-homogeneous variances between groups.

Results

Environmental conditions

Sea surface temperatures were low near the coast (13–14°C) and increased seaward up to 15.5°C

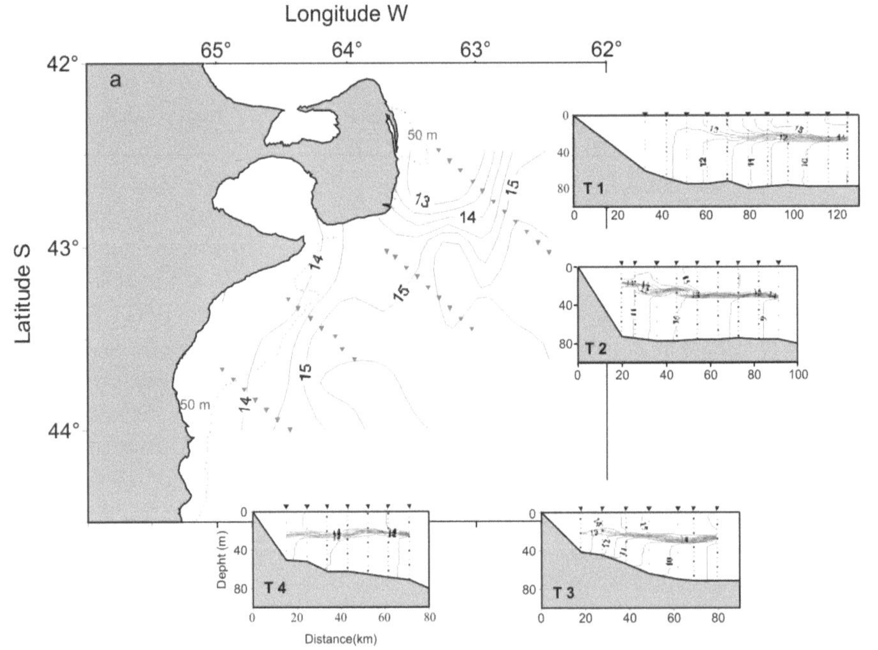

Fig. 2 Sea surface temperature at the tidal front off Península Valdés, Argentina during 6–12 December 1989 and isotherms along the sample transects T1–T4. Isotherms are at 0.5°C intervals

(Fig. 2). A maximum temperature gradient of 0.7°C in 10 km, indicative of the front location, occurred offshore of Península Valdés, and the least steep gradient was observed in the southernmost locations (0.2°C in 10 km). No gradients were observed on transect 2 (T2). Bottom temperatures decreased from 12.5°C near shore to 9°C offshore, with the maximum gradients in the central area (0.66°C in 10 km) and minimum in the north (0.39°C in 10 km) and south (0.19°C in 10 km). During this study, the front was located offshore in the northern area, 68 km from the coastline, and approached as close as 29 km to the coast in the south. On transects 2 and 4, Φ values were higher than 40 J m^{-3}, indicating that the front was located close to the shore. Wind speeds ranged from 2 to 14 m s^{-1} with an average of 8 m s^{-1} for the entire survey area (Fig. 3A). Vertical shear within the water column varied substantially in the surface layers but was always less than ~ 1 s^{-1} deeper than 20 m at all stations (Fig. 3B).

Horizontal biomass patterns

Mnemiopsis leidyi were collected at every station on all transects throughout the study area, and their densities were as high as 138 g ww m^{-3}. Horizontal distribution patterns of *M. leidyi* biomass concentration were significantly related to two features: a north–south gradient and water column stratification. Average biomass concentration of *M. leidyi* increased significantly (K–W $H = 69.2$, df = 3, $N = 255$, $P = 0.000$) in a southerly direction and the southernmost transect (T4) had more than an order of magnitude greater average biomass concentration than the northernmost transect (T1) (Figs. 4, 5). Ctenophore were predominantly small individuals (~ 1 cm) on the northern transect and larger animals on the southern transects (3–14 cm long). Average biomass concentrations were not significantly different between replicates for each transect (K–W $H = 0.045$, df = 1, $N = 255$, $P = 0.832$). The highest distance from shore appeared to influence average biomass concentrations were found offshore; however, the position of a station relative to shore was not in itself a significant factor (K–W $H = 10.175$, df = 5, $N = 255$, $P = 0.070$). Instead, the onshore–offshore pattern was related to the stratification of offshore stations. Water column stratification and the average biomass concentration at a station were significantly positively related (K–W $H = 8.008$, df = 1, $N = 255$, $P = 0.005$). Therefore, strongly stratified stations possessed higher average concentrations of ctenophore biomass, and although some inshore stations were stratified, all the offshore stations were strongly stratified (Fig. 4). The light regime (day or night) during sampling at a station did

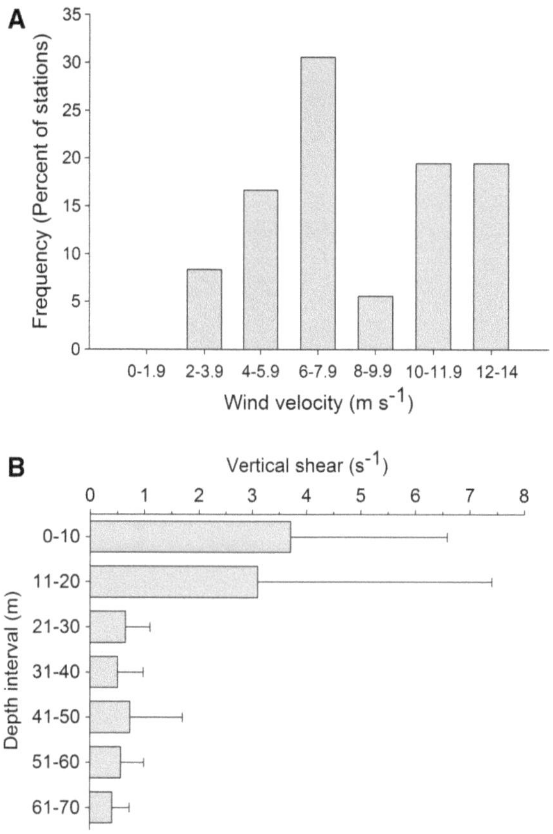

Fig. 3 **A** Distribution of wind velocities ($n = 36$) in the tidal front region off Península Valdés, Argentina. **B** Distribution of vertical shear through the water column in the tidal front region off Península Valdés, Argentina 6–12 December 1989. Histogram represents mean values, and *error bars* are one standard deviation of the mean (only + values shown)

not significantly influence the average biomass concentration at that station (K–W $H = 0.412$, df $= 1$, $N = 255$, $P = 0.521$).

Vertical biomass patterns

The distribution of ctenophore biomass with depth varied widely throughout the study region (Fig. 4). As noted previously, ctenophore biomass was generally greater at stratified than at unstratified stations, but the location of maximum biomass within the water column was not consistently predictable (e.g., Figs. 4, 5B). In order to compare relative vertical positions of ctenophore biomass within the water column, ctenophore concentrations from different depths were grouped by depth, either upper (<20 m) or lower (>20 m). For the stratified regions, this approximated a comparison of ctenophore concentrations in the upper mixed layer with that of the thermocline and below because the thermocline generally was located at a depth between 20 and 30 m (Fig. 4). The ratio of surface to deep concentrations allowed comparison of vertical patterns regardless of differences in absolute biomass or water column stratification.

Vertical distribution patterns were not influenced by the same variables that dominated large-scale horizontal distributions of *M. leidyi* in the study region. Relative vertical distributions (surface or deep) of average ctenophore biomass were not significantly influenced by transect location (K–W $H = 4.213$, df $= 3$, $N = 42$, $P = 0.239$) or water column stratification (K–W $H = 0.098$, df $= 3$, $N = 42$, $P = 0.754$). Light conditions (day or night) did not significantly influence vertical distributions of *M. leidyi* in terms of relative depth (K–W $H = 0.447$, df $= 3$, $N = 42$, $P = 0.504$) or the depth of the biomass maximum (K–W $H = 0.103$, df $= 3$, $N = 42, P = 0.749$). Therefore, there was no evidence of diurnal vertical migration by *M. leidyi* related to light levels. No measured variable was significantly correlated with the depth distribution patterns of *M. leidyi* in the Patagonian shelf region.

Physical factors may constrain rather than directly determine vertical distribution patterns of *M. leidyi* biomass. Vertical shear values found near the surface appear to be linked to the vertical distribution of *M. leidyi* (Fig. 6), although direct correlation between the depth distribution of ctenophore biomass was not significant (Spearman $R = -0.067$, $N = 32$, $P = 0.714$) because biomass vertical distribution patterns were highly variable at low surface vertical shear conditions. However, surface vertical shear constrain *M. leidyi* vertical distributions; at values above 4.0 s^{-1}, the relative proportions of ctenophore biomass above the thermocline were uniformly low (Fig. 6).

Mnemiopsis leidyi responded to wind-induced mixing rapidly (Fig. 7) by altering its vertical distribution within the water column. The potential interaction of surface vertical shear with ctenophore biomass was limited to vertical biomass distributions within a sample location. Surface vertical shear was not significantly linearly correlated with the average (Spearman $R = 0.019$, $N = 32$, $P = 0.920$) or maximum (Spearman $R = 0.005$, $N = 32$, $P = 0.976$) ctenophore biomass concentration at a station or

Fig. 4 Vertical distribution of *Mnemiopsis leidyi* biomass (g ww m^{-3}) along the sample transects T1–T4 (top to bottom, respectively). Each transect was duplicated, with bars over portions of a transect indicating light conditions during sample collection. Isotherms are as shown in Fig. 2

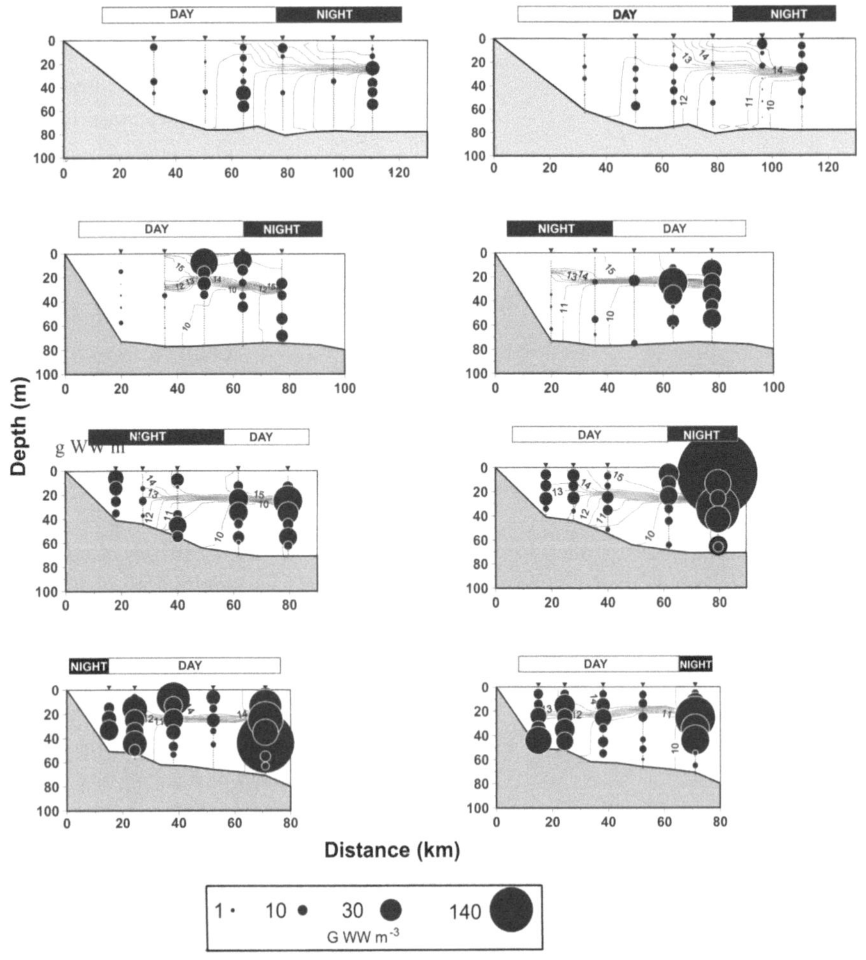

factors affecting horizontal patterns of biomass distribution, such as transect (K–W $H = 4.298$, df = 3, $N = 32$, $P = 0.231$) or water column stratification (K–W $H = 0.086$, df = 3, $N = 32$, $P = 0.769$).

Discussion

The shelf region of southern Patagonia contrasts with the inshore environments typically occupied by *M. leidyi*. Whereas most descriptions of *M. leidyi* distribution patterns over its natural range come from bays or estuaries such as Narragansett Bay (Kremer & Nixon, 1976; Costello et al., 2006), Chesapeake Bay (Purcell & Decker, 2005) and Biscayne Bay (Reeve & Walter, 1978), the southern Patagonian coast is an open water environment, and the maximum concentrations of *M. leidyi* were found tens of kilometres offshore. Yet, *M. leidyi* thrives in this tidal front region. Peak concentrations of ctenophores in the area (138 g ww m^{-3}) exceeded those of the Chesapeake Bay, USA (60 g ww m^{-3}) and were close to the peak values of an invaded region, the Black Sea (184 g ww m^{-3}) (Purcell et al., 2001b). The ctenophore can be so abundant in our study region that it can predominate in the macrozooplankton biomass (Mianzan & Guerrero, 2000) and is detectable acoustically (Alvarez Colombo et al., 2003). The Patagonian shelf environment possesses a diversity of physical conditions that may be instructive for understanding the biology of *M. leidyi*.

Horizontal biomass patterns

Average concentrations of *M. leidyi* biomass were significantly higher in southern stations and in stratified regions. The individual sizes of ctenophores

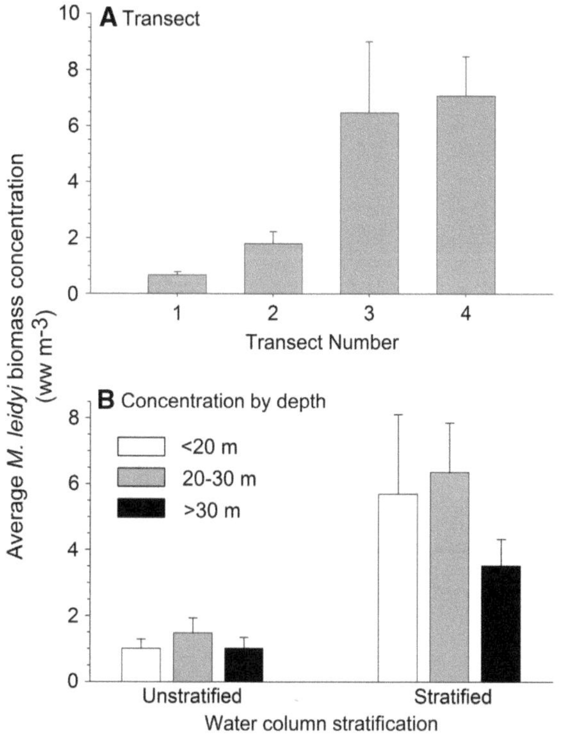

Fig. 5 Biomass distributions of the ctenophore, *Mnemiopsis leidyi*, in the Península Valdés region in December, 1989. Bars represent mean values, error bars ± one standard error of the mean (only + values shown). **A** Horizontal patterns of ctenophore biomass by transect number (as in Fig. 1); **B** Vertical distributions of ctenophore biomass averaged for all transects. Stratified and unstratified stations as illustrated by temperature profiles (as in Fig. 1). Biomass in samples from >20 m depth was designated as in the thermocline (20–30 m) or below the thermocline (>30 m) at stratified stations. Note that these values reflect the individual samples collected at each depth, station, and transect illustrated in Fig. 3

may explain much of the north–south differences in biomass—smaller individuals predominated in the north. The southerly increase of ctenophore biomass probably was not explained by increased food availability. During the field program reported here, micro- and mesozooplankton were obtained from the same stations and depths by means of a multiple opening–closing (0.25-m^2 mouth opening; 64-μm mesh) Multinet (Möhlenkamp, 1996; Viñas & Ramírez, 1996). Zooplankton concentrations were higher in the northern transects where *M. leidyi* biomass was the lowest (Viñas & Ramírez, 1996). This also may be related to dissimilarities in available prey species in the north and south. Zooplankton in the north near Peninsula Valdés was predominated by a small cyclopoid copepod, *Oithona* spp. Baird 1843 (Sabatini & Martos, 2002), which may be less favourable food for adult *Mnemiopsis* than in the south near Isla Escondida, where calanoid copepods predominated.

Inshore–offshore variations in ctenophore abundance occurred within the overall north–south gradient of zooplankton abundance. Microzooplankton (mainly eggs and nauplii of the copepods, *Paracalanus* spp. Giesbrecht 1893, *Acartia tonsa* (Dana, 1848), *Oithona* spp., *Microsetella norvegica* Dana, 1852 and *Euterpina acutifrons* Boeck, 1864) were most abundant at the frontal zone (Viñas & Ramírez, 1996). Mesozooplankton (e.g., mainly small copepods and cladocerans) were also generally more abundant in stratified areas (Möhlenkamp, 1996). Concentrations of *M. leidyi* often were higher in stratified regions (Fig. 4B). The fact that *M. leidyi* biomass distributions and prey availability were not related along a latitudinal gradient but were closely related with the stratification regime suggests that *M. leidyi* biomass distributions are not solely determined by prey availability, but also may be influenced by factors such as water column stratification.

Physical variables often have been related to distribution patterns of *M. leidyi* (Mutlu, 1999; Purcell et al., 2001b; Costello et al., 2006). The influences of tidal currents and mixing, as indicated by barotropic current measurements derived from ADCP data and by outputs from a barotropic model (Tonini et al., 2007), increased from the south to the north and were lower in stratified relative to unstratified stations in the study region. Kinetic energy dissipation rates over most of the region were on the order of 0.1 W m^{-2} but were highly concentrated northeast of Península Valdés, where maximum values exceed 5 W m^{-2} because of the coastline configuration (Glorioso & Simpson, 1994). Barotropic current patterns broadly corresponded inversely with the general north–south distribution patterns of *M. leidyi*, suggesting that increased tidal mixing does not favour accumulation of ctenophore biomass.

Vertical biomass patterns

Vertical distribution patterns of *M. leidyi* were variable in previous reports and within our data. Most commonly, maximum numbers or biomass of *M. leidyi* have been reported to occur above the thermocline (Zaika & Sergeeva, 1991; Zaika &

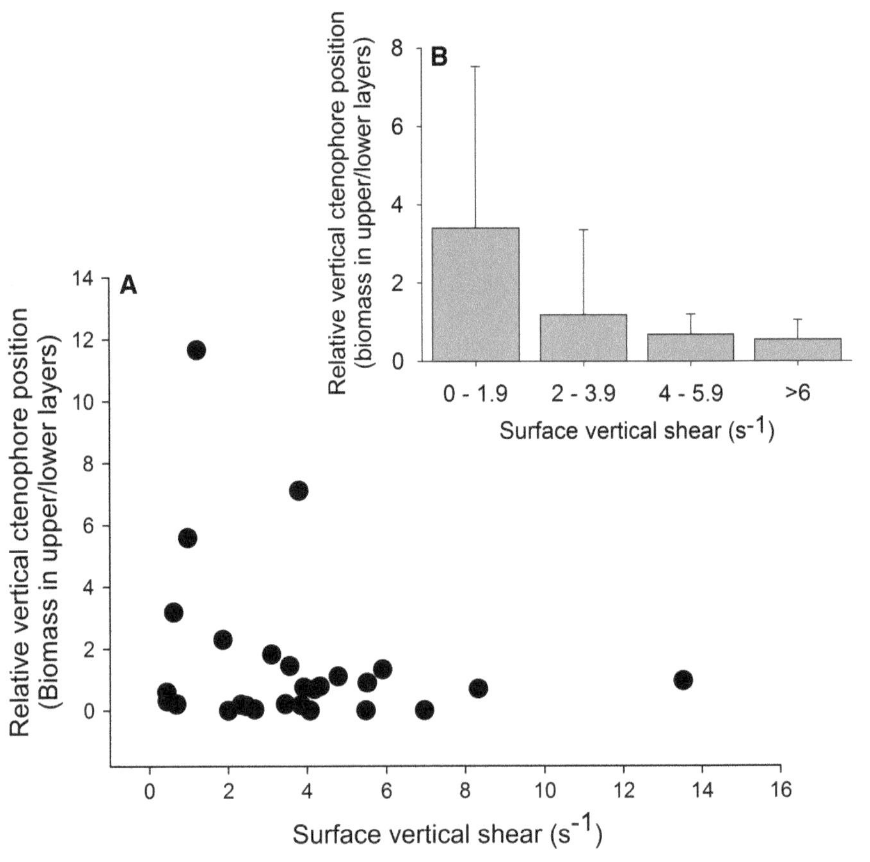

Fig. 6 Relative vertical position of *Mnemiopsis leidyi* biomass versus surface vertical shear in the Península Valdés region during December, 1989. **A** Data for all individual stations, and **B** grouping of stations by depth levels allowing comparison of means and variance by depth. Note that the proportion of ctenophores in the surface layers were always low when surface vertical shear was high (>4.0 s^{-1}); however, when surface vertical shear was low, the vertical positions of ctenophore biomass were more variable

Ivanova, 1992; Mutlu, 1999) or above the pycnocline (Purcell et al., 1994, 2001a). Some evidence indicates that these ctenophores may vertically migrate towards the surface at night (Zaika & Ivanova, 1992) but no consistent pattern has been documented. The advantages for predatory feeding are generally greater in the upper mixed layer where zooplanktonic prey concentrations are frequently steeply high, as in the stratified southern Patagonian shelf region (Möhlenkamp, 1996; Viñas & Ramírez, 1996). *M. leidyi* are capable of vertical migration and can dramatically alter vertical position at one site over a period of less than an hour (Costello & Mianzan, 2003). The relative vertical distribution of the ctenophores across the Patagonian shelf is consistent with other reports indicating high variability. During periods of low vertical shear near the surface, higher biomass concentrations were frequently found in the top 20 m, but when surface levels of vertical shear increased above 4 s^{-1}, ctenophores were found in higher concentrations in or below the thermocline (Fig. 6). Vertical shear in the surface layers may represent a constraint—high levels of vertical shear uniformly coincided with relatively low surface abundances of *M. leidyi*; however, low levels of vertical shear did not correlate with high levels of *M. leidyi* in surface layers, probably because other factors, such as prey availability, influence ctenophore distributions in addition to shear levels. We suggest that low vertical shear levels represent a necessary, but not sufficient, condition for high concentrations of *M. leidyi* to occur in surface waters of the Patagonian shelf region. Downward migrations as an avoidance response to surface mixing in choppy seas caused by high winds have also been described for *M. leidyi* in the Pamlico River, North Carolina, USA (Miller, 1974). In that case, *M. leidyi* appeared to prefer surface layers, but would descend to deeper water where surface winds were high and resurface in areas not exposed to the wind. No similar protection from wind exposure was available for comparison along the Patagonian shelf.

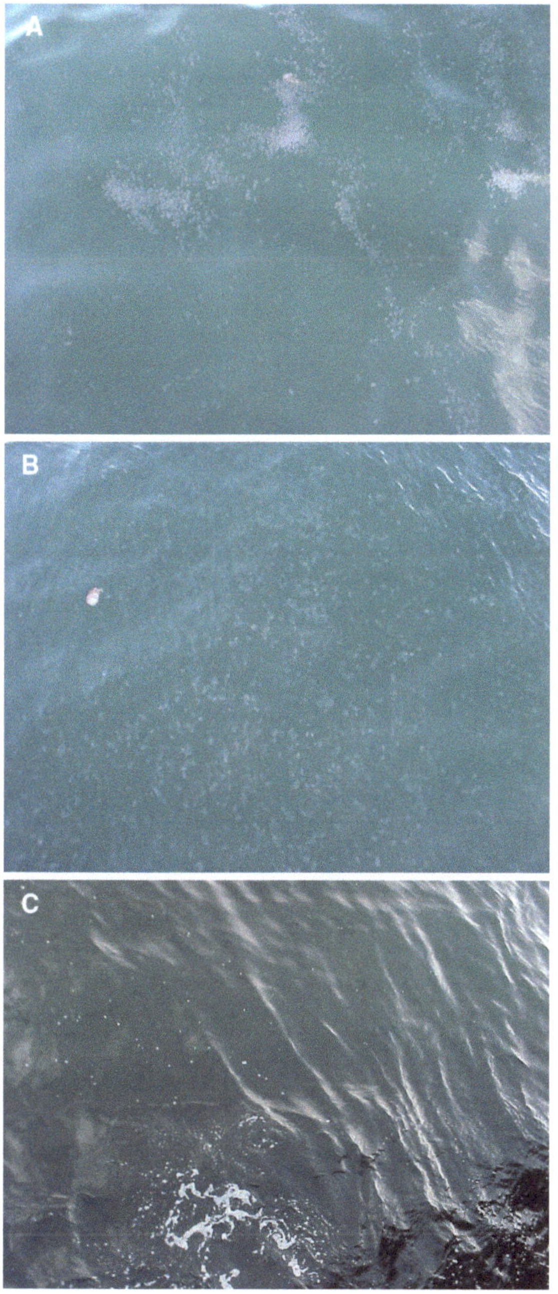

Fig. 7 Alteration of *Mnemiopsis leidyi* surface distributions following the onset of wind activity in the Península Valdés region during February, 2005. The images represent the same station at (**A**, time = 18:33) in very low wind speeds (<1 m s^{-1}) through increasing wind (**B**, time = 18:55) to higher wind speeds, (**C**, time = 19:15) of >5.0 m s^{-1}. Note that during this period of less than 40 min, dense aggregations (**A**) of *Mnemiopsis leidyi* dispersed (**B**) and disappeared from the sea surface (**C**)

Framework for interpreting distributional patterns of *M. leidyi*

Patterns of horizontal and vertical biomass distribution indicate that *M. leidyi* distributions may be influenced simultaneously by a variety of interacting factors. In a system such as the Patagonian shelf, where salinity is relatively constant and prey availability is high (Viñas & Ramírez, 1996; Sabatini & Martos, 2002), factors related to physical mixing may strongly influence ctenophore distribution patterns. Wind stress produces turbulence at the sea surface (Longhurst, 1998) and in high- and mid-latitudes, such as near Peninsula Valdés, the upper ocean usually is well mixed by sustained moderate and high winds (Lozovatsky et al., 2006). In our study, ctenophore vertical distributions appear to be constrained by surface mixing at certain stations. Ambient mixing levels may be important for *M. leidyi* because their prey capture depends upon hydrodynamically mediated processes. Lobate stage *M. leidyi* generates weak feeding currents for prey capture (Waggett & Costello, 1999) that may be disrupted by ambient turbulence. In addition, *M. leidyi* is very sensitive to hydromechanical disturbances created by copepod prey, and capture efficiency is much greater when prey are detected before actual physical contact (Costello et al., 1999). Vertical shear associated with turbulence can be of benefit to some medusae that aggregate at vertical discontinuities (Rakow & Graham, 2006), but turbulence can also interfere with planktonic consumers' abilities to detect prey remotely (Saiz et al., 1992; Saiz & Kiørboe, 1995; Visser, 2007). Zooplankton species that use hydromechanical cues for prey capture or predator avoidance also avoid surface waters during periods of high surface turbulence (Visser et al., 2001). Visser et al. (2009) argue that turbulence increases encounters with predators, as well as reducing feeding effectiveness, and is, therefore, generally unfavourable for most planktonic species. Hence, although not often attributed to *M. leidyi*, avoidance of hydromechanically energetic waters is common among other zooplanktons. Stratified systems may be favourable for *M. leidyi* because stratification can provide a refuge from mixing below the thermocline. In these stratified regions, barotropic models indicate that surface, along-shore velocities can be an order of magnitude higher than the bottom velocities (Tonini et al., 2007). We envision *M. leidyi* as opportunistically utilizing favourable prey availability in the upper mixed layer when wind-generated mixing levels are low enough not to interfere with the ctenophore's feeding currents, sensory capacities, or predation risk. Conversely, as vertical shear levels increase due to tidal or wind mixing, *M. leidyi* can vertically migrate to depths where mixing is low, such as below the thermocline. Because average wind conditions are relatively high along the Patagonian shelf during the summer, *M. leidyi* biomass would often be concentrated in the thermocline or below. Unstratified regions near shore provide limited refuge from strong mixing and consequently are characterized by low average ctenophore biomass concentrations with insignificant variations between depths.

Surface wind speeds and vertical shear levels are crude indices of the stimuli detected by the ctenophores. The tides and winds may be important driving forces behind large-scale ctenophore distributions, but complete understanding of the ctenophore's responses will require direct measurement of fluid activities at scales that match the physical dimensions of the ctenophores.

Acknowledgements The German Research Association (DFG) and the German Ministry for Research and Technology (BMFT) supported this study by providing the research vessel. We would like to thank Prof. W. Nellen and Dr. J. Alheit for providing travelling and shipboard facilities through the IOC–SARP programme. Dr. Y. Matsuura (Brazil) and Dr. L. Djurfeldt (Sweden) kindly sent us the oceanographic data set from the Meteor Cruise 11/3. This article was partially supported by grants from FONCyT PICT no. 01553, CONICET PIP 5009 and 112-200801-00815, and Inter-American Institute for Global Change Research (IAI) CRN 2076 which is supported by the US National Science Foundation (Grant GEO-0452325) to HWM, and US NSF (OCE-9103309 and OCE-9820172) to JHC. The US NSF Grant DEB99-78106 to Prof. Daphne Fautin enabled HWM's participation at the 7th International Conference on Coelenterate Biology. Victor Zaika (Institute of Biology of the Southern Seas, Ukraine) generously translated some of the literature from Russian and exchanged valuable ideas, whose help is particularly acknowledged. E. Palma (UNS, Argentina) provided valuable suggestions and outputs from a barotropic model for the study region. Prof. S. Uye (Japan) kindly reviewed an early version of this article. The help of Dr. W. Ekau (ZTE, Germany), Dr. M. Erhlich (INIDEP), Lic. O. Pin (INAPE, Uruguay), MSc. I. D'apremont Lima (FURG. Brazil), and the scientists and crew of the R/V 'Meteor' is deeply appreciated. This is INIDEP Scientific Contribution no. 1583.

References

Acha, E. M., H. Mianzan, R. A. Guerrero, J. Bava & M. Favero, 2004. Marine fronts at the continental shelves of austral South America. Physical and ecological processes. Journal of Marine Systems 44: 83–105.

Alheit, J., J. Ciechomski, L. Djurfeldt, C. Ebel, M. Ehrlich, J. Elgue, G. Mantero, Y. Matsuura, H. Mianzan, W. Nellen, C. Oderbretch, F. Ramírez, R. Sanchez, G. Shaffer & M. Viñas, 1991. SARP studies on Southwest Atlantic anchovy, *Engraulis anchoita* off Argentina, Uruguay and Brazil. ICES Annual Meeting, France 46(V): 1–32.

Alvarez Colombo, G., H. Mianzan & A. Madirolas, 2003. Acoustic characterization of gelatinous plankton aggregations: four study cases from the argentine continental shelf. ICES Journal of Marine Science 60: 650–657.

Bakun, A. & R. H. Parrish, 1991. Comparative studies of coastal pelagic fish reproductive habitats: the anchovy (*Engraulis anchoita*) of the southwestern Atlantic. ICES Journal of Marine Science 48: 343–361.

Boersma, M., A. M. Malzahn, W. Greve & J. Javidpour, 2007. The first occurrence of the ctenophore *Mnemiopsis leidyi* in the North Sea. Helgoland Marine Research 61: 153–155.

Carreto, J. I., H. R. Benavides, R. M. Negri & P. D. Glorioso, 1986. Toxic red-tide in the Argentine Sea. Phytoplankton distribution and survival of the toxic dinoflagellate *Gonyaulax excavata* in a frontal area. Journal of Plankton Research 8: 15–28.

Costello, J. H. & H. Mianzan, 2003. Sampling field distributions of *Mnemiopsis leidyi* (Ctenophora, Lobata): planktonic or benthic methods? Journal of Plankton Research 25: 455–459.

Costello, J. H., R. Loftus & R. Waggett, 1999. The influence of prey detection on capture success by the ctenophore *Mnemiopsis leidyi* feeding upon adult *Acartia tonsa* and *Oithona colcarva* copepods. Marine Ecology Progress Series 191: 207–216.

Costello, J. H., B. K. Sullivan, D. J. Gifford, D. Van Keuren & L. J. Sullivan, 2006. Seasonal refugia, shoreward thermal amplification, and metapopulation dynamics of the ctenophore *Mnemiopsis leidyi* in Narragansett Bay, Rhode Island. Limnology and Oceanography 51: 1819–1831.

Faasse, M. A. & K. M. Bayha, 2006. The ctenophore *Mnemiopsis leidyi* A. Agassiz 1865 in coastal waters of the Netherlands: an unrecognized invasion? Aquatic Invasions 1: 270–277.

Fuentes, V., D. Atienza, J. M. Gili & J. E. Purcell, 2009. First records of *Mnemiopsis leidyi* A. Agassiz 1865 off the NW Mediterranean coast of Spain. Aquatic Invasions 4: 671–674.

GESAMP (Joint Group of Experts on the Scientific Aspects of Marine Environmental Protection), 1997. Opportunistic settlers and the problem of the ctenophore *Mnemiopsis leidyi* invasion in the Black Sea. Report Study GESAMP 58: 84 pp.

Glorioso, P., 1987. Temperature distribution related to shelf-sea fronts on the Patagonian shelf. Continental Shelf Research 7: 27–34.

Glorioso, P., 2000. Patagonian shelf 3D tide and surge model. Journal of Marine Systems 24(1–2): 141–151.

Glorioso, P. D. & R. A. Flather, 1995. A barotropic model of the currents off SE South America. Journal of Geophysical Research 100: 13427–13440.

Glorioso, P. D. & J. H. Simpson, 1994. Numerical modelling of the M2 tide on the northern Patagonian Shelf. Continental Shelf Research 1(2–3): 267–278.

Hansson, H. G., 2006. Ctenophores of the Baltic and adjacent Seas – the invader *Mnemiopsis* is here!. Aquatic Invasions 1: 295–298.

Harbison, G. R. & S. P. Volovik, 1994. The ctenophore, *Mnemiopsis leidyi*, in the Black Sea: a holoplanktonic organism transported in the ballast water of ships. In Nonindigenous Estuarine and Marine Organisms (NEMO), Proceedings of the Conference and Workshop, Seattle, WA, April 1993. U. S. Department of Commerce, National Oceanic and Atmospheric Administration, Washington, DC: 25–36.

Ivanov, V., A. M. Kamakin, V. B. Ushivtzev, T. A. Shiganova, O. Zhukova, N. Aladin, S. I. Wilson, G. R. Harbison & H. J. Dumont, 2000. Invasion of the Caspian Sea by the comb jellyfish *Mnemiopsis leidyi* (Ctenophora). Biological Invasions 2: 255–258.

Javidpour, J., U. Sommer & T. Shiganova, 2006. First record of *Mnemiopsis leidyi* A. Agassiz 1865 in the Baltic Sea. Aquatic Invasions 1: 299–302.

Kremer, P., 1975. The ecology of the ctenophore *Mnemiopsis leidyi* in Narragansett Bay. Ph.D. thesis, University of Rhode Island.

Kremer, P., 1994. Patterns of abundance for *Mnemiopsis* in US coastal waters: a comparative overview. ICES Journal of Marine Science 51: 347–354.

Kremer, P. & S. Nixon, 1976. Distribution and abundance of the ctenophore, *Mnemiopsis leidyi* in Narragansett Bay. Estuarine, Coastal and Marine Science 4: 627–639.

Kube, S., L. Postel, C. Honnef & C. B. Augustin, 2007. Mnemiopsis leidyi in the Baltic Sea – distribution and overwintering between autumn 2006 and spring 2007. Aquatic Invasions 2: 137–145.

Longhurst, A., 1998. Ecological Geography of the Sea. Academic Press, San Diego, CA, USA: 398 pp.

Lozovatsky, I. D., E. Roget, H. J. S. Fernando, M. Figuero & S. Shapovalov, 2006. Sheared turbulence in a weakly stratified upper ocean. Deep-Sea Research I 53: 387–407.

Martos, P. & R. P. Sánchez, 1997. Caracterización oceanográfica de regiones frontales en la plataforma Patagónica en relación con áreas de desove y cría de la anchoita (*Engraulis anchoita*). Coloquio Argentino de Oceanografía, IAPSO-IADO (Conicet), Septiembre 1997 Bahía Blanca, Argentina.

Mianzan, H., 1999. Ctenophora. In Boltovskoy, D. (ed.), South Atlantic Zooplankton. Backhuys Publishers, Leiden: 561–573.

Mianzan, H. & R. Guerrero, 2000. Environmental patterns and biomass distribution of gelatinous macrozooplankton. Three study cases in the Southwestern Atlantic. Scientia Marina 64(Suppl. 1): 215–224.

Miller, G. R., 1966. The flux of tidal energy out of the deep oceans. Journal of Geophysical Research 71: 2485–2489.

Miller, R. J., 1974. Distribution and biomass of an estuarine Ctenophore population, *Mnemiopsis leidyi* (A. Agassiz). Chesapeake Science 15: 1–8.

Möhlenkamp, M., 1996. Untersuchungen zur kleinskaligen Verteilung von Mikro- und Mesozooplankton im oberen Pelagial des südamerikanischen Schelfmeeres unter Erprobung eines neuen automatischen Messgerätes. Dissertation zur Erlangung des Doktorgrades am Fachbereich Biologie der Universität Hamburg: 112 pp.

Mutlu, E., 1999. Distribution and abundance of ctenophores and their zooplankton food in the Black Sea. II. *Mnemiopsis leidyi*. Marine Biology 135: 603–613.

Nellen, W., 1990. Working report on Cruise N° 11, Leg 3 of RV METEOR. Institut Hydrobiolische Fischereissenschaftich 9: 158 pp.

Oliveira, O. M. P., 2007. The presence of the ctenophore *Mnemiopsis leidyi* in the Oslofjorden and considerations on the initial invasion pathways to the North and Baltic seas. Aquatic Invasions 2: 185–189.

Piola, A. R. & A. L. Rivas, 1997. Corrientes de la plataforma continental. In Boschi, E. E. (ed.), El Mar Argentino y sus recursos pesqueros. 1. Antecedentes históricos de las exploraciones en el mar y las características ambientales. Publicación Especial INIDEP, Mar del Plata: 119–132.

Purcell, J. E. & M. B. Decker, 2005. Effects of climate on relative predation by scyphomedusae and ctenophores on copepods in Chesapeake Bay during 1987–2000. Limnology and Oceanography 50: 376–387.

Purcell, J. E., D. A. Nemazie, S. E. Dorsey, E. D. Houde & J. C. Gamble, 1994. Predation mortality of bay anchovy (*Anchoa mitchilli*) eggs and larvae due to scyphomedusae and ctenophores in Chesapeake Bay. Marine Ecology Progress Series 114: 47–58.

Purcell, J. E., D. L. Breitburg, M. B. Decker, W. M. Graham, M. J. Youngbluth & K. A. Raskoff, 2001a. Pelagic cnidarians and ctenophores in low dissolved oxygen environments: a review. In Rabalais, N. N. & R. E. Turner (eds), Coastal and Estuarine Studies 58: Coastal Hypoxia: Consequences for Living Resources and Ecosystems. American Geophysical Union, Washington, DC: 77–100.

Purcell, J. E., T. A. Shiganova, M. B. Decker & E. D. Houde, 2001b. The Ctenophore *Mnemiopsis* in native and exotic habitats: U.S. estuaries versus the Black Sea basin. Hydrobiologia 451: 145–176.

Rakow, K. C. & W. M. Graham, 2006. Orientation and swimming mechanics by the scyphomedusa *Aurelia* sp. in shear flow. Limnology and Oceanography 51: 1097–1106.

Rapoza, R., D. Novak & J. H. Costello, 2005. Life-stage dependent, in situ dietary patterns of the lobate ctenophore *Mnemiopsis leidyi* Agassiz 1865. Journal of Plankton Research 27: 951–956.

Reeve, M. R. & M. A. Walter, 1978. Nutritional ecology of ctenophores – a review of recent research. Advances in Marine Biology 15: 249–287.

Sabatini, M. & P. Martos, 2002. Mesozooplankton features in a frontal area off northern Patagonia (Argentina) during spring 1995 and 1998. Scientia Marina 66: 215–232.

Saiz, E. & T. Kiørboe, 1995. Predatory and suspension feeding of the copepod *Acartia tonsa* in turbulent environments. Marine Ecology Progress Series 122: 147–158.

Saiz, E., M. Alcaraz & A. Paffenhoffer, 1992. Effects of small-scale turbulence on feeding rate and gross-growth efficiency of three *Acartia* species (Copepoda: Calanoida). Journal of Plankton Research 14: 1085–1097.

Sánchez, R. P., J. Alheit, P. Martos & M. Pájaro, 1998. The spawning and early life of *Engraulis anchoita* in the tidal front off Patagonia. ICES Annual Meeting, Cascais, Portugal, 16–19 September, ICES C.M. 1998/R:1.

Shiganova, T. & A. Malej, 2009. Native and non-native ctenophores in the Gulf of Trieste, Northern Adriatic Sea. Journal of Plankton Research 31: 61–71.

Shiganova, T. A., A. E. Kideys, A. S. Gucu, U. Niermann & V. S. Khoroshilov, 1998. Changes of species diversity and their abundances in the main components of pelagic community during last decades. In Ivanov, L. & T. Oguz (eds), Ecosystem Modelling and Management Tool for the Black Sea. Kluwer Academic Publishers, Dordrecht: 171–188.

Shiganova, T. A., E. I. Musaeva, Y. V. Bulgakova, Z. A. Mirzoyan & M. L. Martynuk, 2003. Ctenophores invaders *Mnemiopsis leidyi* (A. Agassiz), Beroe ovata Mayer 1912 and their effect on the pelagic ecosystem of the north-eastern Black Sea. Biological Bulletin 2: 225–235.

Simpson, J. H., 1981. The shelf-sea fronts: implications of their existence and behaviour. Philosophical Transactions of the Royal Society of London 302 A: 531–546.

Studenikina, E. I., S. R. Volovik, Z. A. Mirzoyan & G. I. Luts, 1991. Comb Jelly *Mnemiopsis leidyi* in the Azov Sea. Okeanologiya 31: 722–725.

Tonini, M., E. D. Palma & A. Rivas, 2007. Simulación Numérica de la Circulación y Frentes Térmicos en los Golfos Norpatagónicos. Mecánica Computacional XXVI: 3757–3768.

Viñas, M. D. & F. C. Ramírez, 1996. Gut analysis of first-feeding anchovy larvae from the Patagonian spawning areas in relation to food availability. Archive of Fishery and Marine Research 43: 231–256.

Visser, A. W., 2007. Motility of zooplankton: fitness, foraging and predation. Journal of Plankton Research 29: 447–461.

Visser, A. W., H. Saito, E. Saiz & T. Kiørboe, 2001. Observations of copepod feeding and vertical distribution under natural turbulent conditions in the North Sea. Marine Biology 138: 1011–1019.

Visser, A. W., P. Mariani & S. Pigolotti, 2009. Swimming in turbulence: zooplankton fitness in terms of foraging efficiency and predation risk. Journal of Plankton Research 31: 121–133.

Waggett, R. & J. H. Costello, 1999. Different mechanisms used by the lobate ctenophore, *Mnemiopsis leidyi*, to capture nauplii and adult life stages of the copepod *Acartia tonsa*. Journal of Plankton Research 21: 2037–2052.

Zaika, V. & N. I. Ivanova, 1992. The ctenophore *Mnemiopsis mccradyi* in the autumn hyponeuston of the Black Sea. Ecologia Moyra 42: 6–10.

Zaika, V. & N. Sergeeva, 1991. Daily changes in population structure and vertical distribution of ctenophore *Mnemiopsis mccradyi* Mayer (Ctenophora) in the Black Sea. Hydrobiology Journal 27: 15–19 (in Russian).

Hydrobiologia (2010) 645:125–133
DOI 10.1007/s10750-010-0222-y

Response of *Chrysaora quinquecirrha* medusae to low temperature

Margaret A. Sexton · Raleigh R. Hood · Judith Sarkodee-adoo · Amanda M. Liss

Published online: 21 March 2010

Abstract Because of their high abundance in Chesapeake Bay, *Chrysaora quinquecirrha* medusae may be an important reservoir of organic matter. The timing and location of the decomposition of biomass from medusae may have implications for carbon cycling in the bay. Our objective was to identify the cause of *C. quinquecirrha* medusa disappearance to better understand when and where decomposition occurs. A time series of visual surface counts and vertical net hauls in the Choptank River, a tributary of Chesapeake Bay, showed that as temperatures approached 15°C, *C. quinquecirrha* medusae disappeared from the surface, but persisted in net hauls until temperatures reached 10°C. In order to test whether medusae sink upon cooling, we exposed *C. quinquecirrha* medusae to low temperatures in large static tanks and measured their depth and pulsation rates twice daily for at least 6 days. This procedure was repeated three times through the 2008 jellyfish season. On average, individuals exposed to temperatures below 15°C were found deeper and pulsed slower than those in the warmer control tank. This suggests that low temperatures cause the medusae to sink before cooling to the limit of their physiological tolerance and may have implications for the deposition of organic matter associated with the seasonal disappearance of medusae from Chesapeake Bay.

Keywords *Chrysaora quinquecirrha* · Temperature · Depth · Seasonal disappearance

Guest editors: J. E. Purcell & Dror Angel / Jellyfish Blooms: New Problems and Solutions

M. A. Sexton (✉) · R. R. Hood · J. Sarkodee-adoo · A. M. Liss
Horn Point Laboratory, University of Maryland Center for Environmental Science, Box 775, Cambridge, MD 21613, USA
e-mail: msexton@hpl.umces.edu

Introduction

The medusa stage of *Chrysaora quinquecirrha* (Desor, 1984) is seasonally abundant in the mesohaline Chesapeake Bay and its tributaries. It has been shown to affect populations of other gelatinous zooplankton, copepods, and ichthyoplankton (e.g., Feigenbaum & Kelly, 1984; Purcell, 1992; Cowan & Houde, 1993). Feigenbaum & Kelly (1984) suggest that *C. quinquecirrha* influences the trophic structure of the bay through its predation on *Mnemiopsis leidyi* (A. Agassiz, 1865). By means of a control over the population of the voraciously feeding ctenophore, high abundances of *C. quinquecirrha* can positively affect secondary production (Purcell et al., 1994b; Purcell & Decker, 2005). A direct effect on fish populations is medusa predation on fish eggs and larvae, which can account for high percentages of mortality (Purcell et al., 1994a). In addition to important trophic interactions,

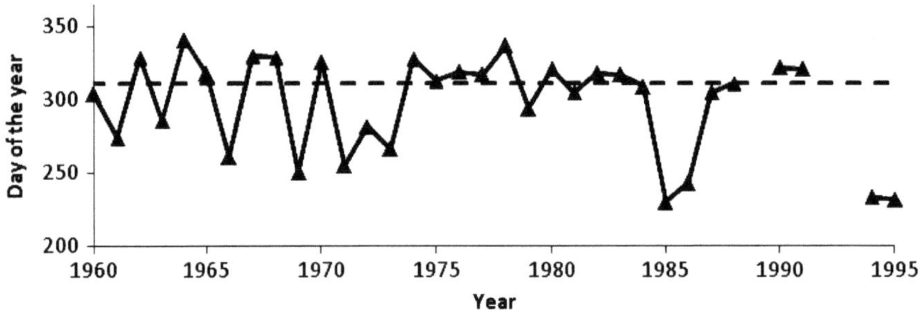

Fig. 1 Day of last occurrence of *Chrysaora quinquecirrha* medusae from visual counts made at the Chesapeake Biological Laboratory on the Patuxent River, Solomons Island, Maryland during1960–1995. Dashed line indicates the median day of last occurrence ($m = 311$, 7 November)

C. quinquecirrha's painful sting has negative influences on recreational activities. For these reasons, it is desirable to understand and predict the occurrence of *C. quinquecirrha*.

Several studies have addressed the environmental factors that determine abundance and distribution of *C. quinquecirrha* medusae in Chesapeake Bay (for example, Cargo & King, 1990; Purcell & Decker, 2005; Breitburg & Fulford, 2006; Decker et al., 2007) and the conditions that cue strobiliation (e.g., Cargo & Schultz, 1967; Cargo & Rabenold, 1980; Purcell et al., 1999); however, the mechanisms of the seasonal disappearance of *C. quinquecirrha* have not been well studied. The day of final occurrence on the Patuxent River, as measured by a time series of average weekly visual counts, usually has been in early November (Fig. 1, median = 311, November 7; D. G. Cargo, unpublished data).

Because of their tendency to form blooms, jellyfish sometimes have important influences on nutrient cycling (Pitt et al., 2009). Several studies have shown that jellyfish can be important to the local carbon cycles. Titelman et al. (2006) identified a shift in the bacterial community when decaying gelatinous matter was the carbon source because bacteria varied in their ability to utilize it. Gelatinous biomass accounted for a large amount of fixed carbon during summer in the mesohaline portion of the York River, a tributary of Chesapeake Bay (Condon & Steinberg, 2008). They suggested that this carbon can be released to the water column, especially during starvation, or to the benthos from gelatinous matter on the sediment. Once on the bottom, Billet et al. (2006) showed that jellyfish carcasses provided a significant input of organic matter to the sea floor, and West et al. (2009) suggested that decomposition of gelatinous biomass can affect sediment nutrient cycling, including causing a significant increase in sediment oxygen demand. Thus, carbon from *C. quinquecirrha* (CQC) may play an important role in carbon cycling in the mesohaline Chesapeake Bay. The time at which *C. quinquecirrha* medusae disappear each year has implications to the timing and location of release of organic matter from gelatinous zooplankton.

We addressed temperature as one possible cause of the annual disappearance of medusae. Gatz et al. (1973) showed that the pulsation rate, the swimming activity of the medusae, decreased with temperature, until pulsation stopped completely at 10°C. This relationship between pulsation rate and temperature may cause *C. quinquecirrha* to sink to the bottom because the negatively buoyant medusa cannot swim as strongly away from the bottom. We compared visual surface counts to vertical net hauls in the Choptank River, a tributary of Chesapeake Bay, in 2005 and 2006, to determine whether the vertical distribution of medusae changes as temperatures approach 15°C. In order to clarify this point further, a large tank experiment was used to determine the effect of low temperature on depth of *C. quinquecirrha* medusae. We hypothesize that *C. quinquecirrha* exposed to temperatures between 10 and 15°C in large tanks will be deeper and pulse slower than *C. quinquecirrha* exposed to warmer temperatures.

Methods

Visual counts and vertical net hauls for *Chrysaora quinquecirrha* medusae were conducted twice daily from the dock at the Horn Point Laboratory,

Cambridge, Maryland, USA on the south side of the Choptank River (38°35.610′ N, 76°7.725′ W). Counts were taken daily at 0700 and 1900 h from June 6, 2005 to September 15, 2005. The count area was defined as the 3 m on the east side of the dock along its entire 61 m length forming a 183 m^2 transect. Consistency in the count area was ensured each day by carrying a 3-m PVC measuring rod with a weighted line on the far end while counting medusae inside the weighted line. Secchi depth measured at the time of each count was used to estimate the depth to which medusae could be seen during the visual count. Densities of *C. quinquecirrha* (medusae m^{-3}) were calculated from the numbers in the area count visually divided by the water volume searched (area × Secchi depth). Immediately after each visual count, a vertical haul from bottom to surface was made with a net (9-m^2 mouth area, 1.6-cm nylon mesh). Water depth was measured at the time of each net haul to calculate volume sampled and density of medusae. On September 16, 2005, the sampling times were adjusted so that the morning count and net haul occurred immediately after sunrise and the evening net haul occurred 20 min before sunset. In subsequent years, observations began on June 1st and followed the sunrise/sunset schedule through the entire season. Counts and net hauls continued on this schedule until no medusae were observed at the surface along the transect or the surrounding area or collected in the net for 10 consecutive days.

Calculations of the importance of carbon from *C. quinquecirrha* (CQC) for medusae relative to other measures of carbon in Chesapeake Bay were made from visual counts and literature values. Two measures of abundance were included: the highest weekly average on the Choptank River during the years 2005–2008, as described above, and the average July–August count on the Patuxent River from Cargo & King (1990). Patuxent River counts were assumed to have a visible depth of 1 m to calculate a density in the count area (medusae m^{-3}). The carbon represented by the densities of *C. quinquecirrha* medusae was calculated using the equation from Purcell & Decker (2005):

$$C = 2.15 * 10^{-4}\ \mathrm{Diam}^{2.903}$$

An average diameter of 33 mm was assumed based on average diameters in late August as reported in Purcell (1992) and used to calculate carbon per individual. This allowed for calculation of the concentration of CQC in the water column, potential CQC flux to the sediment, and CQC deposition rate.

Timing of medusa disappearance in Fig. 1 was from a time series of weekly mean visual counts made at the Chesapeake Biological Laboratory in Solomons, Maryland, USA from 1960 to 1995. Average July–August counts from 1960 to 1986 from this series are published in Cargo & King (1990), but dates of final occurrence were not published. Counts were made by D. G. Cargo with assistance from M. Wiley and H. Millsap until 1991. Wiley continued the counts in 1992 and 1993, and Millsap continued from there in the following years, i.e.,1994 and 1995.

In order to determine whether cold temperatures cause medusae to sink, two 10,000-l tanks were filled with 1-μm filtered Choptank River water. Tanks of 2.3 m depth were chosen to simulate the water depth at the dock where counts and net hauls were made, and where water depth ranged from approximately 1.5 to 3 m depending on tide. We assumed that interaction with the bottom of the tanks would simulate that occurring in situ. One tank was designated the treatment tank, and the other was the control tank. The treatment tank was cooled to 13°C and the control tank was cooled to 16°C. In order to avoid damaging the medusae, the pumps were turned off after initial chilling to the starting temperatures. Temperature was measured twice daily throughout the experiment. The first two trials were terminated after 6 days when the temperature at the bottom of the tanks reached 16°C. The third trial was allowed to continue beyond 6 days despite the increase in temperature. Because changes in light were shown to cause vertical migration in *C. quinquecirrha* (Schuyler & Sullivan, 1997), lights remained off throughout the experiment, and tanks were draped with dark plastic to block out ambient light. Because many of the zooplankton prey of the medusae migrate vertically, food was not introduced to the tanks to eliminate the vertical position of prey as a variable that could influence the vertical position of the medusae.

Chrysaora quinquecirrha medusae were dipped in buckets from the Tred Avon River at Oxford, Maryland, USA immediately before being placed in the tanks and the bell diameter at maximum expansion was measured. The medusae were transported from the river to the laboratory in buckets, and small volumes of water from the chilled tanks were added to the buckets every 0.5 h for 2–3 h to decrease temperature slowly. When the temperatures in the buckets were within 1–2°C of the tank temperatures, 20 medusae were distributed equally between the two tanks to obtain similar size distributions in both tanks and allowed to acclimate 24 h before observations began. Although Gatz et al. (1973) suggested that temperature acclimation to a similar temperature difference occurs within 3 h, Schuyler & Sullivan (1997) reported behavioral changes after the first day of residence in a large tank. Those changes were presumed to be the medusae resuming normal behavior after the stress of capture and transport. For this reason, the conservative acclimation time of at least 24 h was used here.

After 24 h, the depth at which each medusa was swimming was determined by use of a dive light and sounding line. At the same time, the number of swimming pulses in 15 s was counted for each individual. Water temperature was also measured at the surface, 1 m, 2 m, and bottom to calculate a depth-integrated temperature for each tank. These measurements were taken twice daily, and the procedure repeated three times (trials). The first two trials ran for 6 days and the third for 9 days. Although the successive measurements were made over the course of time, they were assumed to be independent because the time between measurements was sufficient for the individuals to travel from top to bottom nearly one hundred times based on a swimming speed of 0.6 cm s^{-1}, which was the most frequent swimming speed observed in the absence of food according to Matanoski et al. (2001). Average depths and pulsation rates observed in the treatment and control tanks were compared using a Wilcoxon two sample test because the distributions of the paired measurements were non-normal. Trends in depth and pulsation with respect to depth-integrated temperature were addressed with least squares regression using S-plus 8.0 statistical software (Sokal & Rohlf, 1995).

Results

Results from the time series of visual counts and vertical net hauls on the Choptank River showed that *C. quinquecirrha* medusae disappeared from the visible surface layer before they disappeared from the entire water column (Fig. 2). Disappearance from

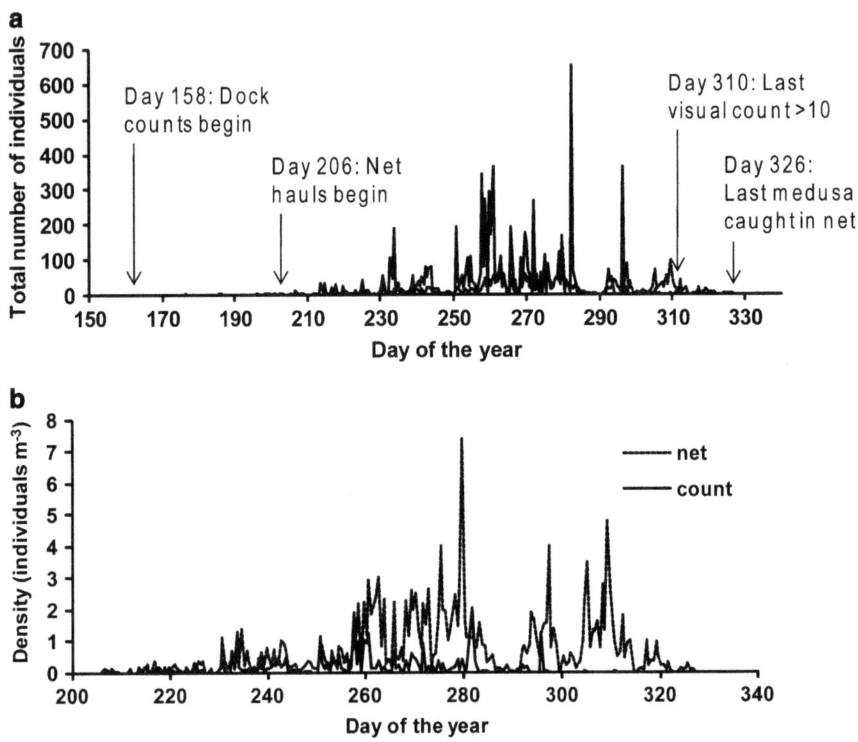

Fig. 2 Abundance (**a**) and density (**b**) of *Chrysaora quinquecirrha* medusae as measured by visual surface counts (*solid*) and vertical net hauls (*dashed*) in 2005 made from the Horn Point Laboratory dock, Cambridge, Maryland

the visible layer coincided with the seasonal decrease in water temperature to 15°C, but complete disappearance from the water column coincided with the decrease in temperature to 10°C (Figs. 2 and 3).

In all the three trials of the tank experiment, medusa average depth was deeper and the average pulsation rate was slower in the cold treatment tank than in the control (Fig. 4). Average pulsation rates ranged from 26 to 36 pulses min^{-1} in the control tank, and from 11 to 28 pulses min^{-1} in the cold treatment. These rates are consistent with those observed by Gatz et al. (1973) in similar temperatures. Results were significantly different according to a one-sided Wilcoxon two sample test with $P < 0.05$ for all trials (Trials 1 and 2, $n = 12$ for both groups; Trial 3, $n = 18$). For depth, $t_s = -4.1312$, -4.130, and -4.411, and for pulsation, $t_s = 3.903$, 4.066, and 2.929 for Trials 1, 2, and 3, respectively.

In the first and second trials, there were no overlaps between the cold treatment and the control for average depth or average pulsation rate (Fig. 5). In the third trial, which lasted 3 days longer than the previous trials, the average depths began similarly to the other trials, but approached one another over time; however, the grand average of all depths over the course of the trial remained significantly deeper in the cold treatment tank than in the control. The relationships of depth and pulsation rate to depth-integrated temperature showed similar patterns in the first and second trials with deeper occurrences and slower pulsation rates in the cold treatments than the controls. In the third trial, where the temperature in the cold treatment tank approached that of the control tank, average depth and pulsation rate increased as temperature increases. Least squares linear regression showed significant relationships between temperature and depth ($r^2 = 0.501$, $P < 0.05$) and pulsation rate ($r^2 = 0.896$, $P < 0.05$) (Fig. 6).

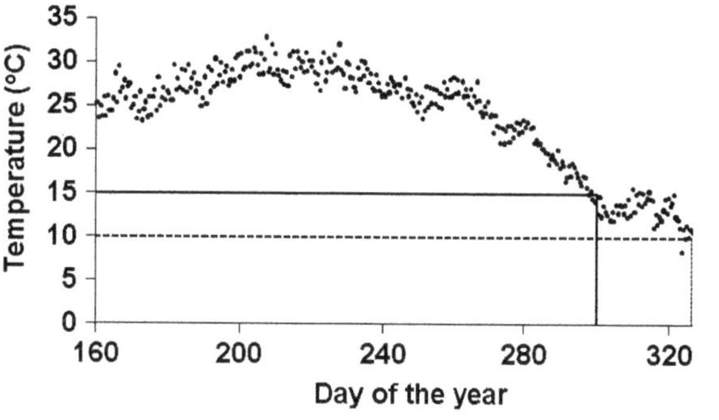

Fig. 3 Time series of water temperature measured at the Horn Point Laboratory dock, Cambridge, Maryland in 2005. Dashed lines indicate the temperatures on the dates of disappearance of *Chrysaora quinquecirrha* medusae from the surface (*solid*) and disappearance from the net hauls (*dashed*) shown in Fig. 2

Discussion

Because the last medusae have been observed most frequently near or after the beginning of November (Fig. 1) when water temperatures are decreasing (Fig. 3), low temperature is a likely cause of the seasonal disappearance of *C. quinquecirrha* in most years. In some years, disappearance occured long before the water temperature began to decrease toward the minimum tolerated by *C. quinquecirrha* medusae. In 7 years of the 35-year time series, the day of final occurrence was at least 50 days earlier than the median day of final occurrence (Fig. 1).

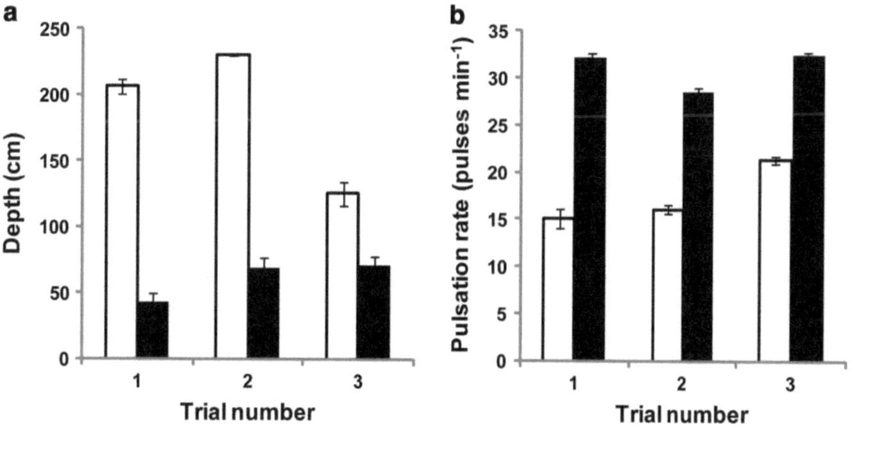

Fig. 4 Average depth (**a**) and pulsation rate (**b**) of *Chrysaora quinquecirrha* medusae for each trial. Average depths and pulsation rates were significantly different in the cold treatment tank (*open bars*) than in the control (*dark bars*) in all trials. Error bars represent standard error

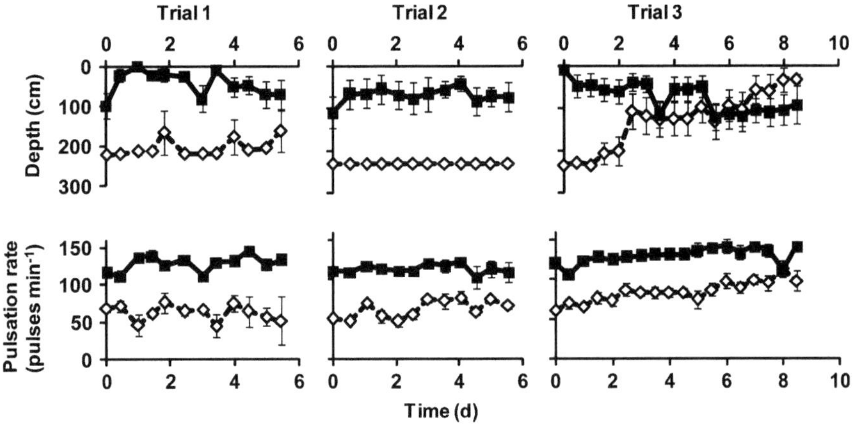

Fig. 5 Time series of average depth (*top row*) and pulsation rate (*bottom row*) of *Chrysaora quinquecirrha* medusae as measured twice daily in the cold treatment (*open symbols*) and the warmer control (*filled symbols*) tanks. Error bars represent standard error

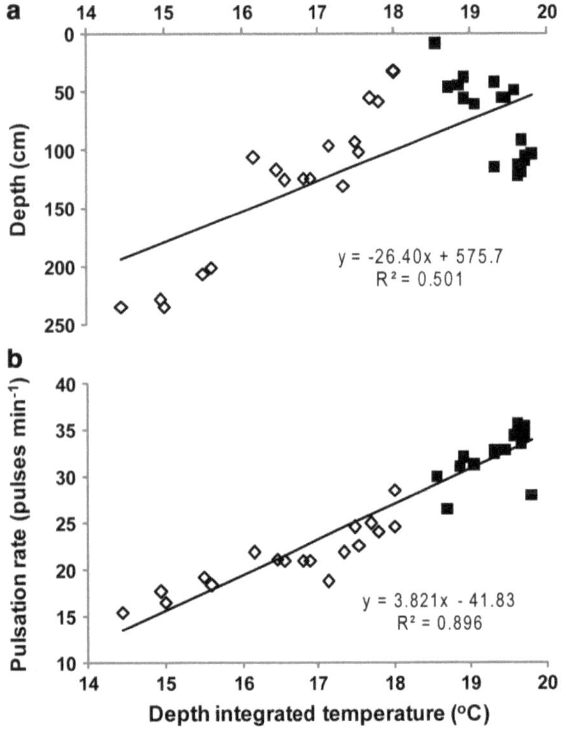

Fig. 6 Least squares linear regression lines calculated for average depth (**a**) and pulsation rate (**b**) of *Chrysaora quinquecirrha* medusae from Trial 3 with respect to depth-integrated temperature. Open points are from the cold treatment and filled points are from the control

Possible mechanisms for these unusually early disappearances include starvation due to low food availability; mortality due to higher than normal rates of disease, parasitism, or predation; or an early cessation of strobiliation accompanied by normal senescence. In addition, unusually low abundance and relatively early disappearance of medusae in 1972 have been attributed in part to Hurricane Agnes (Cargo, 1976); therefore, suboptimal salinity and temperature and physical flushing should also be considered as possible mechanisms. More inquiry is necessary to determine which of these mechanisms may be at work in years with early disappearance. While the reasons for early disappearance in some years are still unclear, results from this study explain the cause of *C. quinquecirrha* medusa's disappearance in most years. Throughout the season, densities found in the net were higher than those measured by the visual counts (Fig. 2b). We interpret this difference to be caused by a non-uniform vertical distribution of *C. quinquecirrha* in the water column. After day 300 when temperatures began to cool below 15°C, densities measured by the net remained as high as in warmer temperatures while those measured by visual counts declined. This indicates that the already vertically stratified distribution had moved farther from the surface at the time of cooling. The experimental results show that temperatures below 15°C cause medusae to reside near the bottom (Figs. 4, 5, 6), as was suggested by the field observations (Figs. 2 and 3). Thus, the living medusae would be deposited on the sediment surface, and the pulsation rate would continue to slow until the temperature reaches 10°C, the limit of their temperature tolerance, as reported by Gatz et al. (1973).

Calculations of the CQC present, annual flux, and deposition rate based on abundance observed in the Choptank and Patuxent rivers showed that medusae contribute organic matter to the tributary carbon cycles (Table 1). Literature values of dissolved organic carbon (Fisher et al., 1998), total annual carbon flux (Kemp et al., 1997), and rate of

Table 1 Carbon from *Chrysaora quinquecirrha* (CQC) expressed as concentration, flux, and deposition rate and as percentages of the dissolved organic carbon (DOC), annual total organic carbon flux (TOC) to the sediment, and deposition from the spring bloom in the Choptank and Patuxent river estuaries of Chesapeake Bay

	Measure of medusa abundance	CQC ($mgC\ m^{-3}$)[c]	% of water column [DOC][d]	Annual CQC flux ($mgC\ m^{-2}$)[e]	% Annual TOC flux[f]	CQC deposition rate ($mgC\ m^{-2}\ d^{-1}$)	% deposition from spring bloom[g]
Choptank[a]	Mean	1.59	0.79	72.51	0.12	3.37	0.66
	Minimum	0.66	0.33	28.36	0.05	1.32	0.25
	Maximum	24.97	1.04	112.37	0.18	5.23	1.02
Patuxent[b]	Mean	1.38	0.69	12.57	0.02	0.58	0.11
	Minimum	$3*10^{-3}$	$1.5*10^{-3}$	0.17	$2*10^{-4}$	$8*10^{-3}$	0.02
	Maximum	0.22	0.11	77.79	0.13	3.62	0.71

[a] Measures of abundance on the Choptank River represent the highest weekly average abundance (no. m^{-3}) from twice daily visual counts at the Horn Point Laboratory dock on the Choptank River each year during 2005–2008. Secchi depth was used to estimate volume sampled

[b] Measures of abundance on the Patuxent River represent the highest average of daily visual counts at the Chesapeake Biological Laboratory on the Patuxent River each year from 1960–1986. Visible depth was assumed to be approximately 1 m to estimate the volume sampled (D. Cargo, unpublished data)

[c] Concentrations of CQC were based on relationships between bell diameter, dry weight, and carbon content from Purcell & Decker (2005) applied to abundance estimates from this study and Cargo & King (1990)

[d] [DOC] of 200 μM was the dissolved organic carbon concentration in Chesapeake Bay at salinities ranging from 10–16 in September 1990 (Fisher et al., 1998)

[e] Flux was calculated from CQC using average depths of each river (Fisher et al., 2006)

[f] Annual TOC flux into the sediment of 61.2 $g\ C\ m^{-2}$ for Chesapeake Bay (Kemp et al., 1997)

[g] Carbon deposition rate from the spring bloom in Chesapeake Bay was calculated to be 0.51 $g\ C\ m^{-2}\ d^{-1}$ by Hagy et al. (2005). Carbon deposition rate of CQC was based on the average observed time for water temperature to drop from 15 to 10°C in the Choptank River

deposition from the spring bloom (Hagy et al., 2005) from the mesohaline portion of Chesapeake Bay were compared with the calculated values (Table 1). Although the total flux from *C. quinquecirrha* deposition may be small relative to the total annual flux of carbon to the sediment, the calculated deposition rate—as much as 1% of deposition from the spring bloom—shows that the end-of-season deposition may represent a sudden pulse of carbon to the sediments. While the in situ observations of the end-of-season disappearance of medusae and the tank experiments suggest that biomass from medusae is deposited on the bottom, the question remains whether this biomass decomposes in place or is further transported along the bottom by currents. We have assumed that CQCremains in the tributaries; however, further study is needed to understand the fate of this carbon once it reaches the sediment surface.

West et al. (2009) showed that the deposition of gelatinous organic matter can double sediment oxygen demand. In Chesapeake Bay and its tributaries where summer hypoxia and anoxia are increasingly common (Kemp et al., 2005), sources of increased oxygen demand are a serious concern. However, if low temperature causes the deposition of organic matter from *C. quinquecirrha* medusae as our results suggest, it occurs late in the year when cool temperatures and reduced stratification result in a well-mixed and oxygenated water column. In fact, because they have few predators, medusae may be a reserve of organic matter that is not respired until late in the season when the threat of anoxia is gone.

The role of jellyfish as predators has been well studied (e.g., Cowan & Houde, 1993; Behrends & Schneider, 1995; Mills, 1995), but because of low apparent removal by predators, the fate of jellyfish biomass is only beginning to be addressed. Excretion from live gelatinous organisms can provide a fraction of the nutrients necessary to fuel primary production (Nemazie et al., 1993; Pitt et al., 2009). In addition to inorganic nutrients, jellyfish release dissolved organic matter to the water, which can fuel bacterial production. Riemann et al. (2006) showed that increased bacterial production coincided with the depth of highest abundance of jellyfish in a Norwegian fjord, presumably as a result of the DOM released by the jellyfish. This suggests that jellyfish are an important link to lower trophic levels (Riemann et al., 2006). Dead jellyfish biomass fueled bacterial production, but not all members of the bacterial community could utilize it, thus the jellyfish played a role in structuring the bacterial community (Titelman et al., 2006; Tinta et al., 2010). Therefore, *C. quinquecirrha* medusa biomass accumulating at the sediment surface at the end of the season may directly increase bacterial production and may also influence the bacterial community composition at that time.

Jellyfish are known for their ability to reach high abundances quickly (Mills, 2001). These blooms can have great effects on the ecosystem through trophic interactions (e.g., Feigenbaum & Kelly, 1984) and nutrient cycling (Pitt et al., 2009). The demise of such blooms can be equally important as nutrients are released through decomposition, as suggested above. In order to understand the role of decomposing gelatinous biomass on nutrient cycling, it is necessary to understand what factors cause the demise of jellyfish blooms. This type of information may lead to the ability to predict when and where decomposing gelatinous biomass will provide nutrients for bacterial production. Anthropogenic activites negatively affect the health of most marine and estuarine systems; therefore, it is likely that some of these factors may cause the death of jellyfish, and their subsequent role in carbon cycling may be affected. For example, Yamamoto et al. (2008) showed that jellyfish carcasses can be an important source of food to benthic scavengers in the Sea of Japan. Because fishermen cut up the jellyfish caught in their nets, they may alter the timing, rate, and location of jellyfish carcasses deposited on the sea floor. Understanding how such factors will continue to affect jellyfish blooms, like that of *C. quinquecirrha* in Chesapeake Bay, may be important to understanding how nutrient cycling will respond to environmental changes.

In summary, the results indicate that low temperature causes medusae to sink in the water column. This information implies that gelatinous organic matter is delivered to the sediment when water temperature cools to 15°C. Although the medusae represent an appreciable amount of carbon, when low temperatures coincide with their demise, biomass deposition is unlikely to contribute to oxygen depletion. The results of this study show that in most years, when medusae disappear as water temperature decreases, the biomass from these organisms may be deposited onto the sediment surface where they will be decomposed.

Acknowledgements This project was funded by grant NA04NOS4780266 from the National Oceanographic and Atmospheric Administration Ecological Forecasting Program to R. R. Hood. The National Science Foundation's Research Experience for Undergraduates Program provided support for J. Sarkodee-Adoo and A. M. Liss during the summers of 2007 and 2008 respectively. We gratefully acknowledge Laura Murray, Court Stevenson, and the Horn Point Laboratory for the use of equipment necessary to this experiment. Comments from the editor and two anonymous reviewers were responsible for significant improvements to earlier drafts of this work.

References

Behrends, G. & G. Schneider, 1995. Impact of *Aurelia aurita* medusae (Cnidaria, Scyphozoa) on the standing stock and community composition of mesozooplankton in the Kiel Bight (western Baltic Sea). Marine Ecology Progress Series 127: 39–45.

Billet, D. S. M., B. J. Bett, C. L. Jacobs, I. P. Rouse & B. D. Wigham, 2006. Mass deposition of jellyfish in the deep Arabian Sea. Limnology and Oceanography 51: 2077–2083.

Breitburg, D. L. & R. S. Fulford, 2006. Oyster-sea nettle interdependence and altered control within the Chesapeake Bay ecosystem. Estuaries and Coasts 29: 776–784.

Cargo, D. G., 1976. Some effects of tropical storm Agnes on the sea nettle population in the Chesapeake Bay. In Davis, J. & B. Laird (eds), The Effects of Tropical Storm Agnes on the Chesapeake Bay Estuarine System. Chesapeake Bay Consortium, Inc. Publ. No. 54: 417–424.

Cargo, D. G. & D. R. King, 1990. Forecasting the abundance of the sea nettle, *Chrysaora quinquecirrha*, in the Chesapeake Bay. Estuaries 13: 486–491.

Cargo, D. G. & G. E. Rabenold, 1980. Observations on the asexual reproductive activities of the sessile stages of the sea nettle *Chrysaora quinquecirrha* (Scyphozoa). Estuaries 3: 20–27.

Cargo, D. G. & L. P. Schultz, 1967. Further observations on the biology of the sea nettle and jellyfishes in Chesapeake Bay. Chesapeake Science 8: 209–220.

Condon, R. H. & D. K. Steinberg, 2008. Development, biological regulation, and fate of ctenophore blooms in the York River Estuary, Chesapeake Bay. Marine Ecology Progress Series 369: 153–168.

Cowan, J. H. & E. D. Houde, 1993. Relative predation potentials of scyphomedusae, ctenophores and planktivorous fish on ichthyoplankton in Chesapeake Bay. Marine Ecology Progress Series 95: 55–65.

Decker, M. B., C. W. Brown, R. R. Hood, J. E. Purcell, T. F. Gross, J. C. Matanoski, R. O. Bannon & E. M. Setzler-Hammilton, 2007. Predicting the distribution of the scyphozomedusa *Chrysaora quinquecirrha* in Chesapeake Bay. Marine Ecology Progress Series 329: 99–113.

Feigenbaum, D. & M. Kelly, 1984. Changes in the lower Chesapeake Bay food chain in presence of the sea nettle *Chrysaora quinquecirrha* (Schyphomedusa). Marine Ecology Progress Series 19: 39–47.

Fisher, T. R., J. D. Hagy & E. Rochelle-Newall, 1998. Dissolved and particulate organic carbon in Chesapeake Bay. Estuaries 21: 215–229.

Fisher, T. R., J. D. Hagy, W. R. Boynton & M. R. Williams, 2006. Cultural eutrophication in the Choptank and Patuxent estuaries of Chesapeake Bay. Limnology and Oceanography 51: 435–447.

Gatz, A. J., V. S. Kennedy & J. A. Mihurski, 1973. Effects of temperature on activity and mortality of the scyphozoan medusa, *Chrysaora quinquecirrha*. Chesapeake Science 14: 171–180.

Hagy III, J. D., W. R. Boynton & D. A. Jasinski, 2005. Modelling phytoplankton deposition to Chesapeake Bay sediments during winter-spring: interannual variability in relation to river flow. Estuarine, Coastal and Shelf Science 62: 25–40.

Kemp, W. M., E. M. Smith, M. Marvin-DiPasquale & W. R. Boynton, 1997. Organic carbon balance and net ecosystem metabolism in Chesapeake Bay. Marine Ecology Progress Series 150: 229–248.

Kemp, W. M., W. R. Boynton, J. E. Adolf, D. F. Boesch, W. C. Boicourt, G. Brush, J. C. Cornwell, T. R. Fisher, P. M. Glibert, J. D. Hagy, L. W. Harding, E. D. Houde, D. G. Kimmel, W. D. Miller, R. E. I. Newell, M. R. Roman, E. M. Smith & J. C. Stevenson, 2005. Eutrophication of Chesapeake Bay: historical trends and ecological interactions. Marine Ecology Progress Series 303: 1–29.

Matanoski, J. C., R. R. Hood & J. E. Purcell, 2001. Characterizing the effect of prey on swimming and feeding efficiency of the scyphomedusa *Chrysaora quinquecirrha*. Marine Biology 139: 191–200.

Mills, C. E., 1995. Medusae, siphonophores, and ctenophores as planktivorous predators in changing global ecosystems. ICES Journal of Marine Science 52: 575–581.

Mills, C. E., 2001. Jellyfish blooms: are populations increasing globally in response to changing ocean conditions? Hydrobiologia 451: 55–68.

Nemazie, D. A., J. E. Purcell & P. M. Glibert, 1993. Ammonium excretion by gelatinous zooplankton and their contribution to the ammonium requirements of microplankton in Chesapeake Bay. Marine Biology 116: 451–458.

Pitt, K. A., D. T. Welsh & R. H. Condon, 2009. Influence of jellyfish blooms on carbon, nitrogen and phosphorus cycling and plankton production. Hydrobiologia 616: 133–149.

Purcell, J. E., 1992. Effects of predation by the scyphomedusan *Chrysaora quinquecirrha* on zooplankton populations in Chesapeake Bay. Marine Ecology Progress Series 87: 65–76.

Purcell, J. E. & M. B. Decker, 2005. Effects of climate on relative predation by scyphomedusae and ctenophores on copepods in Chesapeake Bay during 1987–2000. Limnology and Oceanography 50: 376–387.

Purcell, J. E., D. A. Nemazie, S. E. Dorsey, E. D. Houde & J. C. Gamble, 1994a. Predation mortality of bay anchovy *Anchoa mitchelli* eggs and larvae due to scyphomedusae and ctenophores in Chesapeake Bay. Marine Ecology Progress Series 114: 47–58.

Purcell, J. E., J. R. White & M. R. Roman, 1994b. Predation by gelatinous zooplankton and resource limitation as potential controls of *Acartia tonsa* copepod populations in Chesapeake Bay. Limnology and Oceanography 39: 263–278.

Purcell, J. E., J. R. White, D. A. Nemazie & D. A. Wright, 1999. Temperature, salinity and food effects on asexual reproduction and abundance of the scyphozoan *Chrysaora quinquecirrha*. Marine Ecology Progress Series 180: 187–196.

Riemann, L., J. Titelman & U. Bamstedt, 2006. Links between jellyfish and microbes in a jellyfish dominated fjord. Marine Ecology Progress Series 325: 29–42.

Schuyler, Q. & B. K. Sullivan, 1997. Light responses and diel migration of the scyphomedusa *Chrysaora quinquecirrha* in mesocosms. Journal of Plankton Research 19: 1417–1427.

Sokal, R. R. & F. J. Rohlf, 1995. Biometry. W. H. Freeman and Co., New York.

Tinta, T., A. Malej, M. Kos & V. Turk, 2010. Degradation of the Adriatic medusa *Aurelia* sp. by ambient bacteria. Hydrobiologia. doi:10.1007/s10750-010-0223-x.

Titelman, J., L. Riemann, T. A. Sornes, T. Nilsen, P. Griekspoor & U. Bamstedt, 2006. Turnover of dead jellyfish: stimulation and retardation of microbial activity. Marine Ecology Progress Series 325: 43–58.

West, E. J., D. T. Welch & K. A. Pitt, 2009. Influence of decomposing jellyfish on the sediment oxygen demand and nutrient dynamics. Hydrobiologia 616: 151–160.

Yamamoto, J., M. Hirose, T. Ohtani, K. Sugimoto, K. Hirase, N. Shimamoto, T. Shimura, N. Honda, Y. Fujimori & T. Mukai, 2008. Transportation of organic matter to the sea floor by carrion falls of the giant jellyfish *Nemopilema nomurai* in the Sea of Japan. Marine Biology 153: 311–317.

Hydrobiologia (2010) 645:135–152
DOI 10.1007/s10750-010-0240-9

JELLYFISH BLOOMS

Use of respiration rates of scyphozoan jellyfish to estimate their effects on the food web

Jennifer E. Purcell · Verónica Fuentes · Dacha Atienza · Uxue Tilves · Diana Astorga · Masato Kawahara · Graeme C. Hays

Published online: 9 April 2010

Abstract One of the main objectives of research on jellyfish is to determine their effects on the food web. They are voracious consumers that have similar diets to those of zooplanktivorous fish, as well as eating microplankton and ichthyoplankton. Respiration rates (RRs) can be used to estimate the amount of food needed to balance metabolism, and thereby estimate minimum ingestion. We compiled RRs for scyphozoan medusae in three suborders (Semeaostomeae, Rhizostomeae, and Coronatae) to determine if a single regression could relate RRs to mass for diverse scyphomedusan species. Temperature (7–30°C) was not a significant factor. RRs versus wet weight (WW) regressions differed significantly for semeaostome and rhizostome medusae; however, RRs versus carbon mass over five-orders of magnitude did not differ significantly among suborders. RRs were isometric against medusa carbon mass, with data for all species scaling to the power 0.94. The scyphomedusa respiration rate (SRR) regression enables estimation of RR for any scyphomedusa from its carbon mass. The error of the SRR regression was ±72%, which is small in comparison with the 1,000-fold variation in field sampling. This predictive equation (RR in ml O_2 $d^{-1} = 83.37 * g\ C^{0.940}$) can be used to estimate minimum ingestion by scyphomedusae without exhaustive collection of feeding data. In addition, effects of confinement on RRs during incubation of medusae were tested. Large medusae incubated in small container volumes (CV) relative to their size (ratios of CV:WW < 50) had RRs ~one-tenth those of medusae in relatively larger containers. Depleted oxygen during incubation did not depress RRs of the

Guest editors: J. E. Purcell & Dror Angel / Jellyfish Blooms: New Problems and Solutions

J. E. Purcell (✉)
Shannon Point Marine Center, Western Washington University, 1900 Shannon Point Rd, Anacortes, WA 98221, USA
e-mail: purcelj3@wwu.edu

J. E. Purcell
Coastal and Marine Resources Centre, University College Cork, Naval Base, Haulbowline Island, Cobh, Co. Cork, Ireland

V. Fuentes · D. Atienza · U. Tilves
Institut de Ciencies del Mar, CSIC, P. Marítim de la Barcelona 37–49, 08003 Barcelona, Spain

D. Astorga
Instituto de Ciencias Marinas de Andalucía, CSIC, Avda. República, Saharaui no 2, 11519 Puerto Real, Cádiz, Spain

M. Kawahara
Graduate School of Biosphere Sciences, Hiroshima University, 4-4, Kagamiyama 1 Chome, Higashi-Hiroshima 739-8528, Japan

G. C. Hays
Department of Pure and Applied Ecology, Institute of Environmental Sustainability, Swansea University, Swansea SA2 8PP, UK

medusae; however, swimming may have been restricted and respiration reduced in consequence. We briefly review other problems with RR experiments and suggest protocols and limitations for estimating ingestion rates of jellyfish from metabolic rates.

Keywords Jellyfish · Metabolism · Zooplankton · Methods · Review · Aurelia · Rhizostoma · ETS

Introduction

Reports of jellyfish blooms are increasing worldwide (reviewed in Purcell et al., 2007). These blooms often cause problems for humans and economic damage to various industries such as fishing, aquaculture, power production, and tourism. The possibility that jellyfish populations are increasing has lead to much speculation about the causes. It also has emphasized how little is known about jellyfish populations globally. Thus, there is an urgent need for research on jellyfish.

Purcell (2009) advocates large-scale methods in order to address this global issue. In addition to determination of the species and biomasses of jellyfish populations, their effects on ecosystems need to be evaluated. Recognition of similar feeding and physiological characteristics among species can facilitate prediction of jellyfish ecosystem effects in disparate habitats. For example, the feeding rates of four scyphozoan species, *Aurelia aurita* (Linnaeus), *Aurelia labiata* Chamisso & Eysenhardt, *Cyanea capillata* (Linnaeus), and *Chrysaora quinquecirrha* (Desor) increased significantly with body mass and prey density, but not temperature (Purcell, 2009). Respiration rates (RRs) of those species increased with body mass (slope = 0.936), but not temperature (Purcell, 2009). Glazier (2006) reviewed metabolic rates of pelagic animals, including jellyfish, which scaled isometrically (1:1) with mass, in contrast to the 3/4-power scaling rule commonly accepted for benthic animals and vertebrates (Nagy et al., 1999; Nagy, 2005).

The amounts of prey consumed by jellyfish in situ have been estimated from clearance rate experiments in the laboratory (CR), gut content analysis of field specimens with digestion times measured in the laboratory (GCDT), and RR experiments [reviewed in Purcell (1997)]. Each of these methods has disadvantages. CR experiments are fraught with artifacts that compromise the results, including confinement that detrimentally affects the behaviors of the predator and prey, and unnatural prey at extreme densities; usually, rates are underestimated. GCDT studies minimize those artifacts, but are very time-consuming, and are situation-specific. RR experiments presumably minimize laboratory artifacts and are relatively easy to conduct.

Use of RRs to estimate jellyfish energetic requirements also has disadvantages. Differences in RR experimental protocols may affect the results, as common to laboratory studies on jellyfish and zooplankton (ICES, 2000): (1) treatment of temperature differs among experiments, which may be conducted at ambient water temperature or changed quickly to determine a Q_{10}; (2) the amount of food preceding RR incubations ranges from starved to field-collected to heavily fed. The measured rates usually underestimate actual consumption because specimens have been starved for different durations (hours to days) in order to measure a 'basal' or 'standard' rate and growth and reproductive costs generally are not included; and (3) confinement in the incubation containers could depress RRs in two ways: available oxygen could be depleted sufficiently during incubation to lower measured rates, or swimming could be restricted.

The problems with conditions in RR experiments have been addressed previously, to some degree. For example, (1) RRs measured at near-ambient temperatures did not increase with temperature according to Q_{10} predictions; near-ambient temperatures did not significantly affect the RR equation of the four species examined (Purcell, 2009); (2) it has been demonstrated repeatedly that feeding increases RRs (e.g., Møller & Riisgård, 2007); (3) the possible effects of confinement on RRs, however, have not been examined for jellyfish, to our knowledge. Effects of confinement on feeding rates of jellyfish are well documented (reviewed in Purcell, 1997, 2009). Rhizostome, coronate, and cubozoan medusae, in particular, are vigorous swimmers (e.g., Larson, 1991, 1992; D'Ambra et al., 2001; Gordon & Seymour, 2009; Klevjer et al., 2009). Therefore, restriction of swimming in containers could reduce activity and RRs.

The objectives of this study were twofold. First, to determine if RRs of other scyphozoan species fit the equation derived from four species in Purcell (2009), we used published RRs of scyphomedusae and new data for *Rhizostoma pulmo* (Macri) and *Nemopilema nomurai* (Kishinouye) to calculate a predictive regression. Second, to determine if incubation of medusae in small containers relative to their size lowered the measured RRs, we compared RRs of medusae versus the container volume. Then, to evaluate possible causes of reduced RRs in small containers, we tested whether available oxygen was reduced sufficiently to lower RRs. We discuss evidence that swimming, and hence RRs, may be reduced by confinement.

Materials and methods

Regression analyses of scyphomedusa respiration rates

We tested if RRs can be predicted across scyphomedusan species by regression of medusa mass vs. RR measured at ambient temperatures. We used data previously analyzed in Purcell (2009) for the scyphomedusae *A. aurita*, *A. labiata*, *C. quinquecirrha*, and *C. capillata* (Suborder Semeaostomeae) (Table 1). In the present analysis, we include additional data for *A. aurita*, three other semeaostome species, four species in Suborder Rhizostomeae, and two species in Suborder Coronatae (Tables 1, 2). No RRs were found for cubozoan medusae. In two studies (Nemazie et al., 1993; Pitt et al., 2005), metabolism was measured as excretion rates, which we converted to RRs by the O:N atomic ratio of 11.6 (Purcell & Kremer, 1983). RRs were standardized to ml O_2 medusa^{-1} d^{-1} by the conversions 1 ml O_2 = 1.42 mg O_2 = 44.88 μmol O_2 = 89.76 μg atoms O_2. Conditions of the experiments, i.e., temperature (°C), time without food, ranges of medusa wet weight (WW), and incubation container volume (CV) are in Table 2. Medusa mass was standardized to WW and carbon (C) by published conversions in the same study or from the most-similar species; few carbon contents for rhizostome medusa exist (Table 3).

Previously unpublished RR data also are presented for some species. *R. pulmo* medusae were cultured from polyps in the laboratory of the Institut de Ciencies del Mar in Barcelona, Spain. Before the experiments, medusae were maintained in kreisels with natural seawater at 20°C and fed with *Artemia* sp. nauplii daily. RRs of small medusae were measured in acid-washed jars filled with 5-μm-filtered natural seawater and sealed with Parafilm© under the jar lid. Each jar contained one medusa or only filtered water (controls). After 4–7 h incubation, 50-ml subsamples were transferred without mixing into glass-stoppered bottles. Dissolved oxygen concentrations ([DO]) were measured by Winkler titration (Strickland & Parsons, 1972) with a Mettler Toledo DL50 Graphix Titrator. After incubation, bell diameter (mm) of each medusa was measured. Wet weights of some specimens were measured ±1 μg on a Mettler Toledo Balance MX5. Those specimens then were dried at 60°C until the weight stabilized (average of 4 d) and dry weight (DW) measured as above. Carbon (C) and nitrogen (N) contents were measured with a CHN analyzer. Regression of diameter to WW, and the percentages of DW in WW (DW%WW) and of DW in C (C%DW) enabled conversions among size units (Table 3).

RRs of *A. aurita* medusae collected from the Northern Adriatic Sea were measured at the Marine Biology Station in Piran, Slovenia at 16°C, 38 salinity, and dim light (Garcia, data). After 1-h acclimation, medusae were incubated individually in glass containers with 2.5 l of 0.2-μm-filtered seawater. RRs of *R. pulmo* collected from the Mar Menor coastal saline lagoon (western Mediterranean) were measured at Universidad de Alicante, Spain at 24.6–27°C, 45–48 salinity, and dim light. After 3-h acclimation, medusae were incubated individually in glass containers with 9–12 l of 0.2-μm-filtered seawater at ambient temperature and salinity. For both species, identical containers without medusae were used as controls. The jars were sealed with parafilm and incubated 4 h in a thermostatic chamber. Before and after incubation, 125-ml subsamples were transferred without mixing into glass-stoppered bottles. [DO] was measured by Winkler titration using a Metrohm 794 Titoprocessor. Bell diameter and WW of all medusae were measured after incubation.

RRs of small *N. nomurai* were measured at the Graduate School of Biosphere Sciences, Hiroshima University, Hiroshima, Japan. Ephyrae liberated from polyps were cultured in 18-l kreisel tanks at 24 ± 1°C and fed *Artemia* sp. twice daily. Individual medusae first were placed in 500-ml beakers with filtered (Whatman GF/F) seawater for 3 h, and then incubated in 500-ml airtight glass bottles with filtered seawater for 3 h in darkness. Large *N. nomurai* were

Table 1 Respiration rates (RR) of scyphomedusae measured at near-ambient temperatures (given in °C)

Species	T (°C)	N	R^2	Respiration rate equation	Reference
Semeaostomeae					
Aurelia aurita	15	29[d]	0.98	ln RR (μl O_2 h^{-1}) = −0.905 + 0.63 * ln mg DW	Kinoshita et al. (1997)
	15	26	0.98	ln RR (μl O_2 h^{-1}) = −1.634 + 0.93 * ln mg DW	
	30	12	0.98	RR (μl O_2 h^{-1}) = 1.084 * g $WW^{1.059}$	Dawson & Martin (2001)
	20	25	0.95	RR (ml O_2 d^{-1}) = 0.0401 * g $WW^{1.153}$	Uye & Shimauchi (data)
	28	29	0.83	RR (ml O_2 d^{-1}) = 1.396 * g $WW^{0.711}$	
	20, 28			RR (ml d^{-1}) = 0.0765 * $2.8^{(T-20)/10}$ * $WW^{1.038}$ combined	
	15, 20, 24	46	0.88	RR (ml O_2 h^{-1}) = $10^{(0.11T-3.1)}$ * g $DW^{1.093}$	Ishii & Tanaka (2006)
	15[b]	12[d]	–	RR (ml O_2 d^{-1}) = 0.086 ± 0.010	Møller & Riisgård (data)
	15[c]	6[d]	0.95	RR (ml O_2 d^{-1}) = 1.30 * g $WW^{1.074}$	
	15[b]	12	–	RR (ml O_2 d^{-1}) = 0.18 * g $WW^{1.012}$	
	15[c]	6	–	RR (ml O_2 d^{-1}) = 0.26 * g $WW^{0.914}$	
	16	6	0.91	RR (ml O_2 h^{-1}) = 0.0051 * g $WW^{0.649}$	Garcia (data)
Aurelia labiata	10	18	0.98	Log_{10}RR (μl O_2 h^{-1}) = 0.2 * log_{10} mg $DW^{0.92}$	Larson (1987)
	15	26	0.97	Log_{10}RR (μl O_2 h^{-1}) = 0.39 * log_{10} mg $DW^{0.91}$	
	10	10	–	RR (μmol O_2 g WW^{-1} h^{-1}) = 0.297	Rutherford & Thuesen (2005)
Chrysaora quinquecirrha	18–28	109	0.80	RR (ml O_2 h^{-1}) = (11.6) * 0.974 * Log_{10} g DW + 0.021 * T + 0.134	Nemazie et al. (1993)[e]
				RR (ml O_2 d^{-1}) = 0.558 * g $WW^{0.974}$	
Cyanea capillata	10	16	0.98	Log_{10}RR (μl O_2 h^{-1}) = 0.47 * Log_{10} mg $DW^{1.0}$	Larson (1987)
	15	7	0.99	Log_{10}RR (μl O_2 h^{-1}) = 0.72 * Log_{10} mg $DW^{1.04}$	
Cyanea capillata	10	2	–	RR (μmol O_2 g WW^{-1} h^{-1}) = 0.78	Rutherford & Thuesen (2005)
Pelagia noctiluca (Forskål)	21	13	0.71	ln RR (ml O_2 h^{-1}) = 0.95 * ln vol (ml) + 2.41	Morand et al. (1987)
Phacellophora camtschatica Brandt	10	7	–	RR (μmol O_2 g WW^{-1} h^{-1}) = 0.13	Rutherford & Thuesen (2005)
Rhizostomeae					
Cassiopea xamachana R. P. Bigelow[a]	24	41	0.74	RR (μg O_2 mg $prot^{-1}$ d^{-1}) = 607.91 * mg $prot^{-0.26}$	Verde & McCloskey (1998)
	24	41	0.74	RR (ml O_2 d^{-1}) = 0.4365 * g $WW^{0.740}$	
	30	27	0.38	RR (μg O_2 mg $prot^{-1}$ d^{-1}) = 409.75 * mg $prot^{-0.15}$	
	30	27	0.38	RR (ml O_2 d^{-1}) = 0.1658 * g $WW^{0.850}$	
Catostylus mosaicus (Quoy & Gaimard)	25	16	–	RR (μl O_2 kg WW^{-1} d^{-1}) = (11.6) * 1.54 * kg WW	Pitt et al. (2005)[e]
Mastigias sp.[a]	29	22	0.98	RR (mg O_2 d^{-1}) = 19.07 * g $DW^{0.685}$	McCloskey et al. (1994)

Table 1 continued

Species	T (°C)	N	R^2	Respiration rate equation	Reference
Nemopilema nomurai	24	36	0.58	RR (ml O_2 d^{-1}) = 0.781 * g $WW^{0.675}$	Uye & Kawahara (data)
	24	12	0.77	RR (ml O_2 d^{-1}) = 1.491 * g $WW^{0.880}$	
	24	48	0.99	RR (ml O_2 d^{-1}) = 0.549 * g $WW^{1.0}$ combined	
Rhizostoma pulmo	24.5–27	14	0.66	RR (ml O_2 h^{-1}) = 0.025 * g $WW^{0.993}$	Garcia (data)
Rhizostoma pulmo	20	48	0.72	RR (ml O_2 d^{-1}) = 0.287 * g $WW^{0.8661}$	Barcelona (data)
Stomolophus meleagris L. Agassiz	30	71	0.97	RR (ml O_2 g WW^{-1} h^{-1}) = 0.05 * g $WW^{0.99}$	Larson (1991)
Coronatae					
Linuche unguiculata (Swartz)[a]	26	77	–	RR (µg O_2 h^{-1}) = 1.46 * mg $DW0^{0.88}$	Kremer et al. (1990)[f]
Periphylla periphylla (Péron & Lesueur)	6–8	162	0.71	ln RR (µl O_2 mg C^{-1} h^{-1}) = 2.201 − 0.411 * ln mg C =RR (ml O_2 d^{-1}) = 0.6624 * g $WW^{0.589}$	Youngbluth & Båmstedt (2001)

Equations are given in the original units, but were standardized to wet weight (WW), carbon (C), and RR (ml O_2 d^{-1}) by the conversions 1 ml O_2 = 1.42 mg O_2 = 44.88 µmol O_2 = 89.76 µg atoms O_2, before regression analyses

N number of measurements, R^2 coefficient of determination for the equation, *DW* dry weight, *Data* analyzed from data provided by authors

[a] Has symbiotic zooxanthellae (algal/total RR ~3%; McCloskey et al., 1994), [b] Fed, [c] Starved, [d] Ephyrae, [e] Excretion rates were converted to respiration by the O:N atomic ratios of 11.6, [f] Unpublished personal communication

collected by a scoop net around Tsushima Island, Japan. The medusae immediately were transferred into a 2000-l plastic tank containing 10-µm-filtered seawater for 3–5 h. Each medusa was immersed twice for 10 min in a 50-l tank filled with filtered seawater and then incubated in an 80-l airtight tank with filtered seawater at ambient temperature (24 ± 1°C) and salinity (34) for 40 to 100 min in the shade. [DO] at the beginning and end of the experiments was measured by the Winkler method. Bell diameter and WW of all medusae were measured after incubation. Large medusae were dried at 60°C until the weight stabilized (average 4 d) and DW measured. C and N contents were measured with a Yanako MT-5 CHN analyzer.

For regression analyses, RR was regressed against WW in each of the unpublished data sets (Table 1). Next, we calculated one RR each at the minimum and maximum sizes at each experimental temperature from regressions in each study in Table 1. We $\log_{10}$-transformed RR and mass data prior to analysis. Although temperature was not a significant factor in the earlier analysis (Purcell, 2009), we first tested for effects of temperature and mass on medusa RRs in a multiple regression. Then, the log-transformed scyphomedusan RR (ml O_2 $medusa^{-1}$ d^{-1}) data were regressed against medusa mass in WW (g) and in C (g) in simple linear regressions. We then tested for differences between the regressions (RR vs. WW and RR vs. C) for semeaostomes and rhizostomes with analysis of covariance (ANCOVA). Data transformations and regression analyses were made with SigmaStat© 3.5 software, and ANCOVAs were done with Minitab 8.2 Extended software.

Container effects on scyphomedusa respiration rates

First, we tested the relationships of medusa WW, RR, and container volume (CV), using only the minimum and maximum WWs, RRs, and CVs for each study and temperature (Table 2; Fig. 1). Because often ranges were given for CV and WW, we assumed that the smallest and largest medusae were incubated in the smallest and largest containers, respectively. We $\log_{10}$-transformed RR, WW, and CV data prior to analysis with Pearson product moment correlation to determine if WWs, RRs, and CVs were co-correlated using SigmaStat© 3.5 software.

Table 2 Protocols for metabolic experiments on scyphomedusae at near-ambient temperatures (*T* in °C)

Species	*T* (°C)	Time without food	Medusa wet weight (g)	Container volume (ml)	Ratio of container (ml) to medusa (g)	Reference
Semeaostomeae						
Aurelia aurita	14.6	0	0.02–0.04	10	220–598	Olesen et al. (1994)[a]
	15	16–24 h	0.002–0.3	150	500–2500	Kinoshita et al. (1997)[a]
	15	16–24 h	0.34–58.3	150–8000	137–441	
	30	0	~10–1000	870–5400	5.4–87	Dawson & Martin (2001)[a]
	20	2 h	11.8–888.2	5000–15000	17–423	Uye & Shimauchi (2005)
	28	2 h	76.5–1338.2	5000–20000	15–65	
	15	0	0.01	45	5114	Møller & Riisgård (data)
	15	≥2 d	0.01–0.02	16–45	978–2282	
	15	0	0.3–4.7	300	64–1019	
	15	≥2 d	0.7–215	120–10000	86–465	
	15, 20, 24	–	63–479	20000	41–317	Ishii & Tanaka (2006)
	16	1 h	17.6–147.6	2000	13–113	Garcia (data)
Aurelia labiata	10	1–2 d	0.3–168.4	30–2075	12–100	Larson (1987)
	15	1–2 d	0.3–89.5	30–2075	23–100	
	10	1 d	1.8–26.65	100–500	19–56	Rutherford & Thuesen (2005)
Chrysaora quinquecirrha	18–28	<2 h	0.25–115	500–2000	28–2000	Nemazie et al. (1993)
Cyanea capillata	10	1–2 d	0.5–385.7	30–2075	5–60	Larson (1987)
	15	1–2 d	0.3–18.1	30–2075	100–114	
	10	1 d	0.26–1.17	100?	85–385	Rutherford & Thuesen (2005)
Pelagia noctiluca	21	5 d	9–28	1000	36–111	Morand et al. (1987)
Phacellophora camtschatica	10	1 d	0.54–1.85	100?	54–185	Rutherford & Thuesen (2005)
Rhizostomeae						
Cassiopea xamachana	24.2	0	~10–275	658, 1794, or 4657	17–66?	Verde & McCloskey (1998)
	29.9	0	~10–150		31–66?	
Catostylus mosaicus	25	–	2161	60000	28	Pitt et al. (2005)
Mastigias sp.	29	0	2–117	658, 1794, or 4657	33–39?	McCloskey et al. (1994)
Nemopilema nomurai	24	3 h	1.2–9.6	500	52–406	Kawahara & Uye (data)
	24	3–5 h	800–7800	80000	10–100	
Rhizostoma pulmo	24.5–27	3 h	17.6–1213	9000–12000	10–113	Garcia (data)
	20	0	0.07–0.5	125–1200	687–5529	Barcelona (data)

Table 2 continued

Species	T (°C)	Time without food	Medusa wet weight (g)	Container volume (ml)	Ratio of container (ml) to medusa (g)	Reference
Stomolophus meleagris	30	4–24 h	5–1000	700–8400	8.4–140	Larson (1991)
Coronatae						
Linuche unguiculata	26	<2 d	0.1–2.1	60	30–100	Kremer et al. (1990)[a]
Periphylla periphylla	6–8	12–24 h	1.4–150	1000–5000	33–714	Youngbluth & Båmstedt (2001)

Data data provided by authors. Ephyrae ≤0.3 g WW

[a] ≥1 individuals per container

Next, we examined size-specific RRs versus the ratios of CV to medusa WW to determine if CV affected RRs (Table 2). Minimum and maximum CVs (ml) were divided by minimum and maximum medusa WWs (g) to yield a range of ratios of CV to WW (CV:WW ratio) for each study. CV for studies of McCloskey et al. (1994) and Verde & McCloskey (1998) were in McCloskey et al. (1985). Data on WW, CV, and RR were kindly provided by the authors of previous publications, specifically, Uye & Shimauchi (2005), Møller & Riisgård (2007), and Uye (2008). For those data and our unpublished data, carbon-specific RRs of individual medusae were compared with their CV:WW ratios.

We tested if [DO] was substantially depleted during incubation. We used data on CV, incubation time, temperature (°C), and RR to calculate the percentage (%) reduction of DO during incubation. Because we lacked initial [DO] in all but two studies, we assumed that incubation water was saturated at the start of each experiment. Saturation [DO] (in mg l^{-1}; calculated from www.hbuehrer.ch/Rechner/O2satur.html, accessed 18 July 2009) were converted to ml l^{-1} and multiplied by container volume (=available DO). RRs of medusae (Table 1) were multiplied by CVs and divided by the available DO to estimate % DO depletion.

Data are presented as ranges and mean ± standard error. The numbers of measurements (equal to the number of specimens except when multiple medusae were incubated in a container) are in Table 1.

Results

Regression analyses of scyphomedusan respiration rates

Multiple linear regression of log-transformed scyphomedusan RR (ml O_2 medusa^{-1} d^{-1}) data showed that medusa mass ($t_{2,47} = 26.1$, $P < 0.001$), but not temperature ($t_{2,\ 47} = 1.0$, $P = 0.332$) significantly affected RRs; therefore, temperature was not considered further in the analyses.

When medusa mass was expressed as wet weight (WW), both WW and suborder (Semeaostomeae vs. Rhizostomeae) significantly influenced RR (ANCOVA: effect of $\log_{10}$WW, $F_{1,46} = 366.9, P < 0.001$; effect of suborder, $F_{1,46} = 18.9, P < 0.001$). The slopes of both

Table 3 Biometric conversions for scyphomedusae, with ambient salinities

Species	DW%WW	C%DW	C%WW	Salinity	Reference
Aurelia aurita	3.6	3.7	0.133	>30	Uye & Shimauchi (2005)
Aurelia labiata	3.8	4.3	0.16	28–30	Larson (1986)
Cyanea capillata	4.2	12.8	0.55	28–30	Larson (1986)
Chrysaora quinquecirrha	1.8	11.1	0.19	6–12	Purcell (1992)
Chrysaora fuscescens Brandt	3.33[a]	6.07[a]	0.20[a]	31–33	Shenker (1985)
	3.62	7.74	0.28	31–33	
Pelagia noctiluca	ND	11.4	ND	ND	Morand et al. (1987)
Catostylus mosaicus	9.67	ND	ND	ND	Pitt et al. (2005)
Nemopilema nomurai	ND	ND	0.6	ND	Kawahara & Uye (data)
Phyllorhiza punctata von Lendenfeld (no zooxanthellae)	3.8	12	0.46	29–32.5	Graham et al. (2003)
Rhizostoma pulmo	6.3[a]	5.6[a]	0.34[a]	38	Barcelona (data)
Stomolophus meleagris	4.17	ND	ND	ND	Larson (1987)
Linuche unguiculata	4.7	11.8	0.56	~35	Kremer et al. (1990)
Periphylla periphylla	3.24	19.6	0.64	33.2	Youngbluth & Båmstedt (2001)

DW dry weight, *WW* wet weight, *C* carbon, *ND* no data

[a] immature medusae

regressions were ~1 (Fig. 1A). RRs of coronate medusae coincided with those of rhizostomes, except for the largest *Periphylla periphylla*, which may have been a consequence of confinement (below).

When medusa mass was expressed as carbon (C), RRs of species in the 3 suborders scaled together vs. mass with a slope ~1 (Fig. 1B; slope = 0.939; $R^2 = 0.926$). Some data for *A. aurita* from Garcia (data) and *Phacellophora camtschatica* fell below the prediction lines (Fig. 1B). A few other data (*Cassiopea xamachana*, an upside-down-jellyfish, and the largest *P. periphylla*) were lower than others in their suborders. The points beyond the prediction limits were removed for re-analysis of the regression. The revised regression yielded a higher R^2 value and slope (RR in ml O_2 $d^{-1} = 83.37 * g\ C^{0.940}$; $R^2 = 0.943$; $F_{1,1} = 826.31$; $P < 0.001$). ANCOVA analysis showed a strong effect of mass in g C ($F_{1,46} = 720$, $P < 0.001$), but no effect of suborder ($F_{1,46} = 0.000$, $P = 0.98$); thus a regression equation combining the suborders was justified.

Young medusae (ephyrae) may have higher RRs and different scaling than larger medusae (Kinoshita et al., 1997; Møller & Riisgård, 2007). RRs for *A. aurita* ephyrae against mass (converted to by C%DW = 7.1; Schneider, 1988) were slightly above the 95% confidence interval, but within the prediction error of the SRR regression (Fig. 1B). When ephyrae were included in the SRR regression, it yielded a higher R^2 value but less-steep slope (RR in ml O_2 $d^{-1} = 83.37 * g\ C^{0.919}$; $R^2 = 0.953$; $F_{1,54} = 1106.57$; $P < 0.001$). Because RRs for ephyrae seldom have been measured (except Morand et al., 1987), they were not included in further analyses.

The reliability of predicting RRs is of key importance for use of the scyphomedusan respiration rate (SRR) regression. The Standard Error of the estimate was 0.294. We calculated the coefficient of variation from the ratios of the individual RRs calculated from the original equations (Table 1) to RRs calculated from individual carbon weights entered in the SRR regression. The means and standard deviations of those ratios then were calculated. The coefficient of variation = standard deviation/mean × 100; thus, the error of the SRR regression = 72.2%.

We used respiration regressions that were not used to develop the SRR regression to test how well RRs of the same-sized medusae compared between their original regression and the SRR regression (Table 4). With the exception of large *C. capillata* medusae in Mangum et al. (1972), which were incubated in small containers (CV:WW ratio = 20) and may have had depressed RRs, the SRR regression underestimated some original RRs by 2.1–37.5% and overestimated others by 27–67%.

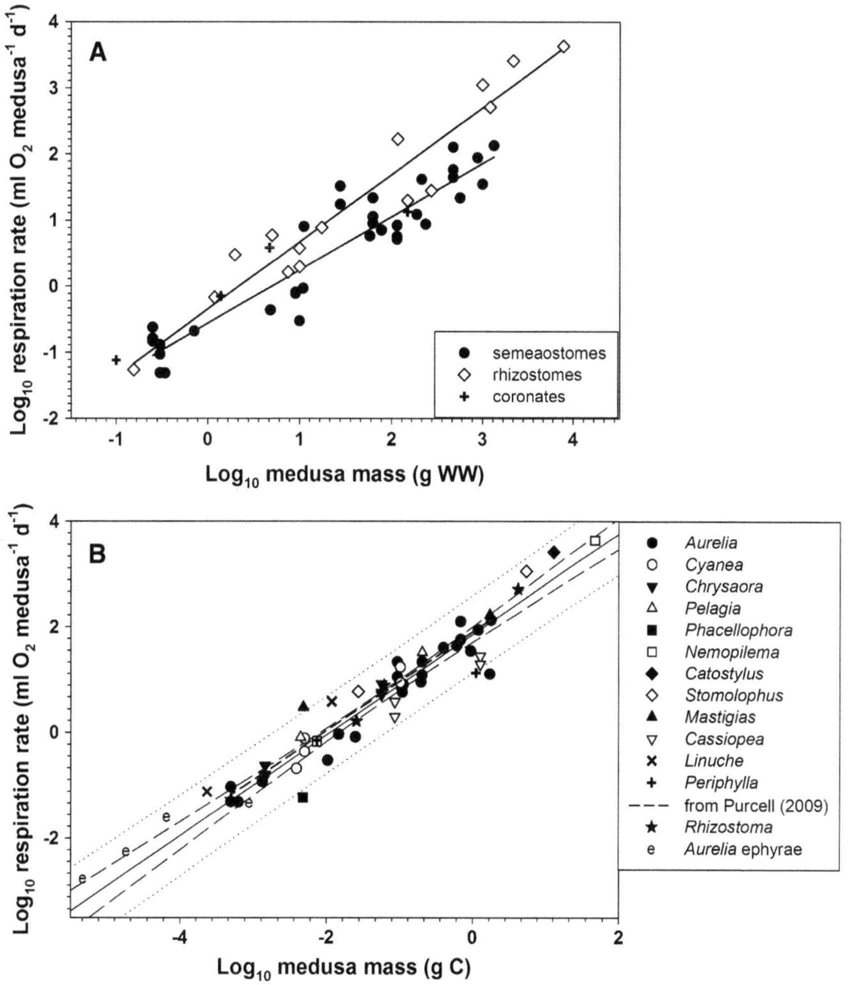

Fig. 1 Log–log plots of scyphomedusan respiration rates (RR) in ml O_2 d^{-1} measured at ambient temperatures versus mass from studies in Table 1. One point (mean) or two points (minimum and maximum mass) are plotted per study at each temperature. **A** RR versus wet weight (WW) in grams by suborder. Regression equations are: semeaostomes $RR = 6.054 * g\ WW^{1.09}$; rhizostomes $RR = 2.698 * g\ WW^{0.917}$; **B** RR versus carbon weight (**C**) in grams by species. *Solid lines* linear regression RR in ml O_2 $d^{-1} = 83.368 * g\ C^{0.940}$, *long dashes* 95% confidence intervals, *dotted lines* prediction errors, *short dashes* semeaostome regression from Purcell (2009). Points below the prediction errors in **B** were omitted from **A**

Container effects on scyphomedusa respiration rates

Medusa WWs were positively correlated with CVs (Fig. 2), which reflects selection of larger containers for larger specimens. All parameters (WW, RR, CV) were co-correlated (Table 5). Correlation of RR with WW is meaningful biologically; however, the positive correlation of RR with CV is not meaningful biologically and should be interpreted to result from increasing WW. The line for a constant ratio of CV to WW of 100:1 shows that most small medusae (<10 g WW) were incubated in containers over 100 times their size (1 ml water weighs 1 g). In contrast, most large medusae (>10 g WW) were incubated in containers less than 100 times their size. This also is apparent from CV:WW ratios in Table 2.

Because RRs and CVs both were correlated with medusa WWs, we examined C-weight-specific RRs vs. the CV:WW ratios from data on individual medusae. C-specific (C-sp) RRs were much higher in small medusae (top right of Fig. 3A), which generally were incubated in containers of greater relative volume than were large medusae. At CV:WW ratios less than ~100:1, incubations of large *A. aurita* and *R. pulmo* medusae at low CV:WW ratios had much lower C-sp RRs (bottom left of Fig. 3A) than did those species at higher CV:WW ratios. Some of those data points were identified as low RRs previously in Fig. 1.

We examined data for individual medusae to determine if dissolved oxygen concentrations [DO] were depleted during incubation. *A. aurita* and *N. nomurai* medusae consumed <1% of the DO

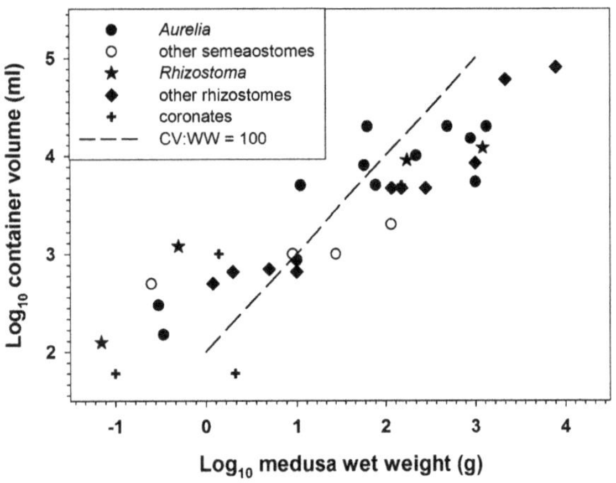

Fig. 2 Incubation-container volume (CV) vs. scyphomedusa wet weight (WW). One point (mean) or two points (minimum and maximum mass) are plotted per study at each temperature. A CV:WW ratio = 100 is shown by the *dashed line*

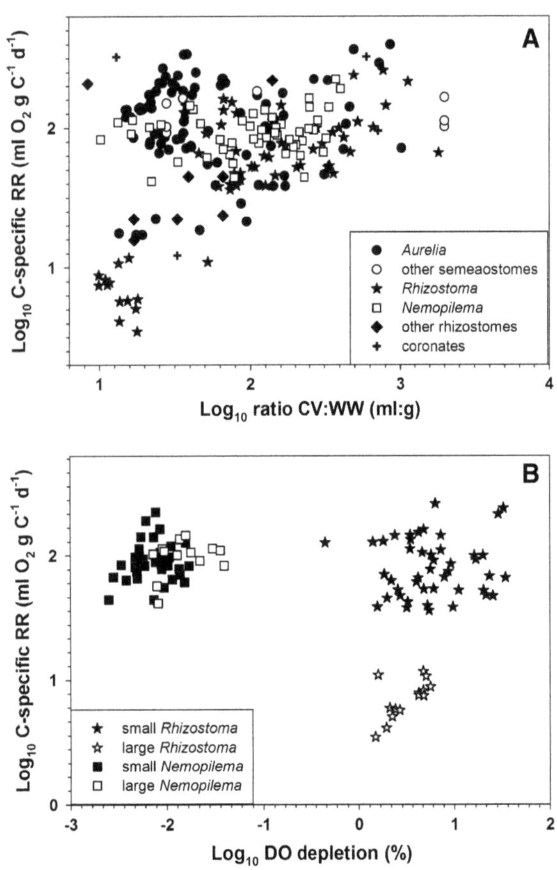

Fig. 3 Carbon-specific respiration rates of scyphomedusae vs. the ratios of container volume in ml to wet weight in grams (CV:WW) (**A**) and vs. the % dissolved oxygen (DO) was depleted during incubation (**B**). Points are data for individual medusae from studies cited as (data) in Table 1

during incubation, and their C-sp RRs were consistent across the range of % DO depletion (Fig. 3B). Small (<5 g WW) *R. pulmo* depleted DO by 0.4–34.2% (mean 7.6 ± 1.0%) in experiments designed to test CV effects on RR, and their C-sp RRs were consistent across the range of % DO depletion (Fig. 3B). Large (17.6–1213 g WW) *R. pulmo* depleted DO by 1.5–5.0% (mean 3.1 ± 0.3%), which was generally less than by small specimens; however, the C-sp RRs of large *R. pulmo* were an order of magnitude lower than of the small specimens (Fig. 3B). Small and large *R. pulmo* had significantly different % DO depletion (one-way ANOVA, $F_{1,57} = 5.67$, $P = 0.021$) and C-sp RRs (one-way ANOVA, $F_{1,57} = 260.47$, $P < 0.001$).

Because RR regressions with slopes of considerably less than 1 (Table 1) suggested the possibility of depressed RRs of large medusae, we also examined % DO depletion in those experiments. Although the slopes of the RR regressions for small (0.675) and large (0.880) *N. nomurai*, *A. aurita* medusae at 28°C (0.717; Uye & Shimauchi, data) and 16°C (0.646, Garcia, data) were <1, % DO depletion was <0.1%. We lacked data for individual medusae of two other species, *C. xamachana* (0.74 and 0.85) and *Mastigias* sp. (0.675) that have symbiotic zooxanthellae; DO production from algal photosynthesis often exceeded DO consumption by medusae, and DO depletion was <1%. The RR slope of *P. periphylla* was 0.589 and DO depletion was <0.03%, assuming the longest incubation time (13 h). Thus, analysis of the [DO] changes in the incubation containers eliminated the possibility that low RRs might have resulted from depleted DO during incubation.

Although DO depletion could not explain depressed RRs in any of the above experiments, long incubation could otherwise detrimentally affect the health of the medusae. For *C. xamachana* (24 h), *Mastigias* sp. (24 h), and large *P. Periphylla* (13 h), long incubations could have contributed to depressing their RRs.

Discussion

Scaling of scyphomedusa respiration rates

Available data from 14 scyphozoan species in three suborders showed that RRs scaled with body mass (g C) over five-orders of magnitude. The SRR

Table 4 Original respiration rates (RR in ml O_2 medusa^{-1} d^{-1}) from studies not used to develop the combined scyphomedusa respiration rate (SRR) regression, RRs calculated from the SRR regression (for all species), and the percentage difference (% dif) between those RRs

Species	T (°C)	WW (g)	Respiration rate equation	Respiration rate (RR in ml O_2 medusa^{-1} d^{-1})			Reference
				Original	Combined scyphomedusae		
				RR	RR	% Difference	
Aurelia aurita	15	0.3	RR (μl O_2 d^{-1}) = 410.89 * mg DW$^{0.86}$	0.08	0.05	−37.5	Frandsen & Riisgård (1997)
		47.2		6.59	6.19	−6.1	
Cyanea capillata	5	0.74	RR (μl O_2 g WW^{-1} h^{-1}) = 90	0.48	0.47	−2.1	Mangum et al. (1972)
		193.6		19.2	87.8	357.3[a]	
Cyanea capillata	15	0.15	RR (μl O_2 g WW^{-1} h^{-1}) = 15.8	0.06	0.10	66.7	Krüger (1968)
		70		26.54	33.76	27.9	
Chrysaora hysoscella (Linnaeus)	15	25	RR (μl O_2 g WW^{-1} h^{-1}) = 6.6	3.96	5.04	27.3	Krüger (1968)
		500		79.2	114.4	44.4	
Rhizostoma octopus (Linnaeus)	15	0.1	RR (μl O_2 g WW^{-1} h^{-1}) = 13.5	0.032	0.046	43.8	Krüger (1968)
		1000		324	264.1	−18.5	
Rhizostoma pulmo	16, 20	0.74	RR (ml O_2 g DW^{-1} h^{-1}) = 0.4	0.30	0.25	−16.7	Vernon (1989) and Yakovleva (1964)

Minimum and maximum sizes from each study were used in the calculations. Dry weights (DW) were converted to wet weights (WW) according to Table 3

[a] Low container volume to WW ratio of 20

[b] RRs reported in Larson (1987); original sizes were not given so small size from Mangum et al. (1972) was used

Table 5 Pearson Product Moment Correlations of container volume (CV in ml), and scyphomedusa wet weight (WW in g), with respiration rate (RR in ml O_2 medusa^{-1} d^{-1})

Pair of variables	Pearson's correlation	
	R	*P*
RR vs. WW	0.909	2.1×10^{-21}
RR vs. CV	0.820	1.8×10^{-12}
WW vs. CV	0.830	5.4×10^{-13}

R correlation coefficient, *P* probability that the factors were not correlated. Data from Tables 1 and 2

regression may enable prediction of RRs of any scyphomedusan species from data only on body mass. As an example, we estimate ingestion from RRs and field size data for *Chrysaora fuscescens* medusae, for which metabolic data do not exist to our knowledge. Mean swimming bell diameter of medusae at Station 1 (17.4 cm: Suchman et al., 2008) was converted to 1.552 g C from data in Shenker (1985) and entered into the SRR regression to yield 126 ml O_2 consumed medusa^{-1} d^{-1}. Multiplication by the RQ gives a minimum of 101 mg C consumed medusa^{-1} d^{-1} to balance respiration.

Scyphomedusae in three suborders scaled similarly with mass as wet weight (g WW), but the RRs of the vigorously swimming rhizostome medusae generally were higher than RRs of semeaostomes. The carbon concentrations in the tissues of rhizostomes are higher than in semeaostomes and so the RRs vs. carbon mass of both suborders scaled together. Convenient measures of medusa size include swimming bell diameter and wet weight (WW), between which numerous conversions exist. Conversions to carbon (C) mass are available for several species, but few exist for rhizostome medusae (Table 3). One of the unavoidable approximations in our analysis was occasional use of WW and DW to C conversions from different species.

The slope of our predictive SRR regression (0.940) is similar to the previous regression for four semeaostome species (slope = 0.936; Purcell, 2009), the regression lines being indistinguishable and nearly isometric (slope = 1) (Fig. 1). The allometric exponents of $\sim$ 1 for scyphomedusae concur with Glazier (2006), who concluded that RRs scaled with body mass of pelagic invertebrates with exponents of $\sim$1. This differs from most benthic animals and vertebrates, which have allometric exponents closer to 0.75 (i.e., 'Kleiber's law'; Nagy et al., 1999). Several characteristics may explain isometric scaling in pelagic animals. Glazier (2006) speculates that their lack of change in body shape and surface area with age, their continued high production costs throughout life, and the energy required to maintain a suspended lifestyle, whether by swimming or buoyancy, could explain the high energy costs of isometric scaling in pelagic animals. RRs of the jellyfish, *C. xamachana*, had scaling exponents of 0.74 and 0.85, which may reflect its epi-benthic lifestyle. We believe that container effects on large jellyfish could contribute to RR regressions with slopes considerably less than 1 in other species (discussed below).

For allometric equations of metabolic rates versus size in vertebrates, size is in live body mass (e.g., Nagy et al., 1999; Nagy, 2005), which would be WW for jellyfish; however, jellyfish have much higher water contents (95–98%) than do fish (71–85%; Doyle et al., 2007), making direct comparisons of the groups by live mass inequitable. For allometric equations of metabolic rates versus size of aquatic invertebrates, including zooplankton, size often is in DW, which is a poor choice for jellyfish because the residual salt in DWs differs with the salinity of the environment (Hirst & Lucas, 1998), as illustrated by the very low DW%WW of *C. quinquecirrha* medusae from salinities 6–12 as compared with other medusae at higher salinities (Table 3). Therefore, we recommend scaling by carbon mass for gelatinous zooplankton taxa.

Sources of error in RR experiments

Temperature changes

Purcell (2009) showed that RRs of 2 scyphomedusan and 1 ctenophore species did not increase with temperature as calculated from Q_{10}s derived from experiments in which temperatures were manipulated. Specifically, Q_{10}s at ambient temperatures were 1.67 for *Aurelia* spp. at 10–30°C and 1.6 for *C. quinquecirrha* at 18 to 28°C, as compared with Q_{10}s of $\sim$3 in manipulated-temperature experiments. Q_{10}s < 2 means that the animal is able to adjust its metabolic rate with temperature; temperature acclimation in medusae has been reported for a long time

(Mangum et al., 1972). RRs used for the SRR regression were measured at ambient temperatures; however, temperature effects of season could have contributed to variability in the data. We conclude that RRs of scyphomedusae should not be adjusted for temperature by Q_{10}s determined from manipulated temperatures; metabolic rates should be measured at the ambient temperatures.

Differences in prior feeding conditions

Differences among experiments in terms of time without food and pre-feeding contribute to differences in RRs (discussed in Purcell, 2009). Pre-experimental protocols differ from days unfed, use of newly collected specimens to reflect rates in situ, to feeding at high prey concentrations. A few studies explicitly tested the effects of food on RRs (e.g., Møller & Riisgård, 2007); RRs of fed medusae decreased within 2.5 h to the rates of starved medusae. Differences in feeding regimes among the studies used in the present analyses probably contributed to variation in the data used here. Unnaturally, high food levels may increase RRs to levels higher than RRs in situ. We believe that for use of RRs for prey consumption estimates, the metabolic rates should be measured on newly collected specimens; however, that raises the issue of acclimation time, which we discuss below.

Different acclimation and incubation times

To our knowledge, no studies on medusae have evaluated how the durations of acclimation or incubation affect medusae in RR experiments. Often 1 h has been chosen arbitrarily for acclimation. The durations of incubation, from <1–24 h, have been compromises between sufficient time to detect differences in [DO] and not to deplete DO or stress the medusae. Differences in acclimation and incubation times among the studies used in the present analyses probably contributed to variability in the data. We suggest that experiments explicitly testing the effects of acclimation and incubation durations on RRs are needed. With improvements in oxygen electrodes, incubation times now can be shortened. We recommend that the durations of medusa confinement be minimized to minimize DO depletion and reduced health.

Container volume effects

We showed that container volume (CV) probably depressed RRs only of large medusae. Most prior studies, in which CV:WW ratios were 100–5,600, yielded consistent C-sp RRs ($\sim$100 ml O_2 g C^{-1} d^{-1}). Evidence of depressed C-sp RRs ($\sim$10 ml O_2 g C^{-1} d^{-1}) occurred with large medusae (*A. aurita*, *C. capillata*, *R. pulmo*, and *P. periphylla*) in containers <50 times larger than the medusa. We showed that depressed RRs probably did not result from depletion of the available DO, as might have occurred with low CV:WW ratios or long incubations.

We tested the validity of our assumption of saturated [DO] at the beginning of incubation with data for measured initial [DO]. The measured initial [DO] generally was higher than the calculated saturated [DO] (mean 112.3 ± 2.2%) for small *R. pulmo* (Barcelona, data). In contrast, initial [DO] was lower than saturated [DO] (mean 93.6 ± 4.1%) for large *N. nomurai* (Kawahara & Uye, data). Nevertheless, the assumption of saturated [DO] gave reasonable estimates (±10%) of available DO at the beginning of the experiments. Depletion of DO is potentially a problem; large *N. nomurai* medusae had slowed swimming pulses at <50% DO saturation (Kawahara, unpublished data).

Large animals usually have lower mass-specific RRs than do small animals (e.g., ICES, 2000). For example, C-sp RRs of ephyrae were 3.4 times those of small medusae (calculated from Kinoshita et al., 1997 and Møller & Riisgård, data). Our analysis for medusae, however, suggests that low C-sp RRs of large medusae may be due in part to confinement effects. Obviously depressed RRs were seen at CV:WW ratios below 50, but less-obvious reductions may have occurred at higher ratios. The C-sp RRs did not differ between small and large *N. nomurai* differing 100–1,000-fold in WW (Fig. 3), even though the CV:WW ratios were much greater for small (mean 184 ± 16) than for large medusae (mean 37 ± 8). That could be due to variation in the CV:WW ratios contributing to variation in the RRs.

Reduced activity in the containers is another probable cause of the low RRs. The best evidence for this is from Larson (1991), who showed higher RRs of active *Stomolophus meleagris* medusae as compared with inactive medusae with crushed rhopalia (Fig. 4). The relationships of *S. meleagris* RRs

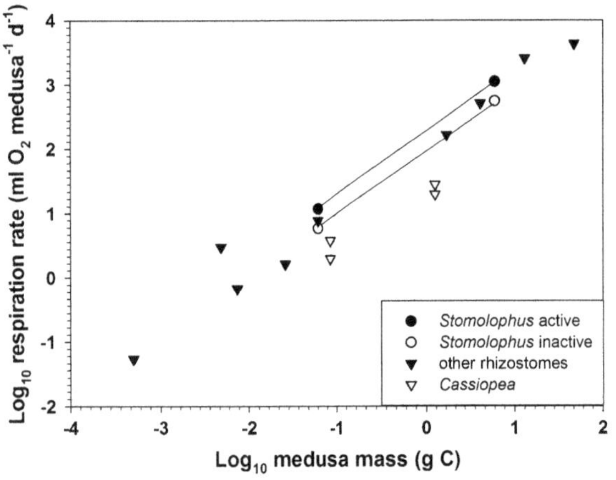

Fig. 4 Effect of activity on respiration rates vs. carbon (C) mass of rhizostome scyphomedusae. *Stomolophus meleagris* medusae were swimming (active) or had crushed rhopalia (inactive) from Larson (1991). *Cassiopea xamachana* medusae rest upside-down on the seabed. One point (mean) or two points (minimum and maximum mass) are plotted per study

versus carbon mass are consistent with other rhizostomes; the RRs of epi-benthic *C. xamachana* medusae were lower than other species in the suborder (Fig. 4). In contrast, confinement of a small hydromedusan species caused increased activity and elevated RRs (Leonard, 1983). We suggest experiments to explicitly test the effects of container size on activity and RRs of medusae.

Minimum rates measured

A weakness of using metabolic rates to estimate ingestion is that requirements for growth and reproduction and losses to respiration are not included and thus underestimate ingestion. Occasionally, growth has been measured in conjunction with RR experiments, in which maximum specific growth rates of ephyrae were 0.2 d^{-1} (Olesen et al., 1994; Møller & Riisgård, 2007). Growth would depend on food availability and be time and location specific. For example, in situ growth rates of *Aurelia* spp. were about 7% WW d^{-1} (Schneider, 1989; Omori et al., 1995; Lucas, 1996; Uye & Shimauchi, 2005), but maximal growth of *C. quinquecirrha* was 60% diameter d^{-1} (~300% WW d^{-1}; Olesen et al., 1996). The higher the growth and reproduction rates, the more the SRR regression would underestimate actual ingestion. Increasing daily metabolic rates over basal rates to account for growth in ingestion estimates would be appropriate.

The variation of data used in the SRR regression is greater than in individual studies because of a combination of the above experimental effects. We hopefully minimized the effects of altered temperatures and varied food conditions by selecting studies at ambient temperatures without added food. Medusae in the various studies experienced different collection and experimental protocols, such as light regime and seawater filtration, which may have affected the results but could not be addressed.

Although we believe, for the above reasons, that the RRs probably were lower than those of free-swimming medusae, especially for large specimens and vigorous swimmers like rhizostomes, we believe that use of the SRR is a reasonable approach for estimating RRs and minimum ingestion of scyphomedusae. The coefficient of variation (72%) of the predictive SRR regression is a low level of variation, given the extreme variation associated with field sampling. Scyphomedusae are renowned for having inhomogeneous distributions (reviewed in Graham et al., 2001); for example, aerial counts of *Rhizostoma octopus* ranged over three-orders of magnitude (Houghton et al., 2006). For all of the above reasons, the SRR regression should be considered to give a conservative estimate of RRs.

Methods to measure field metabolic rates

Due to many possible experimental effects on large, active jellyfish, a method that minimizes confinement effects would be ideal. Many of the blooms of jellyfish around the globe are of very large species, for example *N. nomurai* in Asian waters (up to ~200 kg; Kawahara et al., 2006; Uye, 2008) and *R. octopus* in the Irish Sea (up to 40 kg; Houghton et al., 2007) that are extremely difficult to study because of their great size. Incubation containers to accommodate these jellyfish and maintain a minimum favorable CV:WW ratio >50 should be at least 10 and 2 m^3, respectively. The SRR regression enables estimation of such unwieldy jellyfish; however, the rates used to develop the regression suffer from the above confinement problems. An even better approach would be to eliminate container effects.

Effects of confinement on vertebrate RRs inspired development of the doubly labeled water method

(DLW) to measure the field metabolic rates (FMRs; reviewed in Nagy et al., 1999; Nagy, 2005). DLW has been extensively used on terrestrial species and some marine vertebrates such as sea turtles and pinnipeds; however, the technique can be compromised by high water turnover rates for marine species (Jones et al., 2009) and so probably would be inappropriate for jellyfish. The DLW method is expensive and requires re-capture of the animals for final measurements.

Other methods to measure oxygen consumption are biochemical methods, which involve the determination of the activity of various enzymatic systems, including succinate dehydrogenase, electron transfer system (ETS), lactate dehydrogenase, pyruvate kinase, and citrate synthase (Hernández-León & Ikeda, 2005). Of those indices, ETS activity has been employed most extensively on marine zooplankton (ICES, 2000). The ETS activity is the capacity of a living system to consume oxygen or another electron acceptor and equates respiratory enzyme activity of the ETS with potential oxygen demand (Packard, 1971). The ratio of RR to ETS reflects the fraction of the respiratory capacity that the organism is using (Packard, 1985). The in vivo RR:ETS ratios of marine zooplankton ranged from 0.5 to 1.0, the upper range represented by well-fed specimens (Hernández-León & Ikeda, 2005). Specifically, RR:ETS ratios measured for 5 hydromedusa species, 1 ctenophore, and semi-gelatinous taxa showed that the ratios were about half those of crustacean zooplankton (Owens & King, 1975; King & Packard, 1975; Båmstedt, 2000), presumably because of the high water contents of the gelatinous species.

The ETS method provides good estimates of RRs by the addition of saturated substrate concentrations (Packard, 1971, 1985, Packard & Gómez, 2008). In nature, the cells of organisms may be substrate-limited; therefore, extrapolation of enzyme activities in vitro to those in vivo is not straightforward. In general, ETS measurements reflect the maximum potential oxygen consumption. To closely relate in vitro to in vivo rates, Båmstedt (2000) modified the assay to avoid the addition of substrates and obtained a linear relationship between RR and ETS. Both Packard's and Båmstedt's approaches require calibration of the ETS activity with RRs for each species. Enzyme activity should be standardized by WW (references in Rutherford & Thuesen, 2005). Packard's ETS method requires use of some hazardous chemicals; however, Båmstedt's modification made the method "simple and well suited for field work". Unlike measurements made according to Packard's method, measurements based on Båmstedt's method should be interpreted carefully because it does not measure the potential respiration activity (Packard, personal communication). Application of the ETS method to medusae in nature would minimize the above problems with laboratory RR experiments and enable estimation of the energetic requirements of even the largest species that present the greatest logistic challenges.

Calculation of ingestion from respiration rates

Respiration and excretion are basic physiological processes that are related to body mass, temperature, and activity for all animals. They have been used to estimate the minimum food requirements and ingestion for some gelatinous species (e.g., Ishii & Tanaka, 2006). Because RRs of the scyphozoan species for which respiration data are available are predictable by one equation, it will be easy to estimate minimum energy requirements for scyphomedusae with just data on their size. Ultimately, size data should be in carbon. Size data in the field can be collected as bell diameter if conversions to WW and C are available.

The RRs for medusae and ctenophores can be used to estimate minimum predation rates (e.g., Ishii & Tanaka, 2006). The minimum daily carbon ingestion (MDCI) can be calculated by multiplying the daily RR by the respiratory quotient ($RQ = 0.8$). The MDCI can be converted to numbers of prey ingested from prey carbon mass when the prey types are known (e.g., ICES, 2000). Thus, population estimates of amounts of prey consumed by gelatinous species in situ can be made from the respiration or excretion rate at the mean jellyfish mass, times the jellyfish population density. The effects on the prey populations can be estimated from prey consumption divided by prey densities in situ, in combination with the field data on predator mass and density, prey mass and densities, and temperature. When growth and ingestion rates are relatively low, respiration will approximately equal ingestion and application of the SRR regression to jellyfish population biomass can approximate ingestion and effects of jellyfish on prey populations.

Conclusions

The SRR regression (RR in ml O_2 d^{-1} = 83.37 * g $C^{0.940}$) would allow estimation of RRs and minimum energy demands of scyphomedusae ± 72% from data only on jellyfish mass. Although this method is approximate and should be considered conservative, it is important that gelatinous species be included in ecosystem studies and models that now are conducted on regional to global scales (Pauly et al., 2009). This method offers an alternative to when limited resources and time do not permit exhaustive collection of data on jellyfish ingestion. Use of RRs by species, if available, is preferable to use of the SRR regression.

We briefly summarize recommendations for methods to measure metabolic rates for estimation of ingestion in jellyfish:

- Determine jellyfish densities and sizes, including wet and carbon mass, of each species.
- Report temperature and salinity.
- Use ambient temperature for all experiments.
- Conduct metabolic experiments on newly collected specimens for rates that reflect natural food conditions.
- Incubate jellyfish in volumes at least 100-times greater than the wet mass.
- Do not convert metabolic rates by use of Q_{10} values measured at experimentally manipulated temperatures.
- Calibrate the electron transport system (ETS) method versus RRs to measure field metabolic rates of jellyfish.
- Develop algorithms among taxa that can be used to predict jellyfish effects on large scales.

Acknowledgments This project resulted from a collaboration between the Medusa Project and EcoJel. The Medusa Project was financed by the Agencia Catalana del Agua. EcoJel is a component project of the Interreg 4a Irish-Wales programme, which forms part of the European Regional Development Fund (ERDF). We thank Drs. A. Malej and Gimenez for hosting D. Astorga (funded by P07-RNM-02976) at the Marine Biology Station and Universidad de Alicante, respectively, and Dr. L. Prieto at CSIC at Andalucía for her assistance. Thanks also to Dr. R. M. Callaway for German translation and Dr. P. Kremer and two anonymous reviewers for their constructive comments on the manuscript.

References

Båmstedt, U., 2000. A new method to estimate respiration rate of biological material based on the reduction of tetrazolium violet. Journal of Experimental Marine Biology and Ecology 251: 239–263.

D'Ambra, I., J. H. Costello & F. Bentivegna, 2001. Flow and prey capture by the scyphomedusa *Phyllorhiza punctata* von Lendenfeld, 1884. Hydrobiologia 451(Developments in Hydrobiology 155): 223–227.

Dawson, M. N. & L. E. Martin, 2001. Geographic variation and ecological adaptation in *Aurelia* (Scyphozoa, Semaeostomeae): some implications from molecular phylogenetics. Hydrobiologia 451(Developments in Hydrobiology 155): 259–273.

Doyle, T. K., J. D. R. Houghton, R. McDevitt, J. Davenport & G. C. Hays, 2007. The energy density of jellyfish: Estimates from bomb-calorimetry and proximate-composition. Journal of Experimental Marine Biology and Ecology 242: 239–252.

Frandsen, K. T. & H. U. Riisgård, 1997. Size dependent respiration and growth of jellyfish, *Aurelia aurita*. Sarsia 82: 307–312.

Glazier, D. S., 2006. The ¾-power law is not universal: evolution of isometric, ontogenetic metabolic scaling in pelagic animals. BioScience 56: 325–332.

Gordon, M. R. & J. E. Seymour, 2009. Quantifying movement of the tropical Australian cubozoan *Chironex fleckeri* using acoustic telemetry. Hydrobiologia 616: 87–97.

Graham, W. M., F. Pagès & W. M. Hamner, 2001. A physical context for gelatinous zooplankton aggregations: a review. Hydrobiologia 451: 199–212.

Graham, W. M., D. L. Martin, D. L. Felder, V. L. Asper & H. M. Perry, 2003. Ecological and economic implications of the tropical jellyfish invader, *Phyllorhiza punctata* von Lendenfeld, in the northern Gulf of Mexico. Biological Invasions 5: 53–69.

Hernández-León, S. & T. Ikeda, 2005. Zooplankton respiration. In Del Giorgio, P. & P. J. Williams (eds), Respiration in Aquatic Systems. Oxford University Press, New York: 57–82.

Hirst, A. G. & C. H. Lucas, 1998. Salinity influences body weight quantification in the scyphomedusa *Aurelia aurita*: important implication for body weight determination in gelatinous zooplankton. Marine Ecology Progress Series 165: 259–269.

Houghton, J. D. R., T. K. Doyle, M. W. Wilson, J. Davenport & G. C. Hays, 2006. Jellyfish aggregations and leatherback turtle foraging patterns in a temperate coastal environment. Ecology 87: 1967–1972.

Houghton, J. D. R., T. K. Doyle, M. W. Wilson, J. Davenport & G. C. Hays, 2007. Stranding events provide indirect insights into the seasonality and persistence of jellyfish medusae (Cnidaria: Scyphozoa). Hydrobiologia 589: 1–13.

ICES, 2000. ICES Zooplankton Methodology Manual. Academic Press, London: 705 pp.

Ishii, H. & F. Tanaka, 2006. Respiration rates and metabolic demands of *Aurelia aurita* in Tokyo Bay with special reference to large medusae. Plankton and Benthos Research 1: 64–67.

Jones, T. T., M. Hastings, B. L. Bostrom, R. D. Andrews & D. R. Jones, 2009. Validation of the use of doubly labeled water in the green turtle (*Chelonia mydas*): a word of caution. Journal of Experimental Biology 212: 2635–2644.

Kawahara, M., S. Uye, K. Ohtsu & H. Iizumi, 2006. Unusual population explosion of the giant jellyfish *Nemopilema nomurai* (Scyphozoa: Rhizostomeae) in East Asian waters. Marine Ecology Progress Series 307: 161–173.

King, F. D. & T. T. Packard, 1975. Respiration and the activity of the respiratory electron transport system in marine zooplankton. Limnology and Oceanography 20: 849–854.

Kinoshita, J., J. Hiromi & S. Kadota, 1997. Do respiratory metabolic rates of the scyphomedusa *Aurelia aurita* scale isometrically throughout ontogeny in a sexual generation? Hydrobiologia 347: 51–55.

Klevjer, T. A., S. Kaartvedt & U. Båmstedt, 2009. In situ behaviour and acoustic properties of the deep living jellyfish *Periphylla periphylla*. Journal of Plankton Research 31: 793–803.

Kremer, P., J. Costello, J. Kremer & M. Canino, 1990. Significance of photosynthetic endosymbionts to the carbon budget of the scyphomedusa *Linuche unguiculata*. Limnology and Oceanography 35: 609–624.

Krüger, F., 1968. Stoffwechsel und Wachstum bei Scyphomedusan. Helgoländer wissensaftliche Meeresuntersungen 18: 367–383.

Larson, R. J., 1986. Water content, organic content, and carbon and nitrogen composition of medusae from the northeast Pacific. Journal of Experimental Marine Biology and Ecology 99: 107–120.

Larson, R. J., 1987. Respiration and carbon turnover rates of medusae from the NE Pacific. Comparative Biochemistry and Physiology 87A: 93–100.

Larson, R. J., 1991. Diet, prey selection and daily ration of *Stomolophus meleagris*, a filter-feeding scyphomedusa from the NE Gulf of Mexico. Estuarine and Coastal Shelf Science 32: 511–525.

Larson, R. J., 1992. Riding Langmuir circulations and swimming in circles: A novel form of clustering behavior by the scyphomedusa *Linuche unguiculata*. Marine Biology 112: 229–235.

Leonard, J. L., 1983. The effect of environmental factors on swimming activity in *Sarsia tubulosa* M. Sars (Hydrozoa). Marine Behaviour and Physiology 9: 99–110.

Lucas, C. H., 1996. Population dynamics of *Aurelia aurita* (Scyphozoa) from an isolated brackish lake, with particular reference to sexual reproduction. Journal of Plankton Research 18: 987–1007.

Mangum, C. P., M. J. Oakes & J. M. Shick, 1972. Rate-temperature responses in scyphozoan medusae and polyps. Marine Biology 15: 298–303.

McCloskey, L. R., L. D. Aamodt & W. D. Hazelton, 1985. A computer-controlled respirometer for monitoring production and respiration of symbiotic organisms in situ. In Gabrié, C., et al. (eds), Proceedings of the 5th International Coral Reef Congress 6. Antenne Museum-EPME, Moorea, French Polynesia: 137–142.

McCloskey, L. R., L. Muscatine & F. P. Wilkerson, 1994. Daily photosynthesis, respiration and carbon budgets in a tropical marine jellyfish (*Mastigias* sp.). Marine Biology 119: 13–22.

Møller, L. F. & H. U. Riisgård, 2007. Respiration in the scyphozoan jellyfish *Aurelia aurita* and two hydromedusae (*Sarsia tubulosa* and *Aequorea vitrina*): effect of size, temperature and growth. Marine Ecology Progress Series 330: 149–154.

Morand, P., C. Carré & D. C. Biggs, 1987. Feeding and metabolism of the jellyfish *Pelagia noctiluca* (Scyphomedusae, Semaeostomae). Journal of Plankton Research 9: 651–665.

Nagy, K. A., 2005. Field metabolic rate and body size. Journal of Experimental Biology 208: 1621–1625.

Nagy, K. A., I. A. Girard & T. K. Brown, 1999. Energetics of free-ranging mammals, reptiles, and birds. Annual Review of Nutrition 19: 247–277.

Nemazie, D. A., J. E. Purcell & P. M. Glibert, 1993. Ammonium excretion by gelatinous zooplankton and their contribution to the ammonium requirements of microplankton in Chesapeake Bay. Marine Biology 116: 451–458.

Olesen, N. J., K. Frandsen & H. U. Riisgård, 1994. Population dynamics, growth and energetics of jellyfish *Aurelia aurita* in a shallow fjord. Marine Ecology Progress Series 105: 9–18.

Olesen, N. J., J. E. Purcell & D. K. Stoecker, 1996. Feeding and growth by ephyrae of scyphomedusae *Chrysaora quinquecirrha*. Marine Ecology Progress Series 137: 149–159.

Omori, M., H. Ishii & A. Fujinaga, 1995. Life history strategy of *Aurelia aurita* (Cnidaria, Scyphomedusae) and its impact on the zooplankton community of Tokyo Bay. ICES Journal of Marine Science 52: 597–603.

Owens, T. G. & F. D. King, 1975. The measurement of respiratory electron-transport activity in marine zooplankton. Marine Biology 30: 27–36.

Packard, T. T., 1971. The measurement of respiratory electron transport activity in marine phytoplankton. Journal of Marine Research 29: 235–244.

Packard, T. T., 1985. Measurements of electron transport activity of marine microplankton. In Williams, P. J. & H. W. Jannasch (eds), Advances in Aquatic Microbiology. Academic Press, New York: 207–261.

Packard, T. T. & M. Gómez, 2008. Exploring a first-principles-based model for zooplankton respiration. ICES Journal of Marine Science. doi:10.1093/icesjms/fsn003.

Pauly, D., W. M. Graham, S. Libralato, L. Morissette & M. L. D. Palomares, 2009. Jellyfish in ecosystems, online databases, and ecosystem models. Hydrobiologia 616: 67–85.

Pitt, K. A., K. Koop & D. Rissik, 2005. Contrasting contributions to inorganic nutrient recycling by the co-occurring jellyfishes, *Catostylus mosaicus* and *Phyllorhiza punctata* (Scyphozoa, Rhizostomeae). Journal of Experimental Marine Biology and Ecology 315: 71–86.

Purcell, J. E., 1992. Effects of predation by the scyphomedusan *Chrysaora quinquecirrha* on zooplankton populations in Chesapeake Bay. Marine Ecology Progress Series 87: 65–76.

Purcell, J. E., 1997. Pelagic cnidarians and ctenophores as predators: selective predation, feeding rates and effects on

prey populations. Annales de l'Institut océanographique, Paris 73: 125–137.
Purcell, J. E., 2009. Extension of methods for jellyfish and ctenophore trophic ecology to large-scale research. Hydrobiologia 616: 23–50.
Purcell, J. E. & P. Kremer, 1983. Feeding and metabolism of the siphonophore *Sphaeronectes gracilis*. Journal of Plankton Research 5: 95–106.
Purcell, J. E., S.-I. Uye & W.-T. Lo, 2007. Anthropogenic causes of jellyfish blooms and direct consequences for humans: a review. Marine Ecology Progress Series 350: 153–174.
Rutherford, L. D. Jr., & E. V. Thuesen, 2005. Metabolic performance and survival of medusae in estuarine hypoxia. Marine Ecology Progress Series 294: 189–200.
Schneider, G., 1988. Chemische Zusarnrnensetzung und Biomasseparameter der Ohrenqualle *Aurelia aurita*. Helgolander Meeresuntersungen 42: 319–327.
Schneider, G., 1989. A comparison of carbon based ammonia excretion rates between gelatinous and non-gelatinous zooplankton: implications and consequences. Marine Biology 106: 219–225.
Shenker, J. M., 1985. Carbon content of the neritic scyphomedusa *Chrysaora fuscescens*. Journal of Plankton Research 7: 169–173.
Strickland, J. D. H. & T. R. Parsons, 1972. A practical handbook of seawater analysis. Fisheries Research Board Canada, Bulletin 167: 311 pp.
Suchman, C. L., E. A. Daly, J. E. Keister, W. T. Peterson & R. D. Brodeur, 2008. Feeding patterns and predation potential of scyphomedusae in a highly productive upwelling region. Marine Ecology Progress Series 358: 161–172.
Uye, S., 2008. Blooms of the giant jellyfish *Nemopilema nomurai*: a threat to the fisheries sustainability of the East Asian Marginal Seas. Plankton and Benthos Research 3: 125–131.
Uye, S. & H. Shimauchi, 2005. Population biomass, feeding, respiration and growth rates, and carbon budget of the scyphomedusa *Aurelia aurita* in the Inland Sea of Japan. Journal of Plankton Research 27: 237–248.
Verde, E. A. & L. R. McCloskey, 1998. Production, respiration, and photophysiology of the mangrove jellyfish *Cassiopea xamachana* symbiotic with zooxanthellae: effect of jellyfish size and season. Marine Ecology Progress Series 168: 147–162.
Youngbluth, M. J. & U. Båmstedt, 2001. Distribution, abundance, behavior and metabolism of *Periphylla periphylla*, a mesopelagic coronate medusa in a Norwegian fjord. Hydrobiologia 451: 321–333.

Hydrobiologia (2010) 645:153–165
DOI 10.1007/s10750-010-0221-z

JELLYFISH BLOOMS

Planktonic cnidarian distribution and feeding of *Pelagia noctiluca* in the NW Mediterranean Sea

Ana Sabatés · Francesc Pagès · Dacha Atienza · Verónica Fuentes · Jennifer E. Purcell · Josep-María Gili

Published online: 3 April 2010

Abstract Pelagic cnidarians are important consumers of zooplankton and ichthyoplankton in the world's oceans, and thus harm fisheries as competitors and predators of fish. This study examined the inshore-offshore distribution of pelagic cnidarians and the trophic ecology of *Pelagia noctiluca* ephyrae (<12 mm diameter) and larger medusae in late spring 1995 in the NW Mediterranean Sea. The distribution of pelagic cnidarians was closely related to the presence of the shelf-slope front with most species mainly concentrated close to the front. Meroplanktonic antho- and leptomedusae predominated in coastal waters and more holoplanktonic trachy- and narcomedusae occurred both in shelf and open sea waters. *P. noctiluca* was more abundant than other medusae, including hydromedusae. Siphonophores, particularly *Muggiaea atlantica*, outnumbered medusae at most stations. The diet of *P. noctiluca* ephyrae contained mainly copepods, but ~12% of the prey were fish larvae. *P. noctiluca* exhibited positive prey selection for chaetognaths and mollusc larvae in day and night samples, but fish larvae were positively selected only at night. These differences may be related to the diel vertical distributions of *P. noctiluca* and their prey. Most of the ingested fish larvae belonged to the family Myctophidae, but anchovy and sparid larvae also were found in the gastric pouches. The size of ingested fish larvae increased as ephyra diameter increased; however, in the larger medusae (>12 mm) the number of prey increased with medusa size rather than the size of the larvae. The temporal and spatial co-occurrence of *P. noctiluca* with early life stages of fish suggests that *P. noctiluca* may be an important predator on summer ichthyoplankton.

Keywords Jellyfish · Ichythyoplankton · Fish · Front · Siphonophore · Hydromedusae

Guest editors: J. E. Purcell & Dror Angel / Jellyfish Blooms: New Problems and Solutions

A. Sabatés (✉) · F. Pagès · D. Atienza · V. Fuentes · J.-M. Gili
Institut de Ciències del Mar, CSIC, P. Marítim 37-49, 08003 Barcelona, Spain
e-mail: anas@icm.csic.es

J. E. Purcell
Shannon Point Marine Center, Western Washington University, 1900 Shannon Point Rd, Anacortes, WA 98221, USA

J. E. Purcell
Coastal and Marine Resources Centre, University College Cork, Naval Base, Haulbowline Island, Cobh, Co. Cork, Ireland

Introduction

High concentrations of gelatinous plankton are a natural phenomenon in coastal and oceanic waters. Increased reports about jellyfish suggest that their numbers have increased in recent years (Purcell et al., 2007). This is cause for concern because jellyfish cause serious problems in some economic sectors and in the management of natural resources (Mills, 2001). Modern commercial fishing removes predators and competitors of jellyfish with increasing efficiency enabling many gelatinous zooplankton species to proliferate. When jellyfish occur in large numbers, their predation impact can be very important and may even control the population sizes and compositions of their prey organisms (Behrends & Schneider, 1995). They also may compete with zooplanktivorous fish and fish larvae by eating the same prey items (Purcell & Grover, 1990). Thus, studies of gelatinous zooplankton abundances, population dynamics and feeding ecology are of high priority in all oceans.

The diets of various jellyfish taxa consist primarily of copepods, but may also include meroplankton, ichthyoplankton, and gelatinous plankton (Purcell, 1985, 1997; Purcell & Mills, 1988). There is considerable evidence that scyphomedusae, hydromedusae, siphonophores and ctenophores consume fish eggs and fish larvae (Möller, 1980; Purcell, 1989, 1990; Purcell et al., 1994; reviewed by Purcell & Arai, 2001). Indeed, all jellyfish species tested have shown positive selection for ichthyoplankton (Fancett, 1988; Purcell et al., 1994; Purcell & Sturdevant, 2001), with prey selection being defined as consumption of prey types in disproportion to their abundance in the environment.

In the northwestern Mediterranean, the abundances and distributions of gelatinous zooplankton, especially planktonic cnidarians, have been studied for many years (e.g. Gili et al., 1987a; Dallot et al., 1988). In recent decades, an increase in jellyfish outbreaks in this region has been related to climatic forcing, particularly warming temperatures and dry conditions (Molinero et al., 2005). In the western Mediterranean, pelagic cnidarians are most abundant during spring and summer (Gili et al., 1987a). The distributions of the dominant species are closely related to the hydrographic heterogeneity, the geomorphology of the continental shelf and the presence of mesoscale permanent hydrographic features (Gili et al., 1988).

In the Catalan Sea (NW Mediterranean), the scyphomedusa, *Pelagia noctiluca* (Forsskål), is the most abundant jellyfish species in the summer months and its abundance has increased in recent years (Gili & Pagès, 2005). This species is especially abundant on the shelf-slope where concentrations of plankton occurs (Sabatés et al., 1989). On the shelf-slope, *P. noctiluca* occurs near the surface, especially at night, as observed in other parts of the Mediterranean (review by Mariottini et al., 2008). Their vertical distribution pattern coincides with the nocturnal migration of zooplankton, their main prey (Malej, 1989; Rottini Sandrini & Avian, 1989). *P. noctiluca* is considered a top planktonic predator (Larson, 1987), feeding on almost all types of zooplankton and ichthyoplankton (Giorgi et al., 1991; Zavodnik, 1991; Malej et al., 1993); however, the dietary composition of *P. noctiluca* in the Catalan Sea has never been studied and its predatory effects on zooplankton and ichthyoplankton populations remain unclear.

The Catalan coast is characterized by a permanent shelf-slope front along the shelf edge. Typically the front is defined by salinity differences between waters of the open sea (salinity >38) and the shelf (salinity <38; Font et al., 1988; Salat, 1996). The front runs along the continental slope of the northwestern Mediterranean, reaching a depth of some 400 m where it intersects with the bottom. The associated geostrophic current flows in a strip about 10-km wide at the surface, parallel to the front on its coastal side, and roughly over the 1000-m isobath (García-Ladona et al., 1994). The roles of the front in phytoplankton distribution and primary production (Estrada & Margalef, 1988; Estrada et al., 1999) and in zooplankton distribution and metabolism (Alcaraz et al., 1994; Sáiz et al., 1999) have been documented. The highest densities of fish larvae are located at the edge of the continental shelf in association with the shelf-slope front (Sabatés, 1990). In some cases, the front has been found to delimit shelf and oceanic fish larvae populations and to act as a barrier preventing dispersal of fish larvae to the open sea (Masó et al., 1998).

In this study, we address the distribution of gelatinous zooplankton across the inshore–offshore gradient of the Catalan coast (NW Mediterranean) in relation to mesoscale physical processes, and the trophic ecology of *P. noctiluca* ephyrae (<12 mm diameter) and larger individuals, with particular attention to their predation on fish larvae.

Materials and methods

Sampling was conducted on a transect perpendicular to the coast, from near shore to the slope, in the Catalan Sea (western Mediterranean) during 18–23 June 1995 (FRONTS cruise; Fig. 1). Four stations were sampled in accordance with the general circulation in the Catalan Sea to detect spatial gradients. The first station was in coastal waters (40-m depth), the second was over the shelf (70–80-m depth), the third was over the slope (1000-m depth) near the front, and the last station was in the open sea (>2000-m depth; Fig. 1). This sampling was repeated three times at each station, regardless of time of the day. Thus, a total of 12 samplings were conducted (7 at day and 5 at night).

At each station, vertical profiles of temperature, salinity and chlorophyll fluorescence were obtained with a Neil Brown Mark III-CTD (WOCE standard) equipped with a Sea-Tech fluorometer. Zooplankton and fish larvae were sampled by oblique tows of a 60-cm diameter Bongo net with 500-μm mesh from near-bottom to the surface or from 200 m to the surface where the bottom was deeper than 200 m. The volume of water passing through the net was calculated by means of data obtained from a 2030R General Oceanics Inc. flowmeter mounted in the mouth of the net. Net samples were fixed in a 5% formaldehyde-seawater solution buffered with sodium tetraborate. Earlier on 10–14 June (VARIMED cruise), CTD profiles were made at eight stations on the same transect for physical parameters.

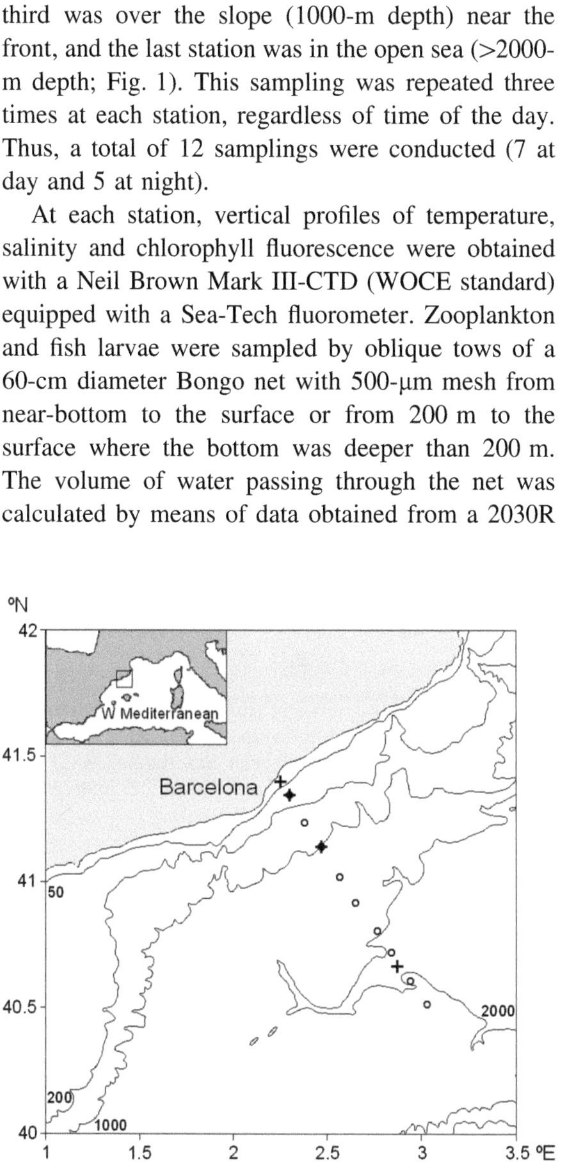

Fig. 1 Map of the northwestern Mediterranean Sea showing with locations of the sampling stations along the coastal–open sea transect during 18–23 June 1995. *Open circles*: CTD casts; *crosses*: zooplankton tows

In the laboratory, medusae, siphonophores and fish larvae were counted and identified to the lowest possible taxonomic level with the aid of a dissecting microscope. For physonect siphonophores, the number of colonies was estimated from the number of nectophores according to Totton (1965); for calycophorans, the number of colonies was determined by the number of anterior nectophores. *P. noctiluca* ephyrae (≤12 mm diameter) were counted. Some larger *P. noctiluca* were occasionally collected but not included in the abundance calculation because the net was inappropriate for collection of large medusae. Nevertheless, the large specimens were included for diet analyses. In addition, the most abundant zooplankton taxa were counted by means of a dissecting microscope from 1/256 to 1/32 aliquots obtained with a Motoda rectangular box splitter (Motoda, 1959). All individuals from each taxon were counted in an aliquot; densities were expressed as the number of individuals per 1000 m^3.

The gastric pouch contents of all *P. noctiluca* ephyrae in the samples were identified, counted, and measured with the aid of a dissecting microscope. We included only partly digested prey to ensure that the prey items had not been captured while in the net. Prior to dissection, the maximum diameter of each jellyfish was measured with an ocular micrometre. The proportion of *P. noctiluca* ephyrae containing food (feeding incidence) was calculated for each sampling station. One-way analysis of variance (ANOVA) was used to assess differences among sampling locations (independent of time of the day) and between day and night (independent of sampling location).

Prey taxon selectivity by *P. noctiluca* ephyrae was estimated by the α index (Chesson, 1978): $\alpha = (r_k/p_k)/\sum_1^n (r_i/p_i)$ where r_k and p_k are the proportions of prey k in the diet and in the field, respectively, and n is the number of prey taxa. Neutral selection would result in a constant $\alpha = 1/n$. This index is density independent and permits the determination of whether prey items were ingested in higher or lower proportion relative to what would be expected owing to their relative biomass in the field (Pearre, 1982). Chesson's α first was calculated for each tow and then averaged by sampling location and by day and night.

To analyse relationships between number and length of captured larval fish and *P. noctiluca* diameter (mm), both ephyrae and large medusae were included. Medusae were grouped into 1-mm size classes, and only classes with more than one prey item were used for further analysis.

Results

Hydrography

The vertical structure of the water column was dominated by thermal stratification, with a well-developed thermocline at around 60 m depth; below this depth temperature remained nearly constant at 14°C (Fig. 2). Vertical salinity distribution along the transect showed the presence of the shelf-slope front, determined by the horizontal gradient at 50–200 m depth (Fig. 2). The marked thermocline prevented the front from reaching the surface. A deep-chlorophyll fluorescence maximum was found near 60 m depth below the thermocline. The highest deep-fluorescence values always occurred on the oceanic side of the frontal system.

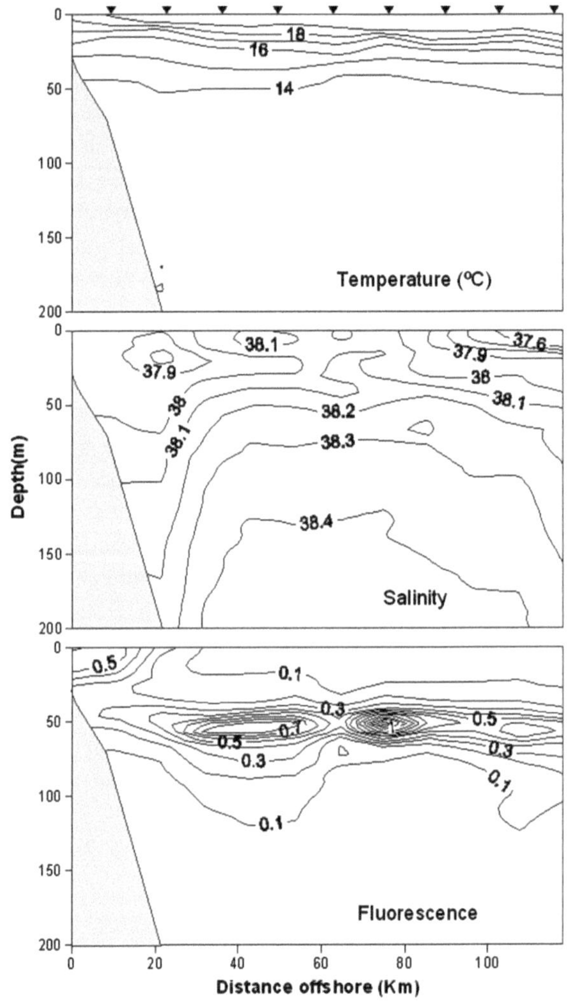

Fig. 2 Vertical distributions of temperature (°C), salinity and fluorescence (arbitrary units) on the coastal–open sea transect in the northwestern Mediterranean Sea during 18–23 June 1995. *Filled triangles* give the locations of hydrographic sampling stations shown in Fig. 1

Pelagic cnidarian distribution

We identified a total of 31 cnidarian species, including 13 siphonophores and 18 medusae (Tables 1, 2). Siphonophores were almost always more abundant than medusae. The highest densities of siphonophores were recorded at the shelf and frontal stations, and medusae were more abundant in the front area.

Among siphonophores, the calycophoran, *Muggiaea atlantica*, predominated at all stations, especially at the shelf and front stations (Fig. 3). The asexual polygastric stage of *M. atlantica* was more abundant on the shelf, while the sexual eudoxid stage peaked at the front area. *Chelophyes appendiculata*, *Lensia conoidea*, *Lensia subtilis* and *Abylopsis tetragona* formed the bulk of the remaining species that showed a wide cross-shelf distribution (Table 1). *C. appendiculata* was absent in coastal waters but occurred at the other stations. *L. conoidea* occurred in low numbers at the coastal and shelf stations and was more abundant in the frontal area and in the open sea. *L. subtilis* occurred at all stations in relatively high numbers, with peak abundances over the shelf.

Most of the medusa species occurred in lower densities than the siphonophores (Table 2). Only the ephyrae of *P. noctiluca* were abundant at all stations, except at the coastal station where they were virtually absent. The highest densities of *P. noctiluca* were observed at the front, but they also were abundant over the shelf and in the open sea. *P. noctiluca* abundance did not differ significantly

Table 1 Mean abundance ± SD of siphonophore colonies per 1000 m^{-3} in a coastal–open sea transect in the NW Mediterranean Sea during 18–23 June 1995

Species	Coast	Shelf	Front	Open sea
Physonectae				
Agalma elegans (Sars, 1846) Fewkes, 1880	–	1.6 ± 2.3	1.9 ± 1.5	1.3 ± 0.9
Agalma sp.	4.1 ± 5.7	–	–	–
Nanomia bijuga (delle Chiaje 1841)	2.3 ± 3.3	1.7 ± 2.4	1.9 ± 1.4	2.5 ± 0.8
Physophora hydrostatica Forskål 1775	–	–	–	0.7 ± 1.0
Calycophorae				
Muggiaea atlantica Cunningham 1892	1797 ± 270	8634 ± 2750	3304 ± 923.3	689.6 ± 160.9
M. atlantica eudoxids	211 ± 160	1728 ± 45.5	3070 ± 1844	248.4 ± 30.9
Lensia subtilis (Chun 1886)	48.5 ± 37	1694 ± 411.2	291 ± 181	298.3 ± 46.5
Lensia conoidea (Keferstein and Ehlers 1860)	5.4 ± 7.7	6.6 ± 9.3	74.8 ± 33.1	59.7 ± 37.4
Abylopsis tetragona (Otto 1823)	2.3 ± 3.3	56.1 ± 29.3	6.5 ± 4.6	37.0 ± 12.9
A. tetragona eudoxids	7.7 ± 6.7	88.9 ± 29.6	54.4 ± 28.1	48.2 ± 10.7
Eudoxoides spiralis (Bigelow 1911)	37 ± 3.7	–	0.7 ± 1.1	9.1 ± 1.1
E. spiralis eudoxids	–	1.6 ± 2.36	–	9.1 ± 10.7
Muggiaea kochi (Will 1844)	–	33.0 ± 28.8	14.6 ± 19.04	16.9 ± 7.3
M. kochi eudoxids	159.6 ± 175.1	20.1 ± 25.2	14.6 ± 20.6	–
Chelophyes appendiculata (Eschscholtz 1829)	–	238.7 ± 126	71.2 ± 8.7	166.4 ± 83.2
Lensia meteori (Leloup 1934)	–	–	10.0 ± 7.1	223.1 ± 13.8
Hippopodius hippopus (Forskål 1776)	–	–	1.1 ± 1.6	0.5 ± 0.8
Unidentified eudoxids	–	366 ± 199	323.1 ± 176.3	332.5 ± 193.6

Table 2 Mean abundances of medusae (individuals 1000 m^{-3} ± standard deviation) in a coastal–open sea transect in the NW Mediterranean Sea during 18–23 June 1995

Species	Coast	Shelf	Front	Open sea
Podocoryne sp.	4.05 ± 7	–	–	–
Leptomedusa	4.1 ± 7.1	–	–	–
Eutima gegenbauri (Haeckel 1864)	2.7 ± 4.7	–	–	–
Leuckartiara octona (Fleming 1823)	2.3 ± 4	–	–	–
Obelia sp.	2.7 ± 4.7	–	1.1 ± 2.0	–
Aglaura hemistoma Perón & Lesueur 1810	2.3 ± 4	112.9 ± 18.8	59.2 ± 39.5	15.1 ± 13.2
Rhopalonema velatum Gegenbaur 1857	13.2 ± 17.3	377.2 ± 288.7	126.9 ± 43.7	99.3 ± 27.7
Solmundella bitentaculata (Quoy & Gaimard 1833)	9.6 ± 10.4	111.1 ± 145.5	68.1 ± 21.3	6.7 ± 5.9
Pelagia noctiluca (Forskål 1775)	–	469.5 ± 457.3	5012.3 ± 7216.8	646.3 ± 459.2
Pandea conica (Quoy & Gaimard 1827)	–	3.3 ± 5.8	4.6 ± 8.0	–
Helgicirrha schulzei Hartlaub 1909	–	1.6 ± 2.9	–	–
Nausithoe punctata Kölliker 1853	–	4.8 ± 4	–	–
Clytia sp.	–	–	8.1 ± 14.0	–
Persa incolorata McCrady 1859	–	–	20.4 ± 19.3	–
Liriope tetraphylla (Chamisso & Eysenhradt 1821)	–	–	1.9 ± 1.7	–
Solmissus flavescens (Kölliker 1853)	–	–	11.6 ± 10.7	–
Rhabdoon singulare Keferstein & Ehlers 1861	–	–	1.5 ± 2.7	–
Velella velella (Linnaeus 1758)	–	–	–	0.6 ± 1.0

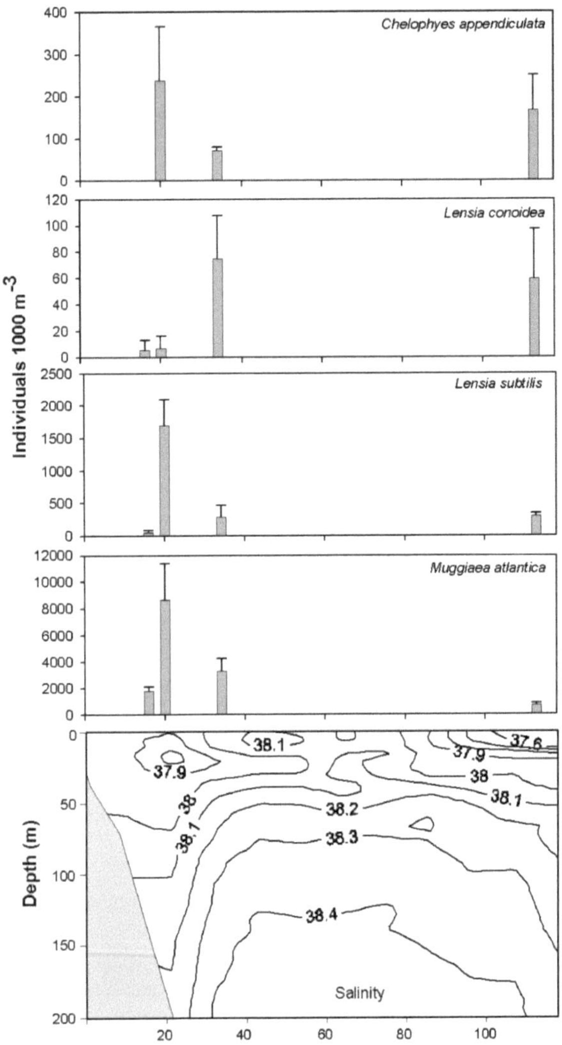

Fig. 3 Salinity distribution (*lower panel*) and abundances (means ± standard deviation) of the siphonophores *Chelophyes appendiculata*, *Lensia conoidea*, *Lensia subtilis* and *Muggiaea atlantica* at the sampling locations on the coastal–open sea transect in the northwestern Mediterranean Sea during 18–23 June 1995

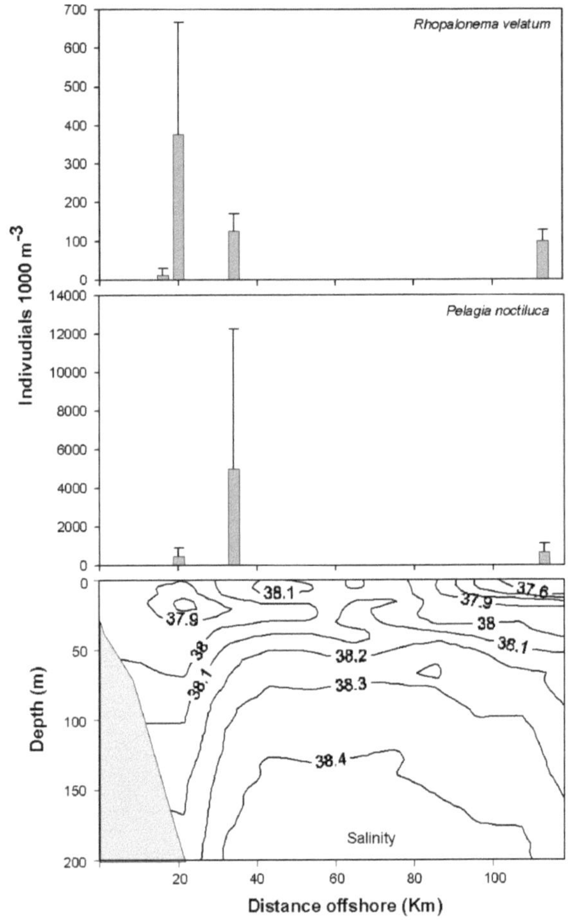

Fig. 4 Salinity distribution (*lower panel*) and abundances (means ± standard deviation) of the medusae *Rhopalonema velatum* and *Pelagia noctiluca* at sampling locations on the coastal–open sea transect in the northwestern Mediterranean Sea during 18–23 June 1995

between day and night tows ($F_{1,8} = 1.22$, $P = 0.31$). *Rhopalonema velatum* was very scarce in the coastal area and showed the highest concentrations over the shelf, with their abundance again decreasing offshore (Fig. 4).

Diet of *Pelagia noctiluca*

A total of 5680 *P. noctiluca* ephyrae were analysed for gastric pouch contents (Table 3). There were significant differences in feeding incidence among the three sampled stations, ranging from 21% over the shelf to 7% in the front ($F_{2,8} = 6.19$, $P = 0.03$). No significant differences in feeding incidence were detected between day (14.0%) and night (14.7%) ($F_{2,8} = 0.01$, $P = 0.92$). The gastric pouches of *P. noctiluca* contained a wide variety of prey, mainly copepods, followed by juvenile euphausiids and fish larvae (Table 3).

Copepods were the most abundant zooplankton group at all sampled stations, with mean densities ranging between 48 and 159 individuals m^{-3} in the shelf and front areas, respectively (Table 4). The next most abundant crustaceans included decapod larvae and juvenile euphausiids. Appendicularians and

Table 3 Number of *Pelagia noctiluca* examined, feeding incidence (percentage of individuals containing prey), diet (percentage number of each prey type) and total number of prey of *Pelagia noctiluca* ephyrae in a coastal–open sea transect in the NW Mediterranean Sea during 18–23 June 1995

	Shelf	Front	Open sea
No. of *P. noctiluca*	145	4400	1135
Feeding incidence (%)	21	7	10
Diet (%)			
Copepoda	72	49	57
Mollusc larvae	8	3	5
Siphonophora	4	1	2
Chaetognatha	4	6	2
Ostracoda	0	1	2
Euphausiacea	0	20	7
Decapoda larvae	0	2	4
Appendicularia	0	4	12
Cladocera	4	3	2
Fish larvae	8	12	6
Total no. of prey	25	224	84

Table 4 Mean abundance of the major zooplankton taxa (individuals m^{-3} ± standard deviation) in a coastal–open sea transect in the NW Mediterranean during 18–23 June 1995

	Shelf	Front	Open sea
Copepoda	48.1 ± 20.9	159.4 ± 226.1	68.4 ± 88.7
Mollusc larvae	0.5 ± 0.2	0.7 ± 0.9	0.6 ± 0.9
Siphonophora	13.0 ± 3.1	7.3 ± 3.4	2.2 ± 0.4
Medusae	1.1 ± 0.2	5.3 ± 7.3	0.8 ± 0.5
Chaetognatha	0.2 ± 0.1	0.7 ± 1.0	0.5 ± 0.6
Ostracoda	0.9 ± 0.9	3.6 ± 4.6	3.9 ± 5.4
Euphasiacea	5.9 ± 1.2	26.4 ± 40.4	7.9 ± 10.2
Decapoda larvae	7.9 ± 3.6	38.0 ± 59.7	2.7 ± 3.5
Appendicularia	5.7 ± 1.8	6.8 ± 8.8	7.3 ± 10.2
Cladocera	5.7 ± 6.6	0.7 ± 0.3	0.2 ± 0.3
Fish larvae	1.0 ± 0.2	0.7 ± 0.1	0.6 ± 0.2

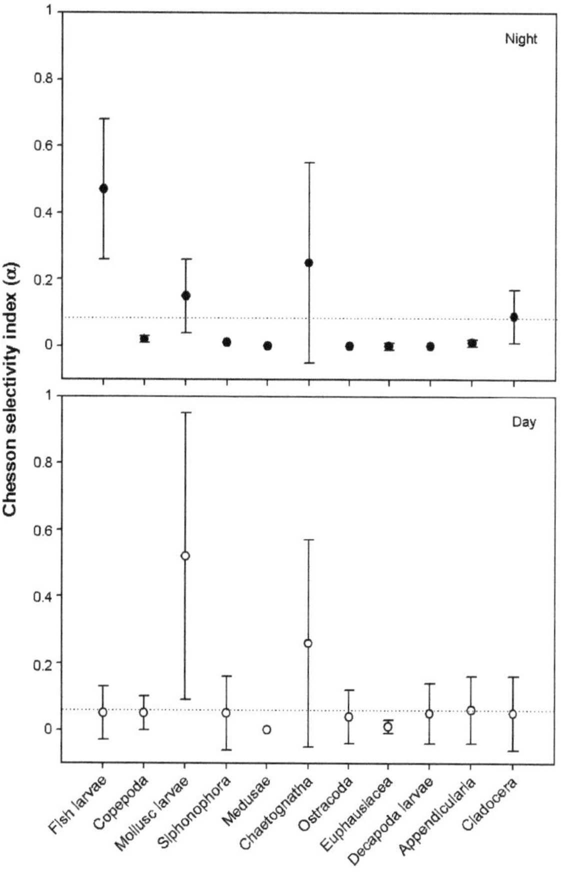

Fig. 5 Mean Chesson's α values (± standard deviation) for the most common prey items in *Pelagia noctiluca* during day and night in the northwestern Mediterranean Sea during 18–23 June 1995. Values above the *dashed line* indicate positive selection

siphonophores were also abundant; yet fish larvae, which were important in the ephyra diet, were clearly a minor group (Table 4). Most zooplankton taxa had highest densities in the front area, where most of the medusae and siphonophores species also occurred.

Selection by *P. noctiluca* ephyrae for certain prey taxa was evident when the gut content data were compared to the potential prey abundance data. At all stations with ephyrae, positive selection was observed for chaetognaths, mollusc larvae and fish larvae. Positive selection was shown for chaetognaths and mollusc larvae in both day and night, but fish larvae were positively selected only at night (Fig. 5). In contrast, selection for copepods was negative or neutral, in spite of the fact that they were the main group of ingested prey (Table 3).

Most species of fish larvae ingested by *P. noctiluca* belonged to the Family Myctophidae, but anchovy, *Engraulis encrasicolus* and Sparidae larvae also were found in the gastric pouches (Table 5). Higher diversity of fish larva prey occurred at the front station, and the species captured most often was *E. encrasicolus* (Table 5).

There was a significant positive correlation ($R^2 = 0.62$, $P < 0.05$; Fig. 6) between mean log

Table 5 Specific composition and relative frequency (% based on total number) of fish larvae found in stomachs of *Pelagia noctiluca* ephyrae during 18–23 June 1995

	Shelf	Front	Open sea
Ceratoscopelus maderensis (Lowe, 1839)	–	21	–
Hygophum benoiti (Cocco, 1838)	50	4	20
Lampanyctus crocodilus (Risso, 1810)	50	8	60
Lampanyctus pusillus (Johnson, 1810)	–	–	20
Myctophum punctatum Rafinesque, 1810	–	4	–
Vinciguerria sp.	–	4	–
Engraulis encrasicolus (Linnaeus, 1758)	–	38	–
Sparidae	–	4	–
Unidentified	–	17	–
Total number	2	26	5

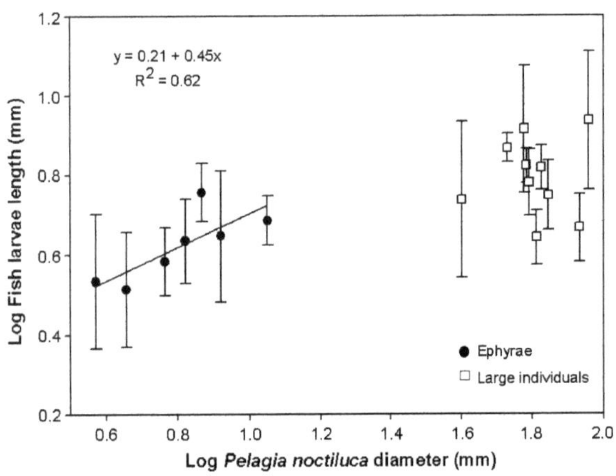

Fig. 6 Mean (± standard deviation) of log-transformed larval fish length (mm) in relation to log mean *Pelagia noctiluca* diameter (mm) in the north western Mediterranean Sea during 18–23 June 1995. *Black circles*: ephyrae (<12 mm diameter); *open squares*: larger individuals; *solid lines*: regressions

standard length of fish larvae ingested by *P. noctiluca* and the mean log diameter of *P. noctiluca* ephyrae (4–12 mm); however, there was no positive correlation between the length of ingested larvae and the diameter of larger *P. noctiluca* (40–90 mm). Nevertheless, the mean number of prey in the gastric pouches was greater for larger medusae (4.96 ± 2.64 SD) than for ephyrae (1.03 ± 0.16 SD), and these differences were statistically significant (Student's t test = −7.923, $P < 0.01$).

Discussion

The distribution of gelatinous zooplankton along the coastal to open sea transect was related to the presence of the shelf-slope front. The pelagic cnidarian community included representatives of all major taxonomic groups and was predominated by the siphonophore, *M. atlantica*. Coastal waters were populated mainly by meroplanktonic species (antho- and leptomedusae) whereas shelf and "offshore" water had more holoplanktonic species (trachy- and narcomedusae).This spatial pattern, characterized by increasing species diversity and abundance with distance from the coast, is a common trend in the northwestern Mediterranean (Gili et al., 1987a; Bouillon et al., 2004). The abundance of several species (Tables 1, 2) agrees with previous studies showing a high abundance of gelatinous carnivores in late spring or early summer in comparison with the rest of the year in the NW Mediterranean (Gili et al., 1987b).

During the last two decades, the neritic siphonophore, *M. atlantica*, and the oceanic scyphomedusan, *P. noctiluca*, have been reported as the most abundant gelatinous predators in the region (Razouls & Thiriot, 1968; Andersen et al., 2001). Moreover, Mackie et al. (1987) reported that *M. atlantica* is the predominant siphonophore species in temperate latitudes of the three major oceans. This species has replaced the previously predominant siphonophore, *Muggiaea kochi*, in the Mediterranean Sea (Riera et al., 1986; Bastić et al., 2007). *P. noctiluca* is the most conspicuous jellyfish in shelf waters of the western Mediterranean (Gili & Pagès, 2005) and in adjacent areas (Zavodnik, 1987). The high densities of most of the gelatinous species observed in this study are in accordance with the high productivity and hydrographic structure in the Mediterranean at the end of spring and early summer (Sabatés et al., 1989). Previous studies in the area corroborate the abundance of neritic species such as *L. subtilis*, *M. atlantica*, *Aglaura hemistoma* and *R. velatum* (Gili et al., 1987b, 1988).

High concentrations of coastal and offshore species of siphonophores and medusae (including hydromedusae) were found close to the shelf/slope front. These high concentrations could be related to increased primary and secondary production in the frontal area (Estrada et al., 1993; Sabatés et al.,

2004), as well as to the convergence associated with the front (Franks, 1992). Sabatés (1990) found high concentrations of shelf and oceanic fish larvae in the frontal area; however, this pattern was subject to considerable spatio-temporal variability, probably due to frontal mesoscale activity (Sabatés et al., 2004). Aggregations of gelatinous organisms around physical discontinuities have been reported in different geographical areas (e.g. Pagès et al., 2001; Lučić et al., 2005). The abundance of various types of gelatinous zooplankton at the front could be explained by interactions between physical and biological mechanisms occurring at local scales (Graham et al., 2001). Coastal hydromedusae would be concentrated at the front as a result of advection because those species do not produce new medusae in open waters far from shore. Siphonophores are holoplanktonic and able to proliferate in open waters at the productive front, thereby establishing large populations where prey are abundant (Purcell, 1982; Graham et al., 2001). In this study, the high concentration of gelatinous predators in the front area could have important trophic consequences for the planktonic communities. Surprisingly, although the abundance of zooplankton prey was highest in the frontal area, the feeding incidence of *P. noctiluca* was lowest there. This observation could be related to the high numbers of *P. noctiluca* and other planktonic predators in the front waters, which could result in low individual food availability. In the same area, Saiz et al. (1999) observed that copepod growth and abundance were uncoupled from phytoplankton abundance, and suggested that predation, rather than advection or physical retention, could explain the copepod distribution and production in the area.

Previous studies on *P. noctiluca* demonstrated that this species is an opportunistic predator that consumes a wide variety of prey (e.g. Malej, 1989). In our study, the diet of *P. noctiluca* ephyrae consisted mainly of copepods, but other prey such as euphausiids, chaetognaths and fish larvae also were eaten. Predation on fish larvae was particularly important in the frontal area, where larvae of both shelf and oceanic species occurred (Sabatés et al., 2007a). Larvae of commercially important fish species, such as anchovy, are often removed in high proportions relative to the other taxa (Purcell, 1989; Purcell et al., 1994). In the Kiel Fjord, the calculated daily predation rate of *Aurelia aurita* medusae was as high as 5% of the herring larvae (Möller, 1984). We were unable to estimate consumption rates of fish larvae in the NW Mediterranean because we lacked the length of time required by medusae to digest fish larvae; nevertheless, consumption of fish larvae clearly is important because they represented up to 12% of total prey captured by *P. noctiluca*.

Very few studies have examined natural prey consumed by ephyrae (Table 6). More microzooplankton prey (ciliates, rotifers and copepod nauplii) were consumed by ephyrae of *Chrysaora quinquecirrha* (Desor) and *A. aurita* (Linneaus) than by *P. noctiluca*, perhaps due to their small sizes relative to *P. noctiluca*. The prey available to the three species also differed because *P. noctiluca* ephyrae were in the open Mediterranean Sea but the others were in shallow estuaries. Even ephyrae only a few millimetres in diameter consumed large prey, including other gelatinous taxa (siphonophores, ctenophores, and hydromedusae) and larval fish.

Day–night variations in feeding could be related to the diel vertical migrations of *P. noctiluca* and their prey. In deep waters, *P. noctiluca* is distributed mostly between the surface and 150 m depth, but during the day it can also be found deeper, between 300 and 500 m (Mariottini et al., 2008). It is not known if ontogenetic changes in depth distribution occur in *P. noctiluca*. In the study area, at the end of spring and beginning of summer, chaetognaths and mollusc larvae are very abundant and concentrate between 60 and 150 m depth (Vives, 1966) and near the bottom in coastal waters (Martin et al., 1997). Fish larvae are mainly distributed in the surface layers above the thermocline (Olivar & Sabatés, 1997; Sabatés, 2004). Anchovy larvae, the most abundant species during spring–summer, are located around 50 m depth during the day and migrate to the surface at night (Olivar et al., 2001; Sabatés et al., 2008). Thus, fish larvae and *P. noctiluca* co-occur in surface waters at night, and our data indicate that prey selection was positive then. Positive selection for ichthyoplankton by other jellyfish species has been reported in other areas as well (Purcell, 1985, 1989; Purcell et al., 1994; Purcell & Sturdevant, 2001).

In this study, the size of ingested larval fish prey increased as ephyra diameter increased, although in the larger medusae, prey size was almost constant in the size range examined. This may be because the abundance of fish larvae in the field as well as their

Table 6 Numbers of ephyrae examined, ephyra diameter, in situ feeding incidence (percentage of individuals containing prey), diets (percentage number of each prey type), and total numbers of prey counted in ephyrae

	Pelagia noctiluca	*Chrysaora quinquecirrha*[a]	*Aurelia aurita*[b]
No. of ephyrae	5680	31	360
Ephyra diameter (mm)	4.0–12.0	3.8 ± 2.9	6.5–11.8
Feeding incidence (%)	8	50	0.2–0.6
	Diet (%)		
Ciliates	NQ	48.4	NQ
Rotifers	NQ	22.4	87.5 +
Copepod nauplii	–	5.0	3.7 −
Copepoda	59.2 −	20.4	4.2 −
Mollusc larvae	4.3 +	0.9	NA
Gelatinous species	Siphonophores 3.7	Ctenophores NQ	Hydromedusae 4.1 +
Chaetognatha	4.7 +	NA	NA
Ostracoda	0.4	NA	NA
Euphausiacea	8.3	NA	NA
Decapoda larvae	0.6	NA	NA
Appendicularia	2.7	NA	NA
Cladocera	2.5	NA	NA
Fish larvae	13.7 +	2.0	NA
Total no. of prey	333	52	1387

NA not available, *NQ* not quantified; + = positive selection; − = negative selection

[a] Olesen et al. (1996)

[b] Sullivan et al. (1997)

susceptibility to predation decline with size (Cowan & Houde, 1992). The maximum size of most fish larvae in the plankton in that area and for the same period is ~10 mm (Sabatés et al., 2007b); therefore, large larvae were very scarce. In addition, digestion is more rapid for small than for large fish larvae, so small larvae would not be as easy to recognize in gut contents as would large larvae. Thus, availability, susceptibility and digestion would explain why large individuals of *P. noctiluca* contained almost a constant larval size range. Similarly, Cowan & Houde (1992) reported size-dependent predation rates on anchovy larvae by *C. quinquecirrha* medusae in controlled experiments. Unlike the larval size, the number of larvae in *P. noctiluca* gastric pouches increased with jellyfish size, presumably because larger medusae can catch a wider range of larval sizes and can effectively fish a larger volume of water than small ephyrae (Madin, 1988).

Our results not only indicate high predation on fish larvae by *P. noctiluca* medusae, which occur in high densities over wide reaches of the oceans (e.g. Doyle et al., 2008), but also demonstrate the potential for competition for zooplankton prey between zooplanktivorous fish (e.g. anchovy) and *P. noctiluca* and other abundant planktonic cnidarians. Most pelagic cnidarians consume mostly copepods (Purcell, 1981, 1997; Purcell & Mills, 1988), and diet overlap between medusae and fishes have been documented (Purcell & Grover, 1990; Purcell & Sturdevant, 2001; Brodeur et al., 2008). Zooplankton predators and prey were concentrated in the front area, but fewer prey items were found in *P. noctiluca*, which suggests that prey might be limiting at the front.

In summary, this study has highlighted how different communities/species of gelatinous zooplankton are found in different habitat types at sea. The highest zooplankton abundances occurred at the frontal area, where some species are concentrated. The diet of the most abundant medusa, *P. noctiluca*, consisted mainly of copepods, but almost 12% of the prey items were fish larvae. Thus, this species may exert important trophic effects on plankton communities, including fish larvae, considering its widespread occurrence. The abundant and diverse pelagic cnidarian fauna of the Mediterranean Sea is of great importance as both predators and competitors of fish populations.

Acknowledgements We greatly appreciate the assistance of the crew of the R/V García del Cid and all the participants during the "FRONTS" cruise. We thank Dr. T. K. Doyle for constructive comments on an earlier version of the manuscript. This study was supported by the projects AMB94-0853 and REN 2002-01339/MAR.

References

Alcaraz, M., E. Saiz & M. Estrada, 1994. Excretion of ammonia by zooplankton and its potential contribution to nitrogen requirements for primary production in the Catalan Sea (NW Mediterranean). Marine Biology 119: 69–76.

Andersen, V., P. Nival, P. Caparroy & A. Gubanova, 2001. Zooplankton community during the transition from spring bloom to oligotrophy in the open NW Mediterranean and effects of wind events. 1. Abundance and specific composition. Journal of Plankton Research 23: 227–242.

Bastić, M., N. Jasprica, M. Carić & D. Lučić, 2007. Annual cycle of the gelatinous invertebrate zooplankton of the eastern South Adriatic coast (NE Mediterranean). Journal of Plankton Research 29: 671–686.

Behrends, G. & G. Schneider, 1995. Impact of *Aurelia aurita* medusae (Cnidaria, Scyphozoa) on the standing stock and community composition of mesozooplankton in the Kiel Bight (western Baltic Sea). Marine Ecology Progress Series 127: 39–45.

Bouillon, J., M. D. Medel, F. Pagès, J. M. Gili, F. Boero & C. Gravili, 2004. Fauna of the Mediterranean Hydrozoa. Scientia Marina 68(Suppl 2): 1–438.

Brodeur, R. D., C. L. Suchman, D. C. Reese, T. W. Miller & E. A. Daly, 2008. Spatial overlap and trophic interactions between pelagic fish and large jellyfish in the northern California Current. Marine Biology 154: 649–659.

Chesson, J., 1978. Measuring preference in selective predation. Ecology 59: 211–215.

Cowan, J. H. & E. D. Houde, 1992. Size-dependent predation on marine fish larvae by ctenophores, scyphomedusae and planktivorous fish. Fisheries Oceanography 1: 113–126.

Dallot, S., J. Goy & C. Carré, 1988. Peuplements de carnivores planctoniques gélatineux et structures productives en Méditerranée occidentale. Oceanologica Acta 10: 193–209.

Doyle, T. K., H. de Hass, D. Cotton, B. Dorshel, V. Cummins, J. D. R. Houghton, J. Davenport & G. C. Hays, 2008. Widespread occurrence of the jellyfish *Pelagia noctiluca* in Irish coastal and shelf waters. Journal of Plankton Research 30: 963–968.

Estrada, M. & R. Margalef, 1988. Supply of nutrients to the Mediterranean photic zone across a persistent front. In Minas, H. J. & P. Nival (eds), Pelagic Mediterranean Oceanography. Oceanologica Acta 9: 133–142.

Estrada, M., C. Marrasé, M. Latasa, E. Berdalet, M. Delgado & T. Riera, 1993. Variability of deep chlorophyll maximum characteristics in the Northwestern Mediterranean. Marine Ecology Progress Series 92: 289–300.

Estrada, M., R. A. Varela, J. Salat, A. Cruzado & E. Arias, 1999. Spatio-temporal variability of the winter phytoplankton distribution across the Catalan and North Balearic fronts (NW Mediterranean). Journal of Plankton Research 21: 1–20.

Fancett, M. S., 1988. Diet and prey selectivity of scyphomedusae from Port Phillip Bay, Australia. Marine Biology 98: 503–509.

Font, J., J. Salat & J. Tintoré, 1988. Permanent features of the circulation of the Catalan Sea. In Minas, H. J. & P. Nival (eds), Pelagic Mediterranean Oceanography. Oceanologica Acta 9: 51–57.

Franks, P. J. S., 1992. Sink or swim: accumulation of biomass at fronts. Marine Ecology Progress Series 82: 1–12.

García-Ladona, E., J. Tintoré, J. M. Pinot, J. Font & M. Manríquez, 1994. Surface circulations and dynamics of the Balearic Sea. In La Violette, P. E. (ed.), Seasonal and Interannual Variability of the Western Mediterranean Sea. AGU, Washington. Coastal and Estuarine Studies 46: 73–91.

Gili, J. M. & F. Pagès, 2005. Les proliferacions de meduses. Bolletí de la Societat d'Història Natural de les Balears 48: 9–22.

Gili, J. M., F. Pagès & F. Vives, 1987a. Distribution and ecology of a population of planktonic cnidarians in the western Mediterranean. In Bouillon, J., F. Boero, F. Cicogna & P. F. S. Cornelius (eds), Modern Trends in the Systematics. Ecology and Evolution of Hydroids and Hydromedusae. Oxford University Press, Oxford: 157–170.

Gili, J. M., F. Pagès & T. Riera, 1987b. Distribución de las especies más frecuentes de sifonóforos calicóforos en la zona norte del Mediterráneo occidental. Investigación Pesquera 51: 323–338.

Gili, J. M., F. Pagès, A. Sabatés & J. D. Ros, 1988. Small-scale distribution of a cnidarian population in the western Mediterranean. Journal of Plankton Research 10: 385–401.

Giorgi, R., M. Avian, S. De Olazabal, L. Rottini-Sandrini, 1991. Feeding of *Pelagia noctiluca* in open sea. Jellyfish Blooms in the Mediterranean. Proceedings of II Workshop on Jellyfish in the Mediterranean Sea. UNEP, Mediterranean Action Plan, Technical Reports Series 47: 102–111

Graham, W. M., F. Pagès & W. M. Hamner, 2001. A physical context for gelatinous zooplankton aggregations: a review. Hydrobiologia 451: 199–212.

Larson, R. J., 1987. Respiration and carbon turnover rates of medusae from the NE Pacific. Comparative Biochemistry and Physiology 87A: 93–100.

Lučić, D., A. Benović, M. Batistić, J. Njire & V. Onofri, 2005. Calycophorae (Siphonophora) in the open waters of the central and southern Adriatic Sea during spring 2002. Journal of the Marine Biological Association of the United Kingdom 85: 495–501.

Mackie, G. O., P. R. Pugh & J. E. Purcell, 1987. Siphonophore Biology. Advances in Marine Biology 24: 97–262.

Madin, L. P., 1988. Feeding behaviour of tentaculate predators: in situ observation and a conceptual model. Bulletin of Marine Science 43: 413–429.

Malej, A., 1989. Behaviour and trophic ecology of the jellyfish *Pelagia noctiluca* (Forsskål, 1775). Journal of Experimental Marine Biology and Ecology 126: 259–270.

Malej, A., J. Faganeli & J. Pezdič, 1993. Stable isotope and biochemical fractionation in the marine pelagic food chain: the jellyfish *Pelagia noctiluca* and net zooplankton. Marine Biology 116: 565–570.

Mariottini, G. L., E. Giacco & L. Pane, 2008. The mauve stinger *Pelagia noctiluca* (Forsskål, 1775). Distribution, ecology, toxicity and epidemiology of stings. A review. Marine Drugs 6: 496–513.

Martin, D., M. Claret, S. Pinedo & R. Sardá, 1997. Vertical and spatial distribution of the near-shore littoral meroplankton off the Bay of Blanes (NW Mediterranean Sea). Journal of Plankton Research 19: 2079–2089.

Masó, M., A. Sabatés & M. P. Olivar, 1998. Short-term physical and biological variability in the shelf-slope region of the NW Mediterranean during the spring transition period. Continental Shelf Research 18: 661–675.

Mills, C. E., 2001. Jellyfish blooms: are populations increasing globally in response to changing ocean conditions? Hydrobiologia 451: 55–68.

Molinero, J. C., F. Ibanez & P. Nival, 2005. North Atlantic climate and northwestern Mediterranean plankton variability. Limnology and Oceanography 50: 1213–1220.

Möller, H., 1980. Scyphomedusae as predators and food competitors of larval fish. Meeresforschung 28: 90–100.

Möller, H., 1984. Reduction of a larval herring population by jellyfish predator. Science 224: 621–622.

Motoda, S., 1959. Devices of simple plankton apparatus. Memoires of the Faculty of Fisheries, Hokkaido University 7: 73–94.

Olesen, N. J., J. E. Purcell & D. K. Stoecker, 1996. Feeding and growth by ephyrae of scyphomedusae *Chrysaora quinquecirrha*. Marine Ecology Progress Series 137: 149–159.

Olivar, M. P. & A. Sabatés, 1997. Vertical distribution of fish larvae in the north-west Mediterranean Sea in spring. Marine Biology 129: 289–300.

Olivar, M. P., J. Salat & I. Palomera, 2001. Comparative study of spatial distribution patterns of the early stages of anchovy and pilchard in the NW Mediterranean Sea. Marine Ecology Progress Series 217: 111–120.

Pagès, F., H. E. González, M. Ramón, M. Sobrazo & J. M. Gili, 2001. Gelatinous zooplankton assemblages associated with water masses in the Humbolt Current System, and potential predatory impact by *Bassia bassensis* (Siphonophora: Calycophorae). Marine Ecology Progress Series 210: 13–24.

Pearre, S., 1982. Estimating prey preference by predators: uses of various indices, and a proposal of another based on X^2. Canadian Journal of Fisheries and Aquatic Sciences 39: 914–923.

Purcell, J. E., 1981. Dietary composition and diel feeding patterns of epipelagic siphonophores. Marine Biology 65: 83–90.

Purcell, J. E., 1982. Feeding and growth in the siphonophore *Muggiaea atlantica*. Journal of Experimental Marine Biology and Ecology 62: 39–54.

Purcell, J. E., 1985. Predation on fish eggs and larvae by pelagic cnidarians and ctenophores. Bulletin of Marine Science 37: 739–755.

Purcell, J. E., 1989. Predation on fish larvae and eggs by the hydromedusa *Aequorea victoria* at a herring spawning ground in British Columbia. Canadian Journal of Fisheries and Aquatic Sciences 46: 1415–1427.

Purcell, J. E., 1990. Soft-bodied zooplankton predators and competitors of larval herring (*Clupea harengus pallasi*) at herring spawning grounds in British Columbia. Canadian Journal of Fisheries and Aquatic Sciences 47: 505–515.

Purcell, J. E., 1997. Pelagic cnidarians and ctenophores as predators: selective predation, feeding rates, and effects on prey populations. Annales de l'Institut océanographique, Paris 73: 125–137.

Purcell, J. E. & M. N. Arai, 2001. Interactions of pelagic cnidarians and ctenophores with fish: a review. Hydrobiologia 451: 27–44.

Purcell, J. E. & J. J. Grover, 1990. Predation and food limitation as causes of mortality in larval herring at a spawning ground in British Columbia. Marine Ecology Progress Series 59: 55–61.

Purcell, J. E. & C. E. Mills, 1988. The correlation of nematocyst types to diets in pelagic Hydrozoa. In Hessinger, D. A. & H. M. Lenhoff (eds), The Biology of Nematocysts. Academic Press, San Diego: 463–485.

Purcell, J. E. & M. V. Sturdevant, 2001. Prey selection and dietary overlap among zooplanktivorous jellyfish and juvenile fishes in Prince William Sound, Alaska. Marine Ecology Progress Series 210: 67–83.

Purcell, J. E., D. A. Nemazie, S. E. Dorsey, E. D. Houde & J. C. Gamble, 1994. Predation mortality of bay anchovy *Anchoa mitchilli* eggs and larvae due to schyphomedusae and ctenophores in Chesapeake Bay. Marine Ecology Progress Series 114: 47–58.

Purcell, J. E., S.-I. Uye & W.-T. Lo, 2007. Anthropogenic causes of jellyfish blooms and direct consequences for humans: a review. Marine Ecology Progress Series 350: 153–174.

Razouls, S. & A. Thiriot, 1968. Le macroplancton de la région de Banyuls-sur-mer (Golfe du Lion). Vie Milieu 19: 133–184.

Riera, T., J. M. Gili & F. Pagès, 1986. Estudio cuantitativo y estacional de dos poblaciones de cnidarios planctonicos frente a las costas de Barcelona (Mediterráneo occidental): Ciclos entre 1966–67; 1982–83. Miscelánea Zoológica 10: 23–32.

Rottini Sandrini, L. & M. Avian, 1989. Feeding mechanism of *Pelagia noctiluca* (Scyphozoa: Semaeostomeae); laboratory and open sea observations. Marine Biology 102: 49–55.

Sabatés, A., 1990. Distribution pattern of larval fish populations in the northwestern Mediterranean. Marine Ecology Progress Series 59: 75–82.

Sabatés, A., 2004. Diel variability of fish larvae distribution during the winter mixing period in the NW Mediterranean. ICES Journal of Marine Science 61: 1243–1252.

Sabatés, A., J. M. Gili & F. Pagès, 1989. Relationship between zooplankton distribution, geographic characteristics and hydrographic patterns off the Catalan coast (western Mediterranean). Marine Biology 103: 153–159.

Sabatés, A., J. Salat & M. Masó, 2004. Spatial heterogeneity of fish larvae across a meandering current in the northwestern Mediterranean. Deep-Sea Research I 51: 545–557.

Sabatés, A., M. P. Olivar, J. Salat, I. Palomera & F. Alemany, 2007a. Physical and biological processes controlling the

distribution of fish larvae in the NW Mediterranean. Progress in Oceanography 74: 355–376.
Sabatés, A., J. Salat, I. Palomera, M. Emelianov, M. L. Fernández de Puelles & M. P. Olivar, 2007b. Advection of anchovy larvae along the Catalan continental slope (NW Mediterranean). Fisheries Oceanography 16: 130–141.
Sabatés, A., N. Zaragoza, C. Grau & J. Salat, 2008. Vertical distribution of early developmental stages in two coexisting clupeoid species, *Sardinella aurita* and *Engraulis encrasicolus*. Marine Ecology Progress Series 364: 169–180.
Saiz, E., A. Calbet, X. Irigoien & M. Alcaraz, 1999. Copepod egg production in the western Mediterranean: response to food availability in oligotrophic environments. Marine Ecology Progress Series 187: 179–189.
Salat, J., 1996. Review of hydrographic environmental factors that may influence anchovy habitats in northwestern Mediterranean. Scientia Marina 60(Suppl. 2): 21–32.
Sullivan, B. K., C. L. Suchman & J. H. Costello, 1997. Mechanics of prey selection by ephyrae of the scyphomedusa *Aurelia aurita*. Marine Biology 130: 213–222.
Totton, A. K., 1965. A Synopsis of the Siphonophorae. British Museum Natural History, London.
Vives, F., 1966. Zooplancton nerítico de las aguas de Castellón (Mediterráneo occidental). Investigación Pesquera 30: 49–166.
Zavodnik, D., 1987. Spatial aggregations of the swarming jellyfish *Pelagia noctiluca* (Scyphozoa). Marine Biology 94: 265–269.
Zavodnik, D., 1991. On the food and feeding in the northern Adriatic of *Pelagia noctiluca* (Scyphozoa). Jellyfish Blooms in the Mediterranean. Proceedings of II Workshop on Jellyfish in the Mediterranean Sea. UNEP, Mediterranean Action Plan, Technical Reports Series 47: 212–216.

Hydrobiologia (2010) 645:167–178
DOI 10.1007/s10750-010-0219-6

Bioenergetics and growth in the ctenophore *Pleurobrachia pileus*

Lene Friis Møller · Javiera Martinez Canon · Peter Tiselius

Published online: 21 March 2010

Abstract Knowledge of how energetic parameters relate to fluctuating factors in the natural habitat is necessary when evaluating the role of gelatinous zooplankton in the carbon flow of coastal waters. In laboratory experiments, we assessed feeding, respiration and growth of the ctenophore, *Pleurobrachia pileus*, and constructed carbon budgets. Clearance rates (F, l d^{-1}) of laboratory-reared *Acartia tonsa* as prey increased as a function of ctenophore polar length (L, mm) as $F = 0.17L^{1.9}$. For ctenophores larger than about 11 mm, clearance rate was depressed in containers of 30–50 l volume. Clearance rates on field-collected prey were highest on the copepod, *Centropages typicus*, intermediate on the cladoceran, *Evadne nordmanni* and low on the copepods, *Acartia clausi* and *Temora longicornis*. Specific growth rates of 8–10 mm *P. pileus* increased with increasing prey concentrations to a maximum of 0.09 d^{-1} attained at prey carbon densities of 40 and 100 μg C l^{-1} of *Artemia salina* and *A. tonsa*, respectively. Weight-specific respiration rates increased hyperbolically with prey concentration. From experiments in which growth, ingestion and respiration were measured simultaneously, a carbon budget was constructed for individuals growing at maximum rates; from the measured parameters, the assimilation efficiency and net growth efficiency were estimated to be 22 and 37%, respectively. We conclude that the predation rates of *P. pileus* depend on ctenophore size, prey species, prey density and experimental container volume. Because the specific growth rates, respiration, assimilation and net growth efficiencies all were affected by food availability, knowledge of the ambient prey field is critical when evaluating the role of *P. pileus* in the carbon flow in coastal waters.

Keywords Feeding · Growth · *Pleurobrachia pileus* · Respiration · Carbon budget · Container effect

Introduction

In estuarine and coastal waters, ctenophores can reach great abundances. They feed at high rates on zooplankton and ichthyoplankton, and thereby may be detrimental to fish populations (Purcell et al., 2001). Ctenophores are well adapted to utilize favourable food conditions, such as patches of zooplankton, with feeding rates not saturating until at very high food levels (Reeve et al., 1978; Greene et al., 1986). Due to their high fecundity and rapid growth potential (Reeve & Walter, 1976; Reeve et al.,

Guest editors: J. E. Purcell & Dror Angel / Jellyfish Blooms: New Problems and Solutions

L. F. Møller (✉) · J. M. Canon · P. Tiselius
Department of Marine Ecology-Kristineberg, University of Gothenburg, Kristineberg 566, 45034 Fiskebackskil, Sweden
e-mail: lene.friismoller@marecol.gu.se

1978), they can quickly increase their abundance and predation rates and thereby act as key organisms in marine pelagic ecosystems by controlling the density of prey (e.g. Frank, 1986; Mutlu et al., 1994; see also Purcell et al., 2001).

Ctenophores have received much attention during the past decades due to the invasion of *Mnemiopsis leidyi* (A. Agassiz) from North America to the Black Sea in the 1980s (Travis, 1993), where it dramatically affected the ecosystem (Kideys, 2002), and subsequently to the Mediterranean, North and Baltic seas (e.g. Riisgård et al., 2007; Fuentes et al., 2009). In its native habitats, *M. leidyi* also greatly affects the pelagic food web (reviewed by Purcell et al., 2001).

The ctenophore *Pleurobrachia pileus* (Müller) also seasonally reaches high densities in coastal waters, with peak abundances reported for autumn (Fraser, 1970), spring/early summer (Greve, 1971; Kuipers et al., 1990), and winter (Schneider, 1987) depending on the geographical location. *Pleurobrachia* spp. are ambush entangling predators (Greene, 1985) that feed on a variety of prey such as fish larvae (Van der Veer & Sadée, 1984; Purcell, 1990) and mesozooplankton, with copepods predominating in the diets (Frank, 1986; Båmstedt, 1998). Even though several studies have investigated feeding in *Pleurobrachia* spp., their feeding effects on different prey types and their roles in the food web are not well understood. The predation effects of *P. pileus* differ depending on the structure of the prey community and the abundance of the ctenophores themselves. Frank (1986) found *P. pileus* 8–13 mm to prefer large zooplankton (>1 mm) and estimated a daily prey removal of 9–23% in 1 year but only 0.7–2.1% in the next. In addition to the patchiness of predator and prey resulting in high variability in the predation effects, the specific prey types also are important to consider when estimating the effects of the *Pleurobrachia* spp. (e.g. Pavez et al., 2006).

The role of *P. pileus* in the carbon flow in coastal waters cannot be predicted solely from feeding rates. Energetic parameters, such as growth and respiration, and how they relate to factors that fluctuate in the natural habitat also are necessary. The objective of this study was to investigate the bioenergetics of the ctenophore, *P. pileus*, to provide data to assess feeding, respiration, and growth, and to make a carbon budget from experiments in which feeding, growth, and respiration were measured simultaneously.

Materials and methods

Collection and maintenance

Pleurobrachia pileus was collected during April–May 2007 in the outer part of the Gullmar Fjord on the west coast of Sweden (58°15′N, 11°26′E). The animals were gently collected with a 500-μm mesh plankton net from the surface waters, transferred to buckets, and brought to the laboratory where they were kept in 100-l containers (15°C, 32 salinity) until experiments were performed. The ctenophores were fed with the copepod *Acartia tonsa* (Dana) for maintenance, but they were starved 2 day prior to each experiment.

Feeding and growth experiments

Pleurobrachia pileus were incubated for 1–3 days when they were fed different prey types (Exp. #1–6). The prey organisms in the experiments were *A. tonsa* (copepodites and adults) and newly hatched nauplii of the brine shrimp, *Artemia salina* (Linnaeus). Both were cultured at 15°C and 32 salinity (S) and the copepods were fed the cryptophyte *Rhodomonas baltica* (Karsten). For experiments with mixed ambient zooplankton prey, the zooplankton was collected with a 250-μm mesh plankton, maintained with gentle air bubbling, and used within 2 days of collection.

For the feeding experiments, prey organisms were counted in sub-samples, and added to containers with a known volume of filtered sea water (15°C, 32 S) to obtain suspensions of known prey concentrations. The polar length (L, mm) of the ctenophores was measured before they were incubated with their prey for 24 h in darkness. After the incubation, the ctenophores were gently removed and the remaining prey organisms were retained on a 200-μm mesh filter and counted in order to estimate the total number of prey organisms ingested by the ctenophores during the feeding period.

Clearance (F, l d^{-1}) was estimated according to (Møller & Riisgård, 2007):

$$F = \frac{V}{n * t} * \ln\left(\frac{C_0}{C_t}\right)$$

where V = volume of water in the container (l), n = number of ctenophores, t = incubation time (d) and C_0 and C_t = prey abundance (prey l^{-1}) at start and end of incubation, respectively.

Ingestion (I, prey d^{-1}) was calculated by multiplying clearance by the average prey abundance (C_m):

$$I = F * C_m$$

and C_m estimated as

$$C_m = e^{\frac{\ln(C_0 * C_t)}{2}}$$

The conditions for Exp #1–6 are outlined in Table 1. Controls consisted of containers with prey organisms, but no ctenophores. The recovery of prey in these controls was always >95% of the initial abundance. Containers were about 1.5 times higher than wide.

Size-dependent feeding experiment (Exp #1)

The effect of ctenophore size on clearance rates was investigated in a series of experiments with ctenophores ranging from 3.2 to 19 mm in polar length. To reduce the effect of container size, three container sizes were used; animals 3.2–8 mm in polar length were incubated in 30-l containers and larger animals were incubated in 50- or 100-l containers. Three *P. pileus* of similar size were transferred to each container with an initial prey abundance of 20 *A. tonsa* l^{-1}.

Prey selection experiments (Exp #2–4)

Clearance may be used as a measure of prey selectivity provided that prey abundances are below saturation levels. Three clearance experiments were carried out with ambient zooplankton communities as prey on three different days. For each experiment, a stock-solution of wild zooplankton was prepared by the same general procedures as above. Only the predominant zooplankton species were counted and the composition was slightly different in the three experiments; very few other species were observed and in very low numbers.

Exp #2 was designed to test the selectivity of two sizes of *Pleurobrachia pileus*. The ctenophores were 7.3 ± 1.2 mm in Exp #2A and 15.0 ± 0.5 mm in Exp #2B. The prey composition was predominated by the copepods, *Acartia clausi* (Giesbrecht) (C_0 = 21 ind l^{-1}), *Centropages typicus* (Kröyer) (C_0 = 6 ind l^{-1}), and *Temora longicornis* (Müller) (C_0 = 4 ind l^{-1}). There were three replicates for each ctenophore size and three ctenophores in each 30-l container.

Exp #3 was slightly modified from Exp #2. The prey composition was similar with the copepods *A. clausi* (C_0 = 12 ind l^{-1}), *C. typicus* (C_0 = 5 ind l^{-1}), *T. longicornis* (C_0 = 3 ind l^{-1}), but in addition the cladoceran, *Evadne nordmanni* (Lovén) (C_0 = 5 ind l^{-1}). Two ctenophores (13 ± 0.7 mm) were placed in each 30-l container (three replicates) and incubations and calculations performed as above.

Exp #4 tested size-selective predation of a range of ctenophore sizes in a single experiment. Prey consisted of a mix of *A. clausi* (C_0 = 12 ind l^{-1}), *C. typicus*

Table 1 *Pleurobrachia pileus*: experimental conditions for experiments (#1–6)

#	Prey type	C (prey l^{-1})	V (l)	t (d)	L (mm)	n	Exp.
1	At	20	30–100	1	3–19	3	F
2A/B	Ac, Ce, T	21, 6, 4	30	1	7/15	3	F
3	Ac, Ce, T, En	12, 5, 3, 5	30	1	13	2	F
4	Ac, Ce, T	12, 5, 4	50–100	1	5–11	3	F
5	Ar	15, 40, 53, 100	30	3	8	3	G, F, R
6	At	7, 16, 33, 59	30	3	10	3	G, F

Prey type: Ar (*Artemia salina*), At (*Acartia tonsa*), Ac (*Acartia clausi*), T (*Temora longicornis*), Ce (*Centropages typicus*), En (*Evadne nordmanni*). *C*: Initial concentration of the different prey types. *V*: incubation volume; *t*: incubation time; *L*: polar length; *n*: number of ctenophores per container. Experiment (Exp.)—F: feeding; G: growth; R: respiration. Mixed wild zooplankton was used in Exp #2–4; only predominant species are mentioned here. All experiments were made at 15°C and 32 salinity

($C_0 = 5$ ind l^{-1}), and *T. longicornis* ($C_0 = 4$ ind l^{-1}). Three similarly sized ctenophores were placed in each 50-l container (ctenophore lengths 5, 7, 9 mm) or 100-l container (ctenophore length = 11 mm).

Growth experiments (Exp #5–6)

Growth experiments were performed with moderate-sized ctenophores offered two prey types in different concentrations to obtain a spectrum of specific growth rates. The experiments lasted for 3 days and the prey consisted either of newly hatched *A. salina* nauplii (Exp #5) or late stage copepodites and adults of *A. tonsa* (Exp #6). *A. salina* at abundances of $C_0 = 15, 40, 53$ and 100 ind l^{-1} were incubated with three *P. pileus* (7.6 ± 0.7 mm polar length) in each 30-l container. *A. tonsa* at abundances of $C_0 = 7, 16, 33$ and 59 ind l^{-1} were incubated with 3 larger *P. pileus* (10.4 ± 0.9 mm) in each 30-l container. To estimate growth, the length of each ctenophore was measured at the start of each experiment and after 3 days. Once each day, the ctenophores were transferred to a newly prepared container with sea water and the initial abundance of prey. The prey remaining in the incubation water were retained on a 200-μm filter and counted.

At the end of the experiment, the ctenophore carbon content was estimated from published conversion factors and the increase in weight translated to specific growth rate (μ, d^{-1}):

$$\mu = \frac{\ln\left(\frac{W_{c,t}}{W_{c,0}}\right)}{t}$$

where $W_{c,0}$ and $W_{c,t}$ are the body carbon contents on day 0 and day t, respectively.

Ingestion (I, prey d^{-1}) and clearance (F, l d^{-1}) were calculated as in Exp #5 and 6. In addition, the respiration rate was measured immediately after termination of Exp #5.

Respiration measurements

Respiration, measured as oxygen consumption, was recorded by five polygraphic oxygen electrodes connected via an amplifier to a computer with data acquisition hardware (Computer Boards CIO-DAS 802). Each electrode was placed individually in a flow-through chamber connected by a tube to a respiration chamber (volume = 4 ml). One ctenophore was placed in each of three chambers and two had no ctenophore (controls). Oxygen-saturated (air bubbled) filtered sea water was pumped at 0.15 ml min^{-1} through the chambers by means of a 5-channel peristaltic pump. All connecting tubing was gas-impermeable 1-mm inner-diameter Tygon tubing. Prior to measurements, the electrodes were calibrated against oxygen-saturated water (100% air saturation) and a saturated aqueous solution of $Na_2S_2O_4$ (0% oxygen saturation).

After introduction of the ctenophores in the three respiration chambers, they were observed to move relatively undisturbed; however, they were unable to extend their tentacles in the chambers, but this was considered not to affect respiration substantially because they were not feeding in the chambers. Water was circulated through the chambers before measurements began. The measurement of oxygen consumption rates was started by closing the system so that the water was re-circulated through the respiration chambers past the electrodes and back again. The oxygen concentration was measured every 10 s, and every minute the average of 6 consecutive measurements were stored for each chamber. Measurements were run for no longer than 60 min. During the initial 40 min the electrodes were allowed to stabilize and the last 20 min were used for calculation of the respiration rates. The electrode output (V) from both experimental and reference chambers were plotted as a function of time, and oxygen consumption rates were calculated by linear regression. Respiration rate for each animal (R, μl O_2 s^{-1}) was calculated as:

$$R = \frac{(b_{\exp} - b_{\mathrm{ref}})}{E_{100}} * S * \frac{V}{n}$$

where $b_{\exp}$ is the slope (V s^{-1}) of the linear regression from the oxygen decline in the respiration chamber and b_{ref} is the average of the slopes of the linear regressions from the oxygen decline in the two reference chambers. E_{100} is the electrode signal at 100% saturation (volt), S = the oxygen solubility (μl O_2 ml water^{-1}; Green & Carrit, 1967), V = the volume (ml) of water in the respiration chamber plus tubes and n = the number of animals.

The weight-specific respiration is defined as: $R_w = R_t \times W_c^{-1}$, where W_c is the body carbon content.

Bioenergetics

The carbon budget can be expressed as $G = (F \times C \times \text{AE}) - R$, where F = clearance rate, C = prey concentration and AE = assimilation efficiency (Møller & Riisgård, 2007).

The following definitions are used in this study (Møller & Riisgård, 2007): minimum prey concentration (C_{min}) resulting in maximum ctenophore growth:

$$C_{min} = \frac{G_{max} + R}{F * \text{AE}}$$

where G_{max} = maximum growth rate = $\mu_{max} \times W_c$, μ_{max} = maximum specific growth rate and W = ctenophore body mass.

The gross growth efficiency (GGE) is

$$\text{GGE} = \frac{G}{I}$$

and the net growth efficiency (NGE)

$$\text{NGE} = \frac{G}{A} = \frac{G}{G + R}$$

Conversion factors

Dry weight (DW, mg) of *P. pileus* was estimated from polar length (L, mm) according to the equation $\log \text{DW} = -1.54 + 2.65 \log (L)$, and converted to carbon content (W_c) as 4.3% of DW (Reeve & Walter, 1976), and to wet weight (WW) as 3.95% of DW (Hoeger, 1983). The carbon content of one newly hatched *A. salina* nauplius = 1.16 µg C (Szyper, 1989) and 1 *A. tonsa* (copepodite or adult, 725 µm) = 2.5 µg C (Berggreen et al., 1988). The carbon content of prey organisms from the field was estimated from published data (Martinussen & Båmstedt, 1995): *Evadne nordmanni* = 1 µg C, *A. clausi* = 4 µg C, *T. longicornis* = 15 µg C and *C. typicus* = 10 µg C. The amount of oxygen consumed was converted to carbon equivalents using a respiratory quotient RQ = 0.74 for *P. pileus* (Kremer, 1977) resulting in a conversion factor of 1 µl O_2 = 0.4 µg C.

Results

In Exp #1, clearance rate as a function of ctenophore polar length was measured using *A. tonsa* as prey.

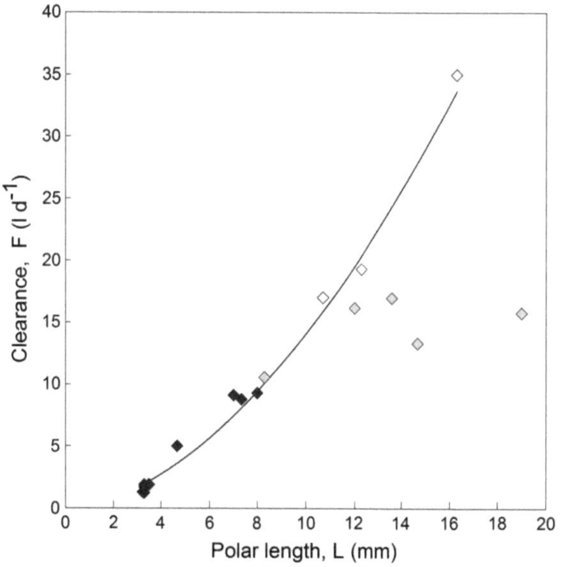

Fig. 1 *Pleurobrachia pileus.* Clearance (F, l d^{-1}) as a function of polar length (L, mm) with *Acartia tonsa* as prey (Exp #1). Nominal prey concentration = 20 ind l^{-1}. Regression line is the power function: $F = 0.2L^{1.9}$, $r^2 = 0.97$. *Filled symbol*: 30 l, *grey symbol*: 50 l and *open symbol*: 100 l. For animals larger than 12 mm, only values from 100 l were used in the regression because the rates for the larger ctenophores were depressed due to too small containers

Clearance (F, l d^{-1}) increased as a function of the polar length (L, mm) as $F = 0.2L^{1.9}$ (Fig. 1). Clearance was depressed for individuals larger than about 12 mm in the smallest (30-l) containers; those results (filled symbols) were not used in the regression.

In Exp #2 and 3, *P. pileus* was incubated with ambient zooplankton predominantly the copepods, *A. clausi*, *C. typicus*, and *T. longicornis* and the cladoceran, *Evadne nordmanni*. Clearance was significantly higher on *C. typicus* than on *A. clausi* and *T. longicornis* (ANCOVA, $F_{3,22} = 4.6$, $P = 0.012$, Table 2), while clearance of *E. nordmanni* (Exp #3 only) did not differ significantly from the others (multiple comparison test, LSD, $0.08 < P < 0.61$). In Exp #4, *P. pileus* of different sizes were incubated with an ambient mixture of *A. clausi*, *C. typicus* and *T. longicornis*. *C. typicus* and *A. clausi* were cleared at similar rates, and both were significantly higher than clearance on *T. longicornis* (ANCOVA, $F_{2,8} = 11.45$, $P = 0.004$, Fig. 2A). The results from Exp #4, shown together with those from Exp #2–3, indicate that clearance was reduced in 30-l containers for the larger ctenophores in Exp #2–3 (Fig. 2B).

Table 2 *Pleurobrachia pileus*: clearance (mean ± SD) of different prey types from experiments #2A ($n = 2$), #2B ($n = 3$), and #3 ($n = 3$)

Prey type	Clearance, F (l d^{-1})		
Experiment:	2A	2B	3
L (mm):	7.3 ± 1.2	15.0 ± 0.5	13.2 ± 0.7
Acartia clausi	5.4 ± 1.4	9.5 ± 0.6	9.8 ± 7.7
Centropages typicus	9.9 ± 2.8	17.0 ± 2.9	21.7 ± 4.4
Temora longicornis	4.3 ± 3.0	10.7 ± 1.2	11.6 ± 5.8
Evadne nordmanni	–	–	15.8 ± 10.8

Sizes of *P. pileus* (L, mm) were 7.3 ± 1.2, 15.0 ± 0.5, and 13.2 ± 0.7, respectively. All tanks were of 30-l volume. The results above are plotted together with results from Exp #4 in Fig. 2B

Growth experiments used *P. pileus* fed different concentrations of *A. salina* (Exp #5, 7.6-mm ctenophores) or *Acartia tonsa* (Exp #6, 10.4-mm ctenophores). The specific growth rates increased with increasing prey concentrations, and the maximum specific growth rate (0.09 d^{-1}) was achieved at prey concentrations of 40 μg C l^{-1} *A. salina* and 100 μg C l^{-1} *A. tonsa* (Fig. 3A). The gross growth efficiency (GGE) was estimated when positive growth was obtained (Fig. 3B). GGE increased from 4 to 8.5% at 40 μg C l^{-1} *A. salina*, but decreased to 4% at higher prey concentrations. GGE increased from 2.5 to 6% for the range of *A. tonsa* concentrations tested.

During the same experiments, ctenophore ingestion increased with prey concentration (Fig. 4A). Ivlev functions fitted to the data indicated maximum ingestion of 850 and 1609 μg C d^{-1} of ctenophores preying on *A. salina* and *A. tonsa*, respectively, and that saturation prey levels were not reached in this study. At the same time, clearance rates declined slightly (Fig. 4B); for *A. salina* clearance ranged from 5 to 8 l d^{-1} and for *A. tonsa* from 9.5 to 16 l d^{-1}.

Finally, the weight-specific respiration rates increased with prey concentration in *P. pileus* feeding on *A. salina* (Fig. 5). The respiration rates increased from 0.21 μl O_2 μg C^{-1} d^{-1} in starved ctenophores to 0.52 μl O_2 μg C^{-1} d^{-1} in ctenophores with the maximum specific growth rate (0.09 d^{-1}, Fig. 3). Thus, the resulting assimilation efficiency (AE) decreased with increasing prey concentration while the net growth efficiency (NGE) increased (Fig. 6).

Because growth (G), ingestion (I) and respiration (R) were measured simultaneously, a carbon budget could be constructed for individuals growing at maximum rates (Table 3). Values of I, R, G_{max} and C_{min} were measured, whereas AE and NGE were estimated on the basis of those values. The food concentration resulting in maximum specific growth rates was 40 μg C l^{-1} (Fig. 3) implying an AE = 22% and NGE = 37% at maximum growth (Fig. 6; Table 3). The total assimilated carbon was 64.9 μg C, of which 63% went into respiration and 37% into growth.

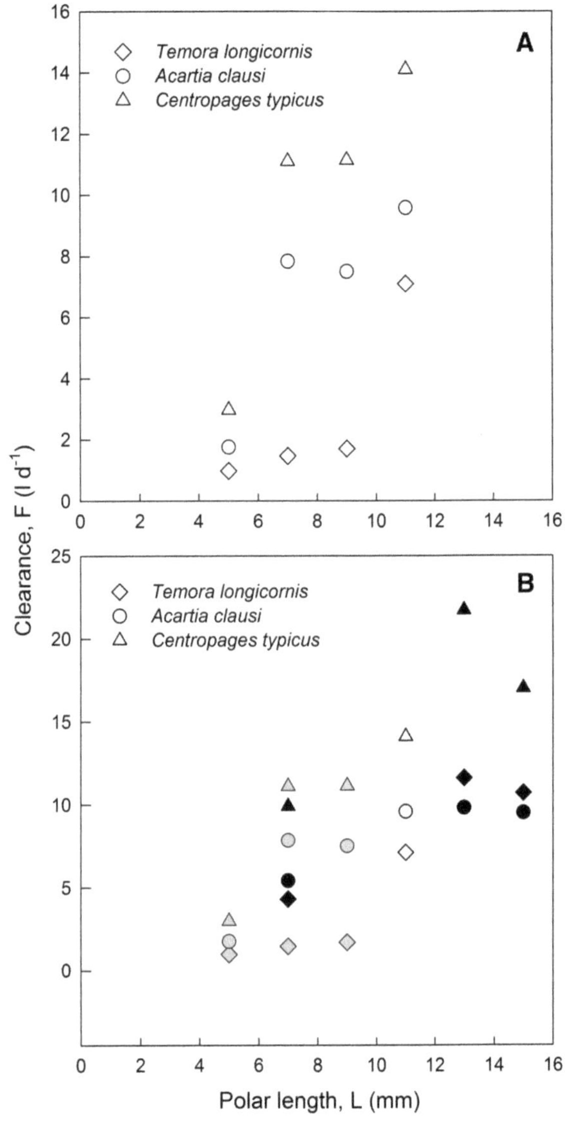

Fig. 2 *Pleurobrachia pileus*. Clearance (F, l d^{-1}) of ctenophores on *Acartia clausi*, *Centropages typicus*, and *Temora longicornis* versus ctenophore polar length (L, mm) in mixtures of all three prey species captured from the field in **A** Exp #4 and in **B** for a combination of Exp #2–4. *Filled symbol*: 30-l, *grey symbol*: 50-l and *open symbol*: 100-l

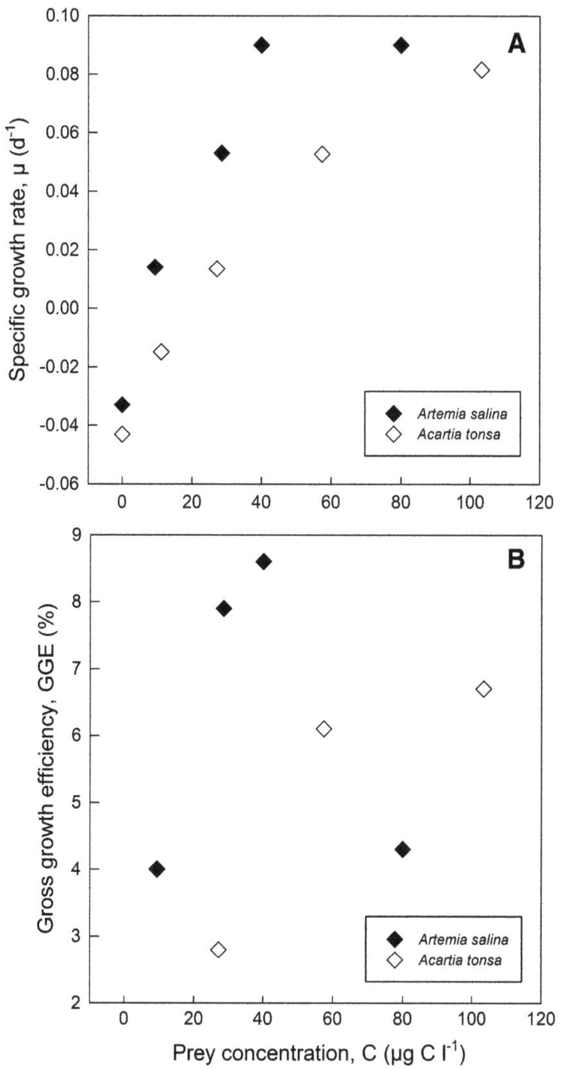

Fig. 3 *Pleurobrachia pileus*. **A** Specific growth rate (μ, d^{-1}) and **B** gross growth efficiency (GGE, %) of ctenophores as a function of prey carbon concentration (C, $\mu g\ C\ l^{-1}$) when preying on *Acartia tonsa* (*open symbols*) and *Artemia salina* (*filled symbols*). Initial size of ctenophores was 7.6 mm for *Artemia salina* and 10.4 mm for *Acartia tonsa*. All tanks were of 30-l volume

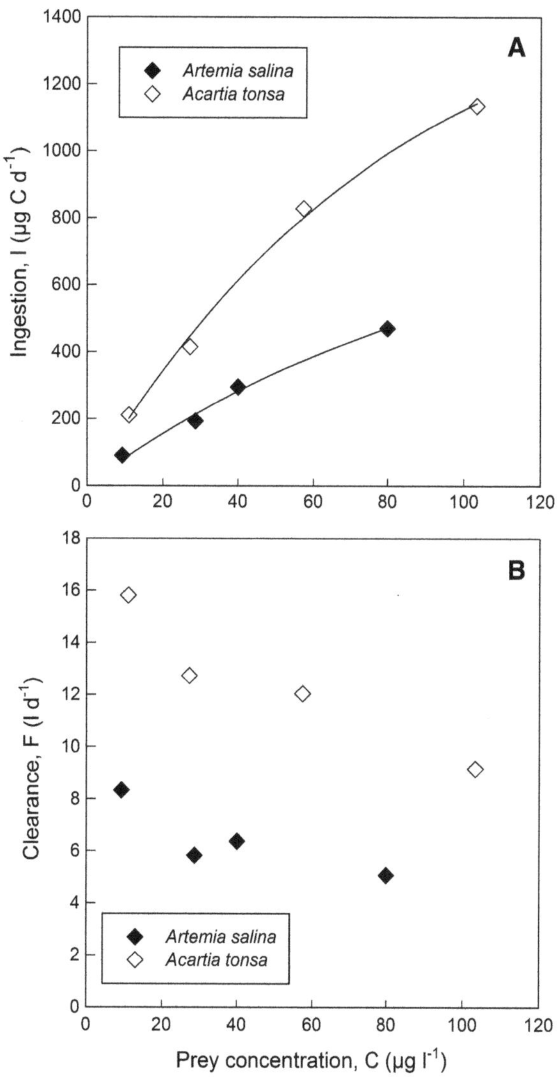

Fig. 4 *Pleurobrachia pileus*. Ingestion (**A**) and clearance (**B**) of ctenophores as a function of prey concentration when preying on *Acartia tonsa* (*open symbols*) and *Artemia salina* (*filled symbols*). Ivlev curve fits: *A. salina*: $I = 850 * (1 - \exp(0.01 * C))$, $R^2 = 0.99$; *A. tonsa*: $I = 1609 * (1 - \exp(0.012 * C))$, $R^2 = 0.99$

Discussion

Feeding

The prediction of predatory effects of ctenophores and other gelatinous species in the field requires knowledge of feeding rates and the factors on which these depend. In feeding experiments the volume of the experimental unit is an important factor to consider (Gibbons & Painting, 1992; Purcell, 2009). *P. pileus* is an ambush entangling predator and tends to retract its tentacles, covering a smaller volume, when they sense the walls of the container (Buecher & Gasser, 1998; Gibbons & Painting, 1992). Clearance rates in *P. pileus* larger than 10 mm depended on container size, and required a container volume of about 40–60 l for maximum rates in 10–15-mm ctenophores (Gibbons & Painting, 1992). Purcell (2009) recently reviewed existing clearance rates of

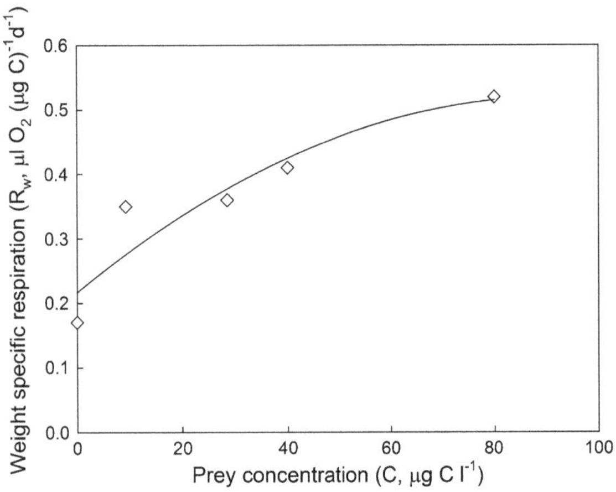

Fig. 5 *Pleurobrachia pileus*. Weight-specific respiration (µl O_2 (µg C)$^{-1}$ d^{-1}) of ctenophores as a function of prey concentration (µg C l^{-1}), when preying on *Artemia salina* (Exp #5)

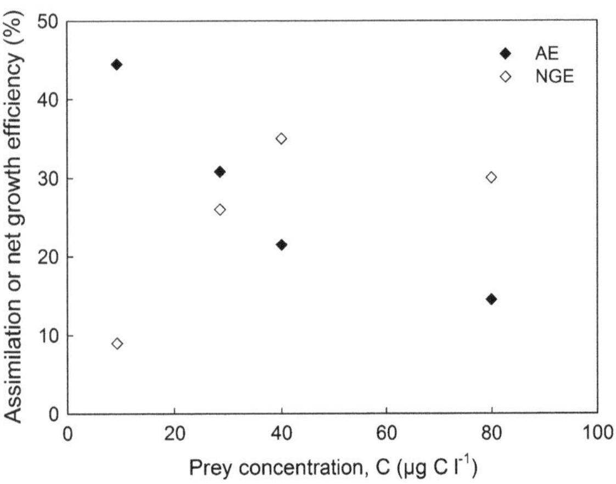

Fig. 6 *Pleurobrachia pileus*. Assimilation efficiency (AE, %) and net growth efficiency (NGE, %) of ctenophores as a function of prey concentration (µg C l^{-1}) when preying on *Artemia salina* (Exp #5). AE = *filled symbols* and *open symbols* = NGE

the lobate ctenophore, *M. leidyi*, and found that rates were reduced in volumes smaller than 20-l and that the ratio of container-volume to ctenophore-volume of should be at least 2500:1. In this study, the objective was to estimate size-dependent clearance in *P. pileus* with *A. tonsa* as prey and the container volumes were selected according to the study by Gibbons & Painting (1992). For individuals larger than 11 mm, the container effect on clearance rates obtained in 30–50-l containers was remarkable when compared with experiments in 100-l containers (Fig. 1). A similar trend was seen for the experiments

Table 3 *Pleurobrachia pileus*: carbon budget for ctenophores growing at maximum rate on a diet of *Artemia salina*

Parameter	Symbol	Unit	Value
Ctenophore carbon content	W_c	µg C	268
Maximum specific growth rate	μ_{max}	d^{-1}	0.09
Clearance	F	l d^{-1}	6.41
Ingestion	I	µg C d^{-1}	295
Respiration	R	µg C d^{-1}	41.2
Growth	G_{max}	µg C d^{-1}	24.1
Assimilation efficiency	AE	%	22
Net growth efficiency	NGE	%	37
Minimum prey concentration	C_{min}	µg C l^{-1}	40

with mixed wild zooplankton, in which clearance seemed depressed for ctenophores larger than 11 mm in 50-l containers (Fig. 2B). After exclusion of the depressed clearance rates of ctenophores larger than 11 mm, the size-dependent clearance rate (F) for *M. leidyi* preying on *A. tonsa* was described by the expression $F = aL^b$, with an exponent close to 2 when related to the polar length (L) (Fig. 1).

For ambush feeders, prey motility is important for the predator–prey encounter rate (Gerritsen & Strickler, 1977). Strong prey selectivity of *Pleurobrachia bachei* (A. Agassiz) occurred in consequence of different swimming speeds and escape responses among different prey types (Greene et al., 1986). Therefore, it is of great importance to quantify feeding rates on the prey species that co-exist with predators in their natural habitats to reliably estimate predation potentials. In this study, the predominant zooplankton species co-existing with *P. pileus* during their peak abundances in the Gullmar Fjord were the copepods, *A. clausi*, *C. typicus*, and *T. longicornis*, and the cladoceran, *Evadne nordmanni*. Clearance rates measured simultaneously on different prey species showed that clearance on *C. typicus* was generally 2–3 times higher than on *A. clausi* and *T. longicornis* (Table 2). All experiments on prey selectivity were performed at relatively high prey concentrations (120–150 µg C l^{-1}); however, saturation did not occur until at even higher prey concentrations (Fig. 4A).

The different clearance rates on different prey depend on the behaviours of both the predator and prey (e.g. Purcell, 1997). *P. pileus* deploys its tentacles by swimming slowly while extending its two tentacles and their tentillae. The curtain of tentillae remains motionless in the water and prey swim into them and become entangled in the sticky tentillae (Greene et al., 1986). The encounter rate is a product of prey size and swimming speed and detection distance of the predator and prey (Gerritsen & Strickler, 1977). *Acartia clausi* alternates between slow sinking (0.3 mm s^{-1}) and short jumps of several body lengths. *Temora longicornis* move slowly while feeding, but swim rapidly or sink (2.5 mm s^{-1}) between feeding bouts (Tiselius & Jonsson, 1990). *Centropages typicus* is a rapid swimmer (1.9 mm s^{-1}) and acts as a cruising predator or filter feeder most of the time (Tiselius & Jonsson, 1990).The selection for the continuously active prey (*C. typicus*) was evident (Table 2; Fig. 2). Chandy & Greene (1995) found a similar difference in clearance between the continuously swimming *Pseudocalanus newmani* (0.48 l h^{-1}) and the hop-sinking *Acartia longiremis* (0.21 l h^{-1}) by *P. bachei*.

Although several studies have quantified feeding in *Pleurobrachiapileus*, the rates have probably been underestimated due to small experimental jars. We compared weight-specific clearance rates of 3 size classes (8, 10 and 13 mm) of *P. pileus* (Table 4). The clearances on *A. tonsa* found in our study were comparable, or slightly higher, than in other studies for 8-mm ctenophores. The rates were 3 times higher in our study for 10-mm ctenophores as compared with rates in 1-l containers (Chandy & Greene, 1995). The container effect was even more pronounced for larger individuals. Clearance was $\approx$10 times higher on *A. tonsa* in our study compared with values for mixed zooplankton >200 μm in 5-l containers (Båmstedt, 1998).

Comparison among studies is difficult due to effects of different prey types as well as container volume. Clearance of *Pleurobrachia* spp. on *Calanus finmarchicus* (Gunner) was three times higher than on smaller copepod species in the same container volume (Båmstedt, 1998). Clearance on *Pseudocalanus newmani* (Frost) was twice that on *Acartia longiremis* (Lilljeborg) in the same container volume (Greene et al., 1986; Chandy & Greene, 1995).The feeding experiments of Greene et al. (1986) demonstrated the importance of different development stages of the copepods in addition to species when evaluating the predation effect; Clearance increased with developmental stages of *C. pacificus* (Brodsky) up to CV. This is consistent with increased encounter rates with larger and faster prey, even though the escape abilities also increased with prey size.

In our study, ingestion by *P. pileus* increased with prey concentration with no saturation observed in the tested prey concentrations; however, at the same time, clearance rates declined slightly (Fig. 4A, B). Previous laboratory studies showed that the ingestion in *Pleurobrachia* spp. increased directly with prey concentrations even much higher than those in the natural environment (Reeve et al., 1978; Båmstedt, 1998).

Growth

The maximum specific growth rate for 7–10 mm ctenophores was 0.09 d^{-1} at prey concentrations of 40 and 100 μg C l^{-1} of *A. salina* and *A. tonsa*, respectively (Fig. 3). Specific growth rates in relation

Table 4 *Pleurobrachia* spp.: comparison with the literature values of ctenophore clearance rates

L (mm)	F_w (l (g WW)$^{-1}$ d^{-1})	Prey species	V (l)	T (°C)	Reference
8	11.4	*Pseudocalanus* sp.	3.5	12	Greene et al. (1986)
8	7.8	*Acartia clausi*	3.5	12	Greene et al. (1986)
8	11.4	*Centropages brachiatus*	5–20	13–15	Gibbons & Painting (1992)
8	12.6	*Acartia tonsa*	30–50	15	This study
10	14.8	*Pseudocalanus newmani*	1	12	Chandy & Greene (1995)
10	6.4	*Acartia longiremis*	1	12	Chandy & Greene (1995)
10	17.3	*Acartia tonsa*	30–50	15	This study
11–14	10	*Calanus finmarchicus*	5	12	Båmstedt (1998)
11–14	2.6	Mixed >200 μm	5	12	Båmstedt (1998)
13	25.6	*Acartia tonsa*	100	15	This study

L = ctenophore polar length; F_w = wet weight-specific (WW) clearance rates; V = container volume; T = temperature to available prey are important when evaluating if the ctenophores are achieving their growth potential in the field; however, growth rates in our study were obtained only for a narrow size range of ctenophores and would not be representative of all size classes. Specific growth rates of *Pleurobrachia* sp. were 0.47 and 0.09 d^{-1} for size classes of 2.6–6 and 6–10 mm, respectively (Reeve & Walter, 1976), which indicated that specific growth rates decreased with ctenophore size. Their results for the larger ctenophores were in the range of those found in this study; however, the effect of size on growth should be related to maximum growth for each size group and prey concentration, which was not ensured in Reeve & Walter (1976). The trend of decreased specific growth with size also was observed for the ctenophore, *M. leidyi* (in Kremer & Reeve, 1989; Reeve et al., 1989), and for maximum specific growth rates in cnidarians, such as the hydromedusa, *Aequorea vitrina* (Gosse), and the scyphomedusa, *Aurelia aurita* (Linnaeus) (in Møller & Riisgård, 2007).

The GGE of *P. pileus* increased from 4 to 8.5% at 40 μg C l^{-1} *A. salina* and then decreased to 4%, but the GGE increased from 2.5 to 6% when the prey was *A. tonsa* (Fig. 3A). These values are comparable with previous studies; the GGE of *M. leidyi* decreased with increasing food density, always being less than 10%; in the same study, the GGE of *Pleurobrachia* spp. was 11 and 3% at low and high food densities, respectively (Reeve et al., 1978).

Respiration

The weight-specific respiration rate of unfed ctenophores was 0.21 μl O_2 (μg C)$^{-1}$ d^{-1}, corresponding to 56 μl O_2 d^{-1} for a 7.6-mm individual, which is comparable to the study by Hirota (1972) who found values of 65 μl O_2 d^{-1} for similarly sized individuals incubated for 24 h.

When an animal is feeding and growing the respiration increases, a phenomenon designated as the specific dynamic action (SDA) (Grisolia & Kennedy, 1966; Kiørboe et al., 1985).Our study measured respiration as a function of food availability as an integral part of the studies of feeding and growth for *P. pileus* feeding on *A. salina*. The relationship between food concentration and weight-specific respiration rate was hyperbolic (Fig. 5). A similar pattern was observed for the ctenophore *M. leidyi*, feeding on *A. tonsa* (in Kremer, 1982).

The weight-specific respiration rates of *P. pileus* increased three folds from starvation level to where maximum specific growth rates were obtained (Fig. 5). Weight-specific respiration rates of *P. pileus* were 2.5 fold higher in fed than in starved individuals (Gyllenberg & Greve, 1979); weight-specific respiration rates were 2.7 times higher when *P. bachei* were fed 100 instead of 10 copepods l^{-1} (Reeve et al., 1978).

Carbon budget

The different parameters of carbon budgets, including the cost of growth, need to be measured simultaneously and under the same conditions. The present carbon budget was constructed for individuals growing at a maximum rate. The growth experiment (Fig. 3) showed that C_{min} was 40 μg C l^{-1} for *P. pileus* feeding on *A. salina*; because G_{max}, R_t and F were measured simultaneously, the unknown factor is the assimilation efficiency (AE). To balance the budget at the maximum specific growth rate, AE would be about 22%. NGE was 37% at maximum specific growth rates, implying that 63% was used for respiration. AE decreased sharply with increasing prey concentration, while NGE increased and plateaued at prey concentrations of 40 μg C l^{-1} and above (Fig. 6). An AE of 22% is low compared with the high assimilation efficiencies observed in ctenophores (e.g. 72%, Reeve et al., 1989). Others also have reported decreased AE in high prey; for example, AE of *Pleurobrachia* sp. decreased from 74 to 19% when fed 10 and 100 copepods l^{-1}, respectively (Reeve et al., 1978), and AE and food concentration were inverse related for the lobate ctenophore *M. leidyi* (in Kremer & Reeve, 1989), and the jellyfish, *Aurelia aurita* (in Møller & Riisgård, 2007). Two possible reasons for this trend were suggested: (1) underestimation of the actual growth by assuming that the total growth (production) is equal to the growth estimated from an increase in polar length, which does not include loss of dissolved organic carbon (DOC) and allocation to reproduction, or (2) that ctenophores and jellyfish might keep feeding at high rates at high prey concentrations by displacing partially digested material, as in the ctenophore *M. leidyi* (in Reeve et al., 1989). Another explanation is that the specific growth rates might be under-estimated because, as pointed out by Reeve et al. (1989), the carbon content of ctenophores is higher in fed than in staved ctenophores. That would affect calculations of the weight-specific respiration rates. In our study, carbon content was not measured but estimated from the literature values and assumed to be constant.

Conclusion

This study provides vital rates and carbon budgets for a key ctenophore species that will allow a better understanding of their role in carbon cycling. We conclude that predation rates of *P. pileus* depend on their size, the prey species, prey density, and container volume, which are all important factors to consider when using laboratory experiments to estimate predation in the field. Also, bioenergetic parameters such as specific growth rate, gross growth efficiency, respiration, net growth efficiency, and assimilation efficiency are affected by food availability; correct estimates of those parameters are necessary when evaluating the role of ctenophores in the carbon flow in coastal waters.

Acknowledgements This research was supported by grant 272-06-0241 from the Danish Agency for Science, Technology and Innovation and FORMAS project 217-2008-917 to L.F.M, by the BONUS+ project 210-2008-1882 BAZOOCA to P.T. and by the Linneaus-Palme student exchange programme to J.M.C. Thanks are due to P. Thor for lending us the respiration set-up. We greatly appreciate the constructive criticism made by P. Kremer, J. Purcell and an anonymous reviewer.

References

Båmstedt, U., 1998. Trophodynamics of *Pleurobrachia pileus* (Ctenophora, Cydippida) and ctenophore summer occurrence off the Norwegian north-west coast. Sarsia 83: 169–181.

Berggreen, U., B. Hansen & T. Kiørboe, 1988. Food size spectra, ingestion and growth of the copepod *Acartia tonsa* during development: implications for determination of copepod production. Marine Biology 99: 341–352.

Buecher, E. & B. Gasser, 1998. Estimation of predatory impact of *Pleurobrachia rhodopis* (cydippid ctenophore) in the Northwestern Mediterranean Sea: in situ observations and laboratory experiments. Journal of Plankton Research 20: 631–651.

Chandy, S. T. & C. H. Greene, 1995. Estimating the predatory impact of gelatinous zooplankton. Limnology and Oceanography 40: 947–955.

Frank, K. T., 1986. Ecological significance of the ctenophore *Pleurobrachia pileus* off Southwestern Nova Scotia. Canadian Journal of Fisheries and Aquatic Sciences 43: 211–222.

Fraser, J. H., 1970. The ecology of the ctenophore *Pleurobrachia pileus* in Scottish waters. ICES Journal of Marine Science 33: 149–168.

Fuentes, V. L., D. Atienza, J.-M. Gili & J. E. Purcell, 2009. First records of *Mnemiopsis leidyi* A. Agassiz 1865 off the NW Mediterranean coast of Spain. Aquatic Invasions 4: 671–674.

Gerritsen, J. & J. R. Strickler, 1977. Encounter probabilities and community structure in zooplankton: a mathematical model. Journal of the Fisheries Research Board of Canada 34: 73–82.

Gibbons, M. J. & S. J. Painting, 1992. The effects and implications of container volume on clearance rates of the ambush entangling predator *Pleurobrachia pileus* (Ctenophora: Tentaculata). Journal of Experimental Marine Biology and Ecology 163: 199–208.

Green, E. J. & D. E. Carrit, 1967. New tables for oxygen saturation of sea water. Journal of Marine Research 25: 140–147.

Greene, C. H., 1985. Planktivore functional groups and patterns of prey selection in pelagic communities. Journal of Plankton Research 7: 35–40.

Greene, C. H., M. R. Landry & B. C. Monger, 1986. Foraging behaviour and prey selection by the ambush entangling predator *Pleurobrachia bachei*. Ecology 67: 1493–1501.

Greve, W., 1971. Ökologische Untersuchungen an *Pleurobrachia pileus*. 1. Freilanduntersuchungen. Helgoländer Wissenschaftliche Meeresunters 22: 303–325.

Grisolia, S. & J. Kennedy, 1966. On specific dynamic action, turnover, and protein synthesis. Perspectives in Biology and Medicine 9: 578–583.

Gyllenberg, G. & W. Greve, 1979. Studies on oxygen uptake in ctenophores. Annales Zoologici Fennici 16: 44–49.

Hirota, J., 1972. Laboratory culture and metabolism of the planktonic ctenophore *Pleurobrachia bachei* A. Agassiz. In Takenouti, A. Y. (ed.), Biological Oceanography of the Northern North Pacific Ocean. Idemitsu Shoten, Tokyo: 465–484.

Hoeger, U., 1983. Biochemical composition of ctenophores. Journal of Experimental Marine Biology and Ecology 72: 251–261.

Kideys, A. E., 2002. Fall and rise of the Black Sea ecosystem. Science 297: 1482–1483.

Kiørboe, T., F. Møhlenberg & K. Hamburger, 1985. Bioenergetics of the planktonic copepod *Acartia tonsa*: relation between feeding, egg production and respiration, and composition of specific dynamic action. Marine Ecology Progress Series 26: 85–97.

Kremer, P., 1977. Respiration and excretion by the ctenophore *Mnemiopsis leidyi*. Marine Biology 44: 43–50.

Kremer, P., 1982. Effect of food availability on the metabolism of the ctenophore *Mnemiopsis mccradyi*. Marine Biology 71: 149–156.

Kremer, P. & M. R. Reeve, 1989. Growth dynamics of a ctenophore (*Mnemiopsis*) in relation to variable food supply. II. Carbon budgets and growth model. Journal of Plankton Research 11: 553–574.

Kuipers, B. R., U. Gaedke, L. Enserink & H. Witte, 1990. Effect of ctenophore predation on mesozooplankton during a spring outburst of *Pleurobrachia pileus*. Netherlands Journal of Sea Research 26: 111–124.

Martinussen, M. B. & U. Båmstedt, 1995. Diet, estimated daily food ration and predator impact by the scyphozoan jellyfishes *Aurelia aurita* and *Cyanea capillata*. In Skjoldahl, H. R., C. Hopkins, K. E. Erikstad & H. P. Leinaas (eds.), Ecology of Fjords and Coastal Waters. Elsevier Science Publishers, Amsterdam: 127–145.

Møller, L. F. & H. U. Riisgård, 2007. Feeding, bioenergetics and growth in the common jellyfish *Aurelia aurita* and two hydromedusae (*Sarsia tubulosa* and *Aequorea vitrina*). Marine Ecology Progress Series 346: 167–177.

Mutlu, E., F. Bingel, A. C. Gücü, V. V. Melnikov, U. Niermann, N. A. Ostr & V. E. Zaika, 1994. Distribution of the new invader *Mnemiopsis* sp. and the resident *Aurelia aurita* and *Pleurobrachia pileus* populations in the Black Sea in the years 1991-1993. ICES Journal of Marine Science 51: 407–421.

Pavez, M. A., L. R. Castro & H. E. Gonzalez, 2006. Across-shelf predatory effect of *Pleurobrachia bachei* (Ctenophora) on the small-copepod community in the coastal upwelling zone off northern Chile (23° S). Journal of Plankton Research 28: 115–129.

Purcell, J. E., 1990. Soft-bodied zooplankton predators and competitors of larval herring (*Clupea harengus pallasi*) at herring spawning grounds in British Columbia. Canadian Journal of Fisheries and Aquatic Sciences 47: 505–515.

Purcell, J. E., 1997. Pelagic cnidarians and ctenophores as predators: selective predation, feeding rates and effects on prey populations. Annales de l'Institut Oceanographique, Paris. Nouvelle Serie 73: 125–137.

Purcell, J. E., 2009. Extension of methods for jellyfish and ctenophore trophic ecology to large-scale research. Hydrobiologia 616: 23–50.

Purcell, J. E., T. A. Shiganova, M. B. Decker & E. D. Houde, 2001. The ctenophore *Mnemiopsis* in native and exotic habitats: US estuaries versus the Black Sea basin. Hydrobiologia 451: 145–176.

Reeve, M. R. & M. A. Walter, 1976. A large-scale experiment on the growth and predation potential of ctenophore populations. In Mackie, G. (ed.), Coelenterate Ecology and Behaviour. Plenum Press, New York: 187–199.

Reeve, M. R., M. A. Walter & T. Ikeda, 1978. Laboratory studies of ingestion and food utilization in lobate and tentaculate ctenophores. Limnology and Oceanography 3: 740–751.

Reeve, M. R., M. A. Syms & P. Kremer, 1989. Growth dynamics of a ctenophore (*Mnemiopsis*) in relation to variable food supply. I. Carbon biomass, feeding, egg production, growth and assimilation efficiency. Journal of Plankton Research 11: 535–552.

Riisgård, H. U., L. Bøttiger, C. V. Madsen & J. E. Purcell, 2007. Invasive ctenophore *Mnemiopsis leidyi* in Limfjorden (Denmark) in late summer 2007 – assessment of abundance and predation effects. Aquatic Invasions 2: 395–401. doi:10.3391/ai.2007.2.4.8.

Schneider, G., 1987. Role of advection in the distribution and abundance of *Pleurobrachia pileus* in Kiel Bight. Marine Ecology Progress Series 41: 99–102.

Szyper, J. P., 1989. Nutritional depletion of the aquaculture feed organisms *Euterpina acutifrons*, *Artemia* sp. and *Brachionus plicatilis* during starvation. Journal of the World Aquaculture Society 20: 162–169.

Tiselius, P. & P. R. Jonsson, 1990. Foraging behaviour of six calanoid copepods: observations and hydrodynamic analysis. Marine Ecology Progress Series 66: 23–33.

Travis, J., 1993. Invader threatens Black, Azov seas. Science 262: 1366–1367.

Van der Veer, H. W. & C. F. M. Sadée, 1984. Seasonal occurrence of the ctenophore *Pleurobrachia pileus* in the western Dutch Wadden Sea. Marine Biology 79: 219–227.

Hydrobiologia (2010) 645:179–191
DOI 10.1007/s10750-010-0223-x

Degradation of the Adriatic medusa *Aurelia* sp. by ambient bacteria

Tinkara Tinta · Alenka Malej · Maja Kos · Valentina Turk

Published online: 26 March 2010

Abstract The decomposition of jellyfish after major bloom events results in the release of large amounts of nutrients, which can significantly alter nutrient and oxygen dynamics in the surrounding environment. The response of the ambient bacterial community to decomposing jellyfish biomass was evaluated in two marine ecosystems, the Gulf of Trieste (northern Adriatic Sea) and Big Lake (Mljet Island, southern Adriatic Sea). The major difference between these two ecosystems is that *Aurelia* sp. medusae occur throughout the year in the oligotrophic Big Lake, whereas in the mesotrophic Gulf of Trieste, they occur only seasonally and often as blooms. Addition of homogenized jellyfish to enclosed bottles containing ambient water from each of these systems triggered considerable changes in the bacterial community dynamics and in the nutrient regime. The high concentrations of protein, dissolved organic phosphorous (DOP), and PO_4^{3-} immediately after homogenate addition stimulated increase in bacterial abundance and production rate, coupled with NH_4^+ accumulation in both ecosystems. Our preliminary results of the bacterial community structure, as determined with denaturing gradient gel electrophoresis, indicated differences in the bacterial community response between the two ecosystems. Despite divergence in the bacterial community responses to jellyfish homogenate, increased bacterial biomass and growth rates in both distinctive marine systems indicate potentially significant effects of decaying jellyfish blooms on microbial plankton.

Keywords *Aurelia* sp. · Decomposition · Nutrients · Bacterioplankton · Jellyfish

Introduction

The material from dead organisms (detritus) plays an important role in the cycling of organic matter in the marine environment (e.g., Alldredge, 1972, 1976, 2005). Bacterial decomposition of phytoplankton and numerous other organic matter sources has been demonstrated in field and laboratory experiments (e.g., Caron et al., 1982; Alldredge & Youngbluth, 1985; Alldredge et al., 1986; Simon et al., 1990; Passow & Alldredge, 1994; Passow et al., 1994). Gelatinous macrozooplankton often are regarded as nutritionally poor food sources due to their high water content and predation on them often disregarded; however, predation on jellyfish by fish and large marine animals, such as sunfish and leatherback turtles (reviewed in Arai, 2005; Houghton et al., 2006), indicates that these organisms are good organic sources and potential substrates for bacteria. In particular, during the decomposition of blooms,

Guest editors: J. E. Purcell & Dror Angel / Jellyfish Blooms: New Problems and Solutions

T. Tinta · A. Malej · M. Kos · V. Turk (✉)
National Institute of Biology, Marine Biology Station, Fornace 41, 6330 Piran, Slovenia
e-mail: turk@mbss.org

jellyfish carcasses can serve as major carbon sources to the environment (Billett et al., 2006; Yamamoto et al., 2008). The total organic content of jellyfish is 1–2% of wet weight (Larson, 1986; Clarke et al., 1992) and generally consists of low carbohydrate (7 ± 5%), intermediate lipid (22 ± 12%), and high protein (72 ± 14%) contents (reviewed in Pitt et al., 2009).

The few studies that have addressed the fate of dead jellyfish (Titelman et al., 2006; West et al., 2009) indicated rapid decomposition and nutrient release, whether in the water column or on the sediment surface. Scavengers tend to enhance the decomposition of carcasses by increasing the available surface area, but most of the organic matter is ultimately processed by bacteria. Bacteria hydrolyze and utilize virtually all natural polymers by means of extracellular enzymes. The turnover rate of soluble proteins by bacteria in seawater is only hours (Hollibaugh & Azam, 1983; Hoppe et al., 1988) and similar to the turnover rate of free amino acids in solution (Keil & Kirchman, 1993). The end products of protein degradation, such as dissolved primary amines and free amino acids, contribute to the dissolved organic and inorganic nitrogen (e.g., ammonium) pools (Kremer, 1977). Degradation of lipids (phospholipid components of cell membranes) and nucleic acids (DNA and RNA) contributes to the dissolved organic phosphorous pool (DOP) (Kremer, 1975; Båmstedt & Skjoldal, 1980; Arai, 1997). Thus, the release of large amounts of nutrients into the marine environment from jellyfish bloom decomposition potentially causes substantial changes in substrate quality and quantity, and could affect the composition and activity of the bacterial community (Martinez et al., 1996; Pinhassi et al., 1999; Schafer et al., 2001; Carlson et al., 2002). This has been the subject of only one previous study (Titelman et al., 2006).

We have conducted incubation experiments to evaluate the responses of native bacterial communities to the addition of dead jellyfish biomass. Homogenized tissue of *Aurelia* sp. was used as the jellyfish substrate. This genus of scyphomedusae occurs worldwide in numerous coastal and shelf sea environments (Lucas, 2001) where it frequently forms large blooms (Hamner & Dawson, 2009). We examined the degradation of *Aurelia* sp. and its effects on the ambient bacterial community during spring of 2008 and 2009 in the Gulf of Trieste (northern Adriatic Sea), where they often form blooms in the winter–spring period (Kogovšek et al., this volume) and in Veliko jezero (Big Lake, Mljet Island, southern Adriatic Sea), where they are abundant throughout the year (Benović et al., 2000). Effects of jellyfish tissue on bacterial abundance and production, and on the bacterial community structure, were monitored concurrently with changes in protein and nutrient concentrations.

Materials and methods

Sampling and locations

Experiments were performed in the Bay of Piran (Gulf of Trieste, northern Adriatic Sea) and in the marine lake Veliko jezero (Big Lake) at Mljet Island (southern Adriatic Sea). Experiments were conducted at both locations during spring of 2008 and 2009. The two sets of results from each location were treated as replicates for the site. Seawater samples were collected with a Niskin sampler at 3-m depth at the station BF (45°32.804N; 13°33.034E) in the Bay of Piran, and at 3- and 30-m depths at station Buza (42°46,477N; 17°23.304E) in Big Lake. Before each sampling, seawater temperature and salinity profiles were determined with a CTD fine-scale probe (Microstructure Profiler MSS90, Sea & Sun Technology GmbH). The water temperatures in spring 2008 and 2009 were 14 and 11°C, respectively, in the Bay of Piran (3-m depth), between 17 and 19°C in Big Lake at 3-m depth, and ~10°C at 30-m depth in both years. The ambient ammonium concentration varied in the spring 2008 and 2009 between 0.5 and 0.82 μM in the Bay of Piran and between 0.19 and 0.43 μM in Big Lake. The ambient orthophosphate concentration was low and similar in both years in both ecosystems, between 0.1 and 0.17 μM at 3-m depth, and 0.03 and 0.04 μM at 30-m depth (Table 1).

Immediately after sampling, seawater was filtered through a 200-μm mesh net and subsequently through GF/F filters (Whatman) to yield a filtrate containing microorganisms <0.8 μm in size. The filtered seawater was collected in 8-l acid-washed, autoclaved polycarbonate bottles (Nalgene) and was used for the jellyfish decomposition experiments as described

Table 1 Summary of jellyfish degradation experiments with details of the incubations and ambient seawater temperature (*T*), ammonium (NH_4^+), and orthophosphate (PO_4^{3-}) concentrations

Exp. code	Incubation time	Depth (m)	*T* (°C)	Inoculate (g l^{-1})	NH_4^+ (μM)	PO_4^{3-} (μM)
Piran 2008	17 Apr–5 May 2008	3	14	30.0	0.50	0.10
Piran 2009	24 Mar–15 Apr 2009	3	11	12.5	0.82	0.17
Mljet 2008	8 May–12 May 2008	3	17	15.0	0.30	0.11
		30	10	15.0	0.19	0.03
Mljet 2009	15 May–22 May 2009	3	19	5.5	0.32	0.15
		30	11	3.1	0.43	0.04

below. At the same time, subsamples of filtrate were frozen at −20°C for nutrient analysis.

Jellyfish decomposition experiments

Aurelia sp. medusae were collected by dip net from the surface during bloom events (April 2008 and March 2009) in the Bay of Piran and by divers in Big Lake. Bell diameter and wet weight were recorded for each individual. Bell diameters ranged from 6.3 to 18.5 cm, and a mean wet weight was 15 g. In each experiment, 3–5 jellyfish of different sizes were selected and their whole bodies homogenized with an Ultra-Turrax TP 18/10 (Janke & Kundel®) at 20,000 rpm for several minutes. The whole tissue was homogenized to enhance its degradation by the ambient bacterial community in the sampled sea water. Sea water samples were inoculated with different concentrations of the homogenate (from 3 to 30 g l^{-1}, Table 1). The volume of added homogenate was recalculated according to the average organic content of *Aurelia* sp. and approximate jellyfish concentration during the bloom (Alvarez-Colombo et al., 2008; Turk et al., 2008). Part of the homogenate was frozen and stored for CHNS elemental analysis.

All experiments were carried out in 8-l polycarbonate bottles with GF/F pre-filtered sea water, as described above. A measured weight of homogenized jellyfish tissue was added to each experimental bottle (Table 1). Bottles with filtered sea water without jellyfish homogenate served as controls. The bottles were incubated in situ at each study site to provide ambient conditions of irradiance and water temperature. Locations, dates, durations, and depths of the incubations are in Table 1.

The same sampling procedure was followed at both locations throughout the experiments. Bottles were sampled immediately after inoculation (T_0), and subsamples for nutrient and bacterioplankton analysis were taken on days 1, 2, 4, 6, 8, and 18 in 2008 and on days 1, 2, 3, 6, 9, and 22 in 2009 Bay of Piran experiments. During the experiments in Big Lake, the subsamples were taken daily. Prior to sampling, the enclosed fluid was mixed gently. Subsamples were taken by pouring 500 ml of seawater into acid-washed, autoclaved flasks. Some air was introduced into incubation bottles during subsampling.

Subsamples (5 ml) for bacterial carbon production measurements were incubated immediately after sampling. 300-ml subsamples were filtered (GF/F Whatman) and kept frozen for nutrient analysis. 30 ml of seawater was fixed for bacterial counts and stored at −4°C. At the beginning and end of the experiments, sea water from the bottles was filtered in triplicate onto 0.2-μm polyethersulfone membrane filters (PALL Inc.) for bacterial community analysis. Those filters were stored at −80°C for DNA extraction.

Bacterial abundance and carbon production

Bacterial abundance was determined from formaldehyde-fixed (<0.2 μm pre-filtered, 2% final concentration) seawater samples. The samples were filtered onto 0.2-μm black polycarbonate filters (Millipore), and bacteria were stained with DAPI (4′,6-diamino-2-phenylindole, 1 μg ml^{-1}, final, Sigma) (Porter & Feig, 1980). From each sample, more than 200 bacteria or at least 10 fields per filter were counted using an Olympus BX51 epifluorescence microscope. Pictures were taken using an Olympus Microscope Digital DP70 camera and analyzed with computer software Olympus DP-Soft BX-51. A carbon-to-cell ratio of 19.8 fg C $bacterium^{-1}$ (Lee & Fuhrman, 1987) was used to estimate bacterial biomass.

Bacterial carbon production (BCP) was measured by ^{3}H-leucine incorporation (Kirchman et al., 1985). Each sample was incubated in triplicate with ^{3}H-leucine (20 nM, final, Amersham) for 1 h. Samples with trichloroacetic acid (TCA) added to 5% final concentration prior to addition of ^{3}H-leucine served as controls. Incubation was stopped by adding TCA (5% final concentration) to all samples. The samples were washed according to the centrifugation protocol (Smith & Azam, 1992). Radioactivity was measured using a liquid scintillation counter (Canberra Packard and TriCarb Liquid Scintillation Analyzer, model 2500 TR). BCP was calculated as described by Simon & Azam (1989). Bacterial community growth rate, μ, was calculated as the ratio of BCP to bacterial biomass (B).

Bacterial community structure

DNA was extracted from the filters as described in Boström et al. (2004), with slight modifications. DNA was precipitated at −20°C for 1 h, with 0.1 volume of ammonium acetate (3 M NaAc, pH 5.2) and 0.6 volume of isopropanol; 5-µl baker's yeast tRNA (stock 9.7 mg ml^{-1}) was added as co-precipitant. The pellet was washed with 70% ice-cold ethanol and air dried. Precipitated DNA was re-suspended in 0.02-µm pre-filtered, autoclaved H_2O, and kept at −20°C. Bacterial 16S rDNA was amplified using a universal primer complementary to position 907–927 (Muyzer & Smalla, 1998) and a bacterial primer complementary to position 341–358, with a 40-bp GC-clamp (Muyzer et al., 1993). The PCR reaction mix contained 1× reaction buffer (Tris KCl), 2 mM $MgCl_2$, 0.2 mM dNTP, 1 µM of each primer, Taq polymerase (5 U µl^{-1}, Fermentas), and 1 µl of DNA (50–100 ng). The PCR touchdown protocol of Don et al. (1991) was used; entailing initial denaturation at 94°C for 5 min, followed by a thermal cycler program: denaturation for 1 min at 94°C, primer annealing for 1 min at an initial 65°C, decreasing 1°C every two cycles to a final 50°C, primer extension for 3 min at 72°C. Ten touchdown cycles were run followed by 20 standard cycles (denaturation for 1 min at 94°C, primer annealing for 1 min at 55°C, primer extension for 3 min at 72°C). The last cycle was followed with a 10-min final incubation at an annealing temperature of 72°C. The size and quality of PCR products were confirmed by agarose gel electrophoresis. DNA concentration was measured fluorometrically using PicoGreen quantification reagent (Molecular Probes). PCR products were analyzed by DGGE electrophoresis (C.B.S. Scientific Co.). 60 ng µl^{-1} of PCR product was loaded onto 6% polyacrylamide gels (acrylamide:N,N′-methylbis-acrylamide 37:1) containing a denaturant gradient top to bottom of 29–52% (100% denaturant is defined as 7 M urea and 40% (v/v) formamide). Electrophoresis was run at 150 V for 6 h using 0.5× TAE running buffer at 60°C. The concentration and volume of DNA loaded per lane were constant, which enabled quantitative comparison between bands. A dendrogram was constructed from DGGE banding patterns by the software Quantity One 4.6.2 (Bio-Rad) using Dice coefficient and cluster analysis by the unweighted-pair group method using arithmetic average. DGGE bands were visually detected, and the presence and position of bands were employed in the similarity analysis.

Nutrient analysis

All nutrient analyses were performed on GF/F pre-filtered water samples. Samples were analyzed for total dissolved nitrogen (TDN), ammonium (NH_4^+), nitrite (NO_2^-), nitrate (NO_3^-), total dissolved phosphorus (TDP), and orthophosphate (PO_4^{3-}), using the standard protocols (Grasshoff et al., 1983). Dissolved organic phosphorus (DOP) was calculated as the difference between TDP and PO_4^{3-}. Dissolved oxygen (DO) concentration was measured in triplicate by the standard Winkler method.

Protein concentrations were determined by the standard Bradford method (Bradford, 1976) using the Bio Rad Protein Assay Kit. Seawater samples for protein assays were filtered through GF/F filters and stored at −20°C prior to analysis. For each sample, triplicate 2-ml aliquots were centrifuged for 10 min at 20,000×g at 4°C. OD_{595} was measured versus reagent blanks (0.22-µm pre-filtered seawater) and protein concentrations were determined using a standard curve prepared by dilutions of a bovine serum albumin standard in the range 1–20 µg protein ml^{-1}.

Results

Bacterial and nutrient dynamics

Bacterial colonization of jellyfish micro-particles was observed by microscope within the first 24 h of the experiment, and bacterial numbers increased substantially in the following days. In experiments in the Bay of Piran in 2008, the abundance of bacteria increased from an initial value of 4.1×10^8 to 2.3×10^{10} cells l^{-1} by day 2. At the same location in spring 2009, the bacterial abundance increased from 2.8×10^8 to 2.7×10^{10} cells l^{-1} by day 6, showing a much longer lag phase (Fig. 1A). Bacterial carbon production (BCP), measured only in 2009, showed a pattern similar to that of bacterial abundance. At the beginning of the experiment, BCP was low (0.1 μg C l^{-1} d^{-1}) in the bottles with added jellyfish, but increased by day 3 to 24.7 μg C l^{-1} d^{-1}. After an initial lag period, BCP reached high values on day 6 (90.5 μg C l^{-1} d^{-1}) (Fig. 2A). Bacterial growth rate varied from 0.002 to 1.15 d^{-1}.

In experiments in both years at Big Lake, bacterial abundance peaked in the first 3 days, increasing from 3×10^8 to 2.9×10^9 cells l^{-1} in 2008 and from 8×10^8 to 4.6×10^9 cells l^{-1} in 2009 (Fig. 1B). BCP showed a pattern similar to that for bacterial abundance in 2009. In contrast to the Bay of Piran experiment, BCP peaked at 101.5 μg C l^{-1} d^{-1} in the first 2 days in the jellyfish-inoculated bottle, but then decreased to 8.3 μg C l^{-1} d^{-1} at the end of the experiment (day 7) (Fig. 2B). Bacterial growth rate varied from 0.29 to 1.14 d^{-1}. Changes in bacterial abundance and BCP were much smaller in the control bottles, in both locations (Fig. 2).

The initial dissolved protein concentration, measured immediately after addition of jellyfish homogenate, was 2 μg ml^{-1} in 2008 and 0.7 μg ml^{-1} in 2009 in the Bay of Piran experiment (Fig. 3A, B) and increased in the first 1–2 days of incubation. After this increase, protein concentration declined over the next 24 h by 55% in 2008 and 80% in 2009 (Fig. 3A, B).

In Big Lake, the initial protein concentrations in the bottles with jellyfish was 0.64 μg ml^{-1} in 2008 and 1.07 μg ml^{-1} in 2009 (Fig. 3C, D). After 2–3 days of incubation, it decreased by 30 and 50% in 2008 and 2009, respectively (Fig. 3C, D). The decreases in protein concentrations corresponded to the increase in bacterial abundance and productivity. In the control bottles at both locations, the protein concentrations were consistently low (below 0.35 μg ml^{-1}) and did not change during the experiments.

Ammonium concentration (NH_4^+) increased with time in all of the jellyfish-inoculated bottles. At the Bay of Piran, in the first 5 days of each experiment, ammonium increased continuously from 0.5 to 60 μM in 2008 and from to 0.82 to 100 μM in 2009. At Big Lake in the first 4 days of each experiment, NH_4^+ concentration increased continuously from 0.3 to 38.2 μM in 2008 and from 0.32 to 17 μM in 2009. As different amounts of jellyfish homogenate were added to the bottles, the concentration of ammonium was expressed as relative accumulation per amount of added homogenate (Fig. 4). NH_4^+ concentrations in all control bottles remained low and did not change throughout the experiments.

The concentrations of DOP and PO_4^{3-} were higher in all experimental bottles with jellyfish homogenate than in the control bottles (Fig. 5). In the 2008 Bay of

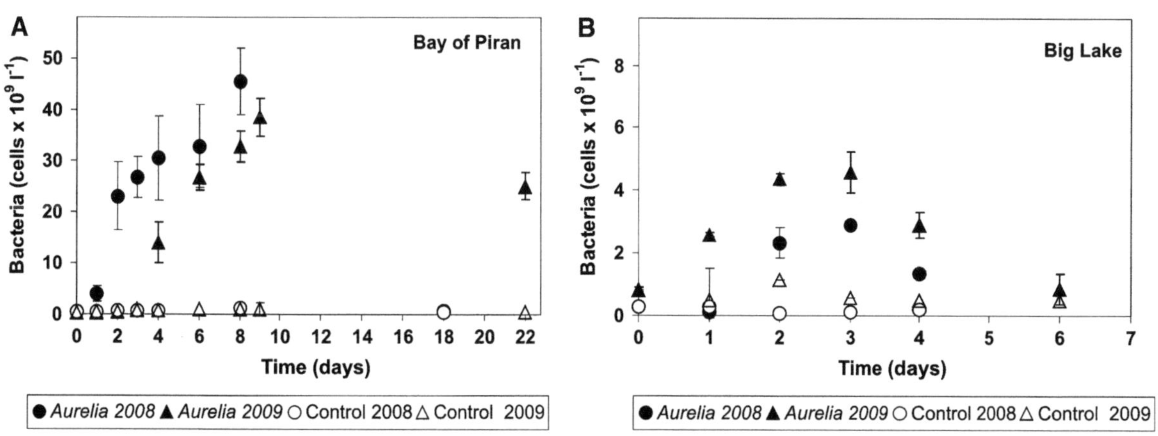

Fig. 1 Bacterial abundance measured during the jellyfish enrichment experiments in the Bay of Piran (**A**) and Big Lake (**B**) in spring 2008 and 2009. Each point represents 30 replicates (mean ± standard deviation)

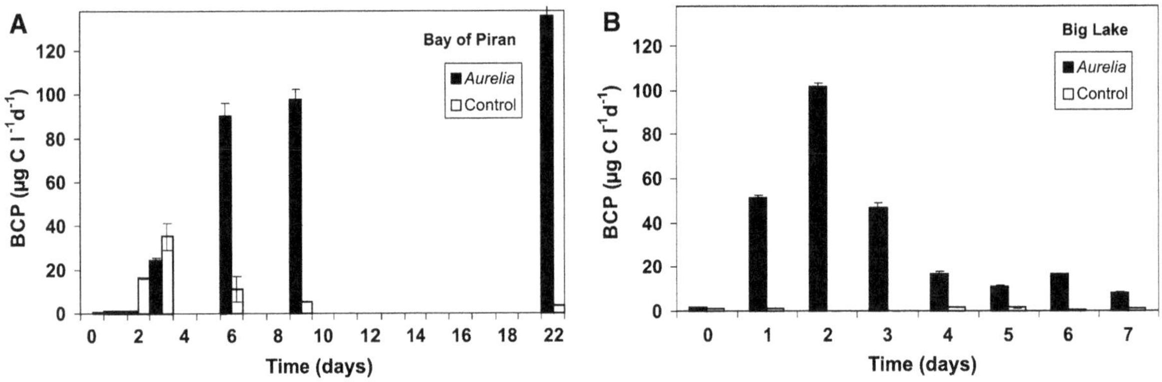

Fig. 2 Bacterial carbon production (BCP) measured during the jellyfish enrichment experiments in the Bay of Piran (**A**) and Big Lake (**B**) in spring 2009. Each point represents 3 replicates (mean ± standard deviation)

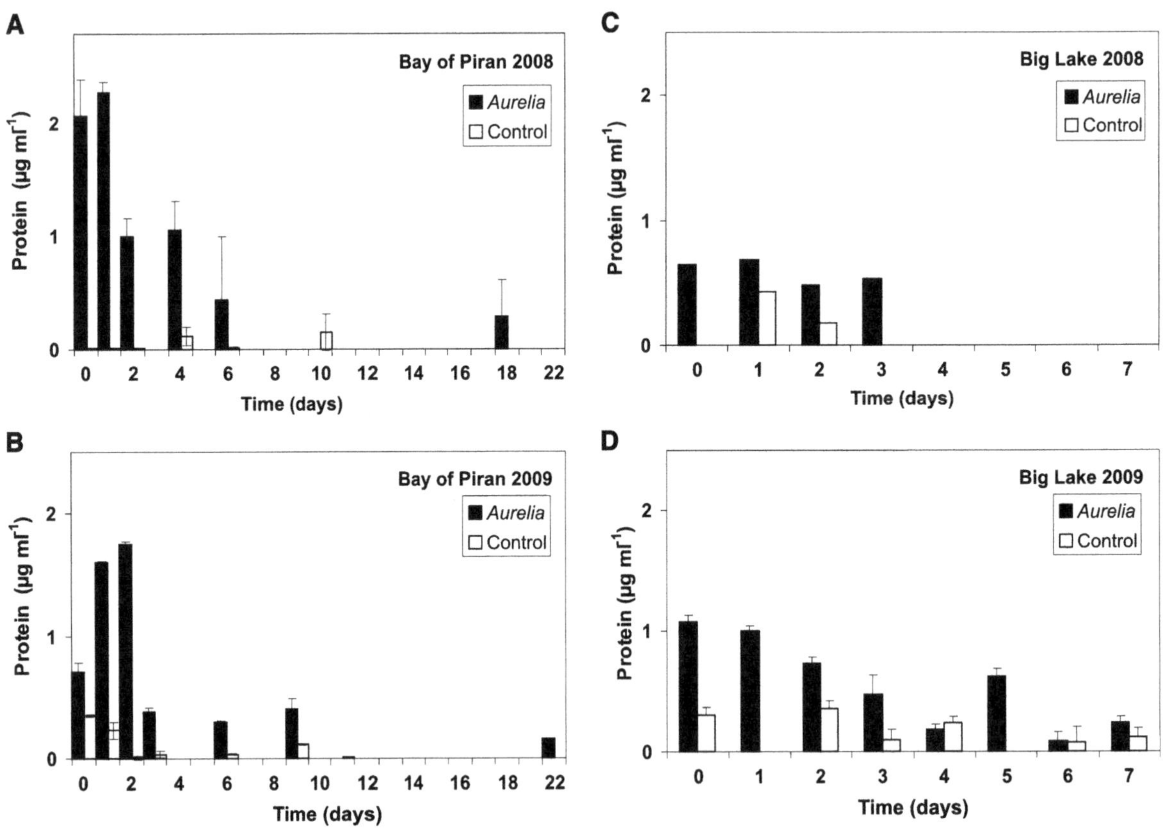

Fig. 3 Dissolved (GF/F filtrate) protein concentration (µg ml^{-1}) measured during the jellyfish enrichment experiments in the Bay of Piran (**A**, **B**) and Big Lake (**C**, **D**) during spring of 2008 and 2009. Each point represents 3 replicates (mean ± standard deviation)

Piran experiment, the initial DOP and PO_4^{3-} concentrations were four to seven times higher than those in the controls (Fig. 5A). After the first 24 h, DOP concentrations decreased continuously to the end of the experiment, while the PO_4^{3-} initially decreased but accumulated at the end of the experiment. A

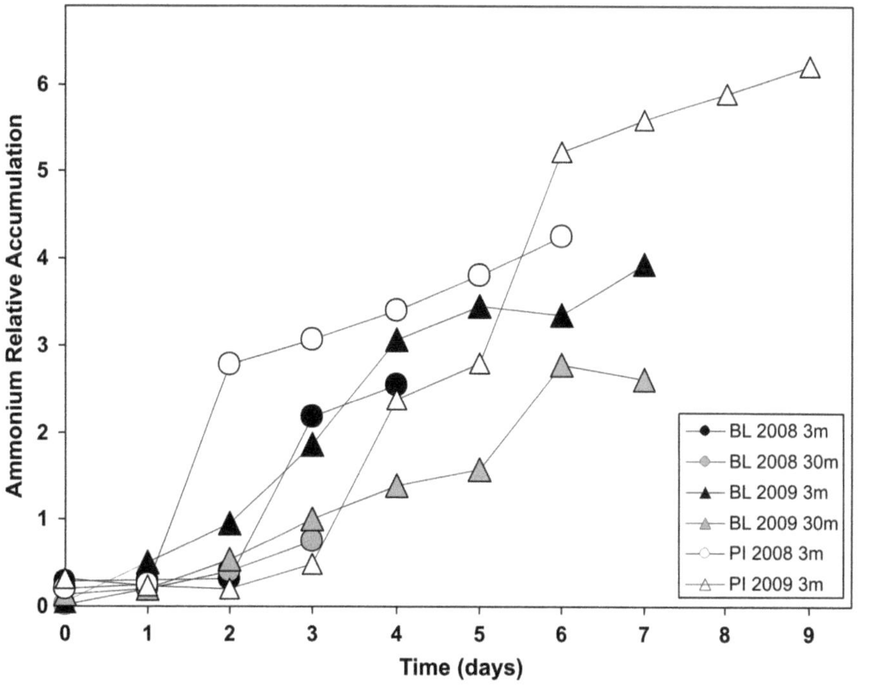

Fig. 4 Concentrations of ammonium, expressed as relative accumulation per amount of added jellyfish homogenate, during the jellyfish enrichment experiments conducted in the Bay of Piran (PI—3 m) and in Big Lake (BL—3 m and 30 m depth) in spring 2008 and 2009

similar pattern was observed in 2009, except that DOP increased sharply until day 2, reaching eight times the concentration in the control and then decreased to below initial level. PO_4^{3-} concentrations doubled during the first 3 days, reaching four times the concentration in the control, and then levelled off until day 9, reaching a much higher concentration on day 22 (Fig. 5B).

In Big Lake experiment in 2009, a similar pattern was observed. After an initial increase, DOP concentrations decreased slightly but steadily, while PO_4^{3-} levels decreased but then rose sharply from days 4 to 7 (Fig. 5D). The 2008 experiment lasted only 4 days, so only the initial increases of DOP and PO_4^{3-} concentrations were recorded, the latter starting to decrease slightly on day 4, which suggested a pattern similar to that observed in 2009 (Fig. 5C).

The dissolved oxygen (DO) concentration was not measured regularly, and the experimental bottles were gently aerated each sampling time; however, a decrease in DO concentration from 5.9 to 3.04 ml l^{-1} was observed in Big Lake experiment in spring 2009.

Bacterial community composition

PCR-DGGE analysis of 16S rDNA was used to track changes in bacterial community composition from the beginning (T_0) to the end of the experiments in the bottles with jellyfish homogenate (T_fA) and controls (T_fC) in 2008 (Fig. 6). The bacterial community composition in the jellyfish-inoculated bottles changed with time, as indicated by differences between the community fingerprints at T_0 and T_f. In the Big Lake experiment, the number of bands in the fingerprints did not change between T_0 and T_f, either in the jellyfish-inoculated bottles (T_fA) or in the control (T_fC). Comparison of the positions of bands in the fingerprints showed that some were conserved in all fingerprints (T_0, T_fA, and T_fC), while other bands either disappeared or appeared in the jellyfish-inoculated bottles and in the control.

In the Bay of Piran experiment, the number of bands decreased between T_0 and T_f, especially in the controls (T_fC) (Fig. 6). Although, some bands were conserved at the same position in all fingerprints, indicating that some bacterial groups were always present, both the band positions and intensities clearly showed that temporal changes in bacterial community composition occurred in the jellyfish-inoculated bottles and in the controls.

A dendrogram based on the presence and/or absence of bands showed that the bacterial communities grouped in two clusters; the jellyfish-inoculated communities (A) formed one cluster and the T_0

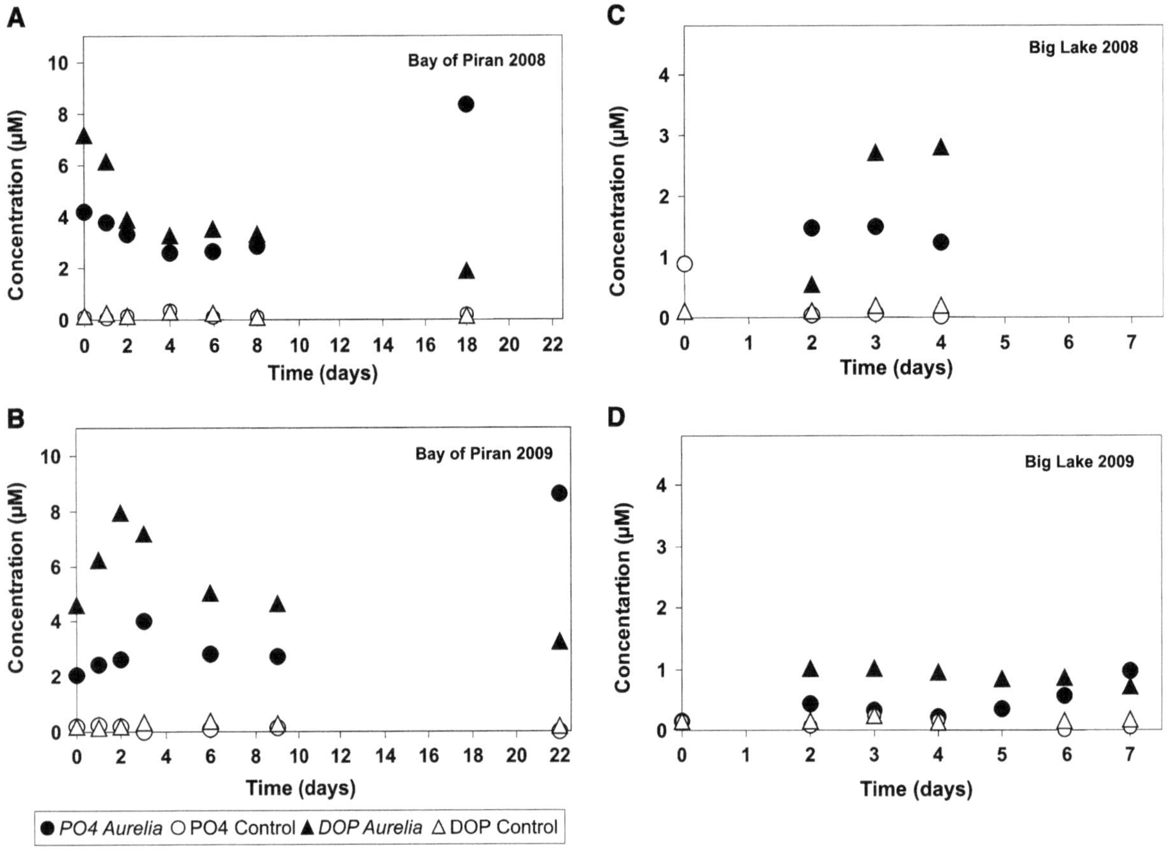

Fig. 5 Dissolved organic phosphorous (DOP) and orthophosphate (PO_4^{3-}) concentrations measured during the jellyfish enrichment experiments in the Bay of Piran (**A**, **B**) and in Big Lake (**C**, **D**) in spring 2008 and 2009 (note the difference in duration between figures)

communities another (Fig. 7). The control communities fit into separate clusters.

Discussion

It has been shown that *Aurelia* sp. exerts direct predatory pressure on mesozooplankton and microzooplankton populations (Stoecker et al., 1987; Sullivan et al., 1994; Purcell & Sturdevant, 2001; Malej et al., 2007; Lo & Chen, 2008). There is also evidence of an indirect cascading effect of *Aurelia* sp. on autotrophic and heterotrophic microbial plankton (Turk et al., 2008). Heterotrophic bacteria, as major consumers of dissolved organic matter (DOM) in marine ecosystems (Azam & Malfatti, 2007; Kirchman, 2008), may benefit most from increased DOM levels as jellyfish decompose. Decomposition of other gelatinous zooplankton, such as salps, ctenophores, and appendicularia, contributes to the general pool of marine snow in the ocean (Caron et al., 1982). Jellyfish play an important role in providing large amounts of carbon and nutrients to the microbial loop by several possible pathways: (a) excretion, (b) mucus production and release (Schneider, 1989; Hansson & Norrman, 1995; Riemann et al., 2006), and (c) decaying biomass (Titelman et al., 2006; West et al., 2009). The decomposition of jellyfish after major bloom events releases large amounts of nutrients, and may significantly alter nutrient and oxygen dynamics in the surrounding environment (Pitt et al., 2009). Depending on the degree of mixing in the water column, decaying jellyfish may also contribute substantially to bottom water hypoxia (West et al., 2009). Moreover, the decomposition of jellyfish can lead to a large flux of particulate carbon to the sediment (Mills, 1995; Billett et al., 2006).

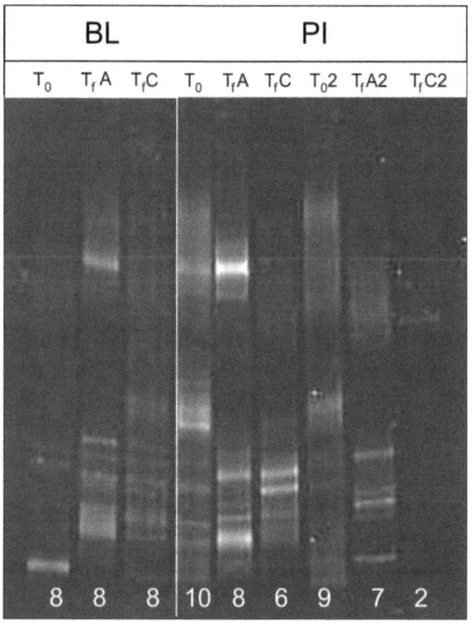

Fig. 6 Temporal changes in bacterial community composition fingerprints analyzed by DGGE during the jellyfish enrichment experiments conducted in the Bay of Piran and Big Lake in spring 2008. First *three lanes* represent the results of the community fingerprints from the experiment in the Big Lake (BL) and next *six lanes* community fingerprints determined during the jellyfish enrichment experiments in the Bay of Piran (PI, two replicates). T_0—at the beginning of the experiments, T_f—at the end of the experiments, *A*—jellyfish-inoculated bottle, *C*—control. The numbers below each lane are the numbers of bands detected

We have evaluated the response of native bacterial communities to a large influx of particulate organic matter in marine ecosystems—the Bay of Piran (northern Adriatic Sea) and Big Lake (Mljet Island, southern Adriatic Sea). The important difference between these ecosystems is the year-round presence of *Aurelia* sp. in Big Lake (Benović et al., 2000) as compared with their periodic occurrence in great abundance in the Gulf of Trieste (Purcell et al., 1999; Kogovšek et al., this volume). Furthermore, Big Lake is an enclosed marine lake, connected to the Adriatic Sea by a long, narrow, shallow channel, but the Bay of Piran is part of the regularly flushed northern Adriatic Sea. The experiments were conducted in both locations during the spring, and ambient temperature, salinity, chlorophyll *a*, nutrient levels, and bacterial abundance were similar, though temperatures were slightly lower and phytoplankton biomass slightly higher in the Bay of Piran experiments (data not shown).

The addition of homogenized jellyfish to sea water triggered considerable changes in bacterial community dynamics and in the nutrient regime at both locations (Figs. 1, 2, 3, 4, 5). Although different concentrations of inoculate were added, similar patterns of bacterial and nutrient dynamics were observed in different years at each location (Table 1). The initially high

Fig. 7 Dendrogram constructed from DGGE banding patterns using the cluster analysis by the unweighted-pair group method using arithmetic average. The number of discernible bands used in the construction of the dendrogram is indicated below each lane in Fig. 6. BL—Big Lake, PI—Bay of Piran, T_0—at the beginning of the experiment, T_f—at the end of the experiment, *A*—jellyfish-inoculated bottle, *C*—control bottle

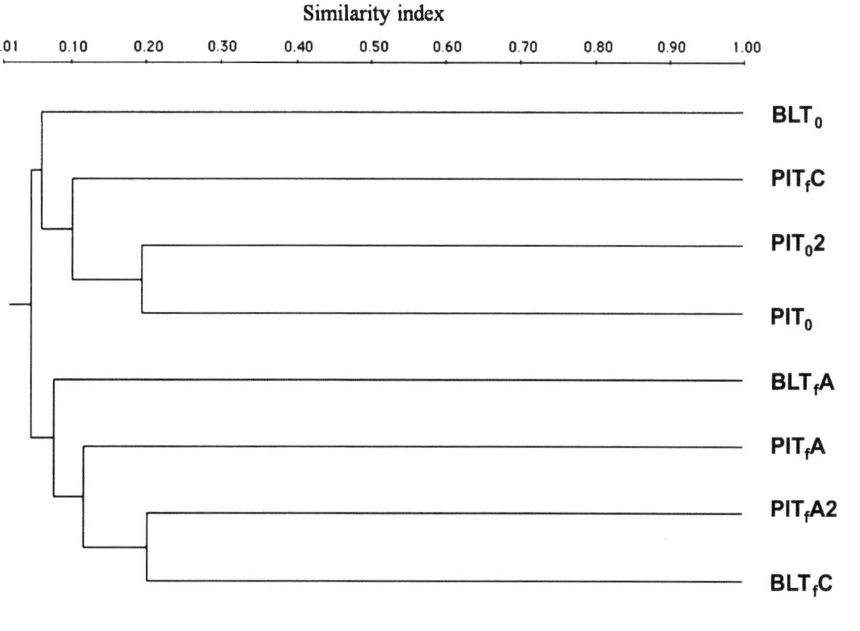

concentrations of protein, DOP, and PO_4^{3-} in the inoculated bottles maintained an increase in bacterial abundance and productivity. In the Bay of Piran experiment, the protein concentration increased in the first 24–48 h due to rapid jellyfish degradation, as also observed by West et al. (2009). This was followed by a steady decrease, presumably due to proteolysis and uptake by bacteria; however, in Big Lake, protein concentration peaked upon addition of jellyfish homogenate to the bottles and decreased rapidly thereafter. The addition of jellyfish homogenate to water from Big Lake triggered a rapid bacterial community response, with increased bacterial abundance and production in 24–48 h (Fig. 1B, 2B). In contrast, the bacterial response lagged in the Bay of Piran experiment, with the greatest increase in bacterial abundance and production after 6 days (Fig. 1A, 2A). One explanation for the difference in the bacterial community response times could be the higher seawater temperature in Big Lake in both years (Table 1). Another explanation is that the native bacterial community in Big Lake may be predisposed toward decomposition of jellyfish tissue because *Aurelia* sp. occur and die in the lake year-round, whereas the bacterial community from the Bay of Piran may need to adapt to the new jellyfish substrate.

Utilization of substrates is believed to be driven by the extent of nutrient or energy limitation in the system (Thingstad et al., 1997; Caron et al., 2000). Given that the two ecosystems have low ambient concentrations of NH_4^+, rapid depletion of proteins followed by an accumulation of ammonium in the incubation bottles is not surprising (Figs. 3, 4). Although PO_4^{3-} increased steadily, the DOP concentration did not change dramatically in Big Lake, suggesting that other sources of phosphorous may have been available to the microbial community (Fig. 5D). Because inorganic and organic phosphorous compounds are the limiting elements for microbial growth in both ecosystems, those available from the degrading jellyfish were rapidly consumed by bacteria, and PO_4^{3-} accumulation was observed only in the experiments with prolonged incubation (>6 days) (Fig. 5A, B).

Protein consumption coupled with NH_4^+ accumulation and oxygen consumption in the jellyfish-inoculated bottles suggests that the proteins were catabolized for energy rather than assimilated in salvage pathways (Cherrier & Bauer, 2004; West et al., 2009). Because dissolved oxygen levels in the incubation bottles decreased with time, the hypoxic conditions may have affected bacterial community composition as well as their metabolic processes. Information from several recently completed bacterial genome sequences shows substantial cellular allocations to amino acid and peptide transport, which suggests that these substrates are important for meeting bacterial nitrogen demands (Giovannoni et al., 2005). Among the nitrogen-containing reduced carbon substrates, dissolved-free amino acids and protein (or dissolved combined amino acids) appear to support the greatest fraction of bacterial growth (Kirchman, 2008).

In addition to the different times of the bacterial response to the addition of jellyfish homogenate, the bacterial community structure dynamics diverged in the two ecosystems. The number of bands in the community fingerprints did not change between T_0 and T_f in the Big Lake experiment. Although, the final bacterial community fingerprint (T_f) appears to differ from that of the T_0 community, both T_fA and T_fC groups appear in the same cluster, which indicates a similar community composition. The similarities between the T_f community fingerprints suggest that the native Big Lake bacterial community is adapted to decomposing jellyfish biomass, which occurs throughout the year in this environment. The rapid response of the bacterial community to the addition of jellyfish homogenate in Big Lake also supports this conclusion. The relative intensities of bands from the T_f communities in Big Lake indicate that certain bacterial groups in the T_fA community proliferate in the presence of jellyfish homogenate, suggesting that some specific groups are better adapted than others to utilize this substrate for their growth.

In contrast, in the Bay of Piran where jellyfish are not always present, the number of bands decreased from T_0 to T_f bacterial community fingerprints (T_fA and T_fC). This decrease in bacterial diversity could be due to the lack of certain nutrients in the bottles, as compared to ambient concentrations, because no supplements other than jellyfish homogenate were added to the enclosed communities. This could explain the large decrease in the number of bands in the T_fC community fingerprints. Surprisingly, the number of bands in T_fA was considerably greater than that in the T_fC fingerprints, and the high intensity of some bands indicates that certain bacterial groups in the T_fA communities were dominant. These results, together with high and constantly increasing productivity of the bacterial community exposed to decaying jellyfish tissue, suggest that jellyfish detritus supports bacterial growth, although only part of the bacterial community was able to utilize this substrate. Changes in bacterial community composition resulting from introduction of a new substrate are not surprising because ectohydrolytic enzyme profiles and activities are highly variable among different bacterial groups (Martinez et al., 1996). Adaptation of the bacterial community to the new substrate apparently took time because the greatest increase in bacterial abundance and productivity was recorded after 6 days (as compared to the rapid response in Big Lake). Another explanation for this result is the difference in the ambient seawater temperature (Table 1). The dendrogram (Fig. 7) shows that jellyfish-inoculated bacterial communities (T_fA) clustered together, indicating similarities in species composition, which may be due to stimulation or inhibition of specific bacterial groups (Titelman et al., 2006).

Conclusions

Addition of jellyfish homogenate to ambient (GF/F filtered) seawater triggered responses in the bacterial communities from two different Adriatic Sea marine ecosystems. The initially high concentrations of protein, DOP, and PO_4^{3-} in the inoculated bottles, resulting from addition of jellyfish homogenate, supported increased bacterial abundance and production, coupled with NH_4^+ accumulation and oxygen consumption in both ecosystems. The response of the bacterial community to the added substrate was more rapid in Big Lake, where jellyfish occur all year, than in the Bay of Piran where jellyfish are not always present. In addition, the bacterial community composition diverged in the two ecosystems. In Big Lake, the final and control community fingerprints were not much different; however, the addition of jellyfish homogenate to Bay of Piran seawater resulted in dramatic changes in bacterial community composition. Our results suggest significant effects of decaying jellyfish blooms on bacterial and nutrient dynamics. Despite divergence in bacterial community responses to jellyfish homogenate, increased bacterial biomass and growth rates in both distinctive marine systems indicate potentially significant effects of decaying jellyfish blooms on microbial plankton.

Acknowledgments This research was financed by the Ministry of Higher Education, Science and Technology of the R Slovenia (P1-0237), and bilateral cooperation with the Ministry of Science, Education and Sport of the Croatia. We are grateful to the Mljet National Park authorities for their hospitality during field work. Also thanks to A. Benović, D. Lučić, V. Onofri, J. Forte, and T. Makovec for their help in field work. We are grateful to anonymous reviewers for their critical and valuable comments on the manuscript.

References

Alldredge, A. L., 1972. Abandoned larvacean houses: a unique food source in the pelagic environment. Science 177: 885–887.

Alldredge, A. L., 1976. Discarded appendicularian houses as sources of food, surface habitats and particulate organic matter in planktonic environments. Limnology and Oceanography 21: 14–23.

Alldredge, A. L., 2005. The contribution of discarded appendicularian houses to the flux of particulate organic carbon from oceanic surface waters. In Gorsky, G., M. J. Youngbluth & D. Deibel (eds), Response of Marine Ecosystems to Global Change. Contemporary Publishing International, Paris, France: 309–326.

Alldredge, A. L. & M. J. Youngbluth, 1985. The significance of macroscopic aggregates (marine snow) as sites of heterotrophic bacterial production in the mesopelagic zone of the subtropical Atlantic. Deep Sea Research 32: 1445–1456.

Alldredge, A. L., J. Cole & D. A. Caron, 1986. Production of heterotrophic bacteria inhabiting organic aggregates (marine snow) from surface waters. Limnology and Oceanography 31: 68–78.

Alvarez-Colombo, G., A. Benović, A. Malej, D. Lučić, T. Makovec, V. Onofri, M. Acha, A. Madriolas & H. Mianzan, 2008. Acoustic survey of a jellyfish-dominated ecosystem (Mljet Island, Croatia). Hydrobiologia 616: 99–111.

Arai, M. N., 1997. A Functional Biology of Scyphozoa. Chapman & Hall, London.

Arai, M. N., 2005. Predation on pelagic coelenterates: a review. Journal of the Marine Biological Association of the United Kingdom 85: 523–536.

Azam, F. & F. Malfatti, 2007. Microbial structuring of marine ecosystems. Nature Reviews Microbiology 5: 782–791.

Båmstedt, U. & H. R. Skjoldal, 1980. RNA concentration of zooplankton: relationship with size and growth. Limnology and Oceanography 25: 304–316.

Benović, A., D. Lučić, V. Onofri, M. Peharda, M. Carić, N. Jasprica & S. Bobanović-Čoli, 2000. Ecological characteristics of the Mljet Island seawater lakes (Southern Adriatic Sea) with special reference to their resident populations of medusae. Scientia Marina 64: 197–206.

Billett, D. S. M., B. J. Bett, C. L. Jacobs, I. P. Rouse & B. D. Wigham, 2006. Mass deposition of jellyfish in the deep Arabian Sea. Limnology and Oceanography 51: 2077–2083.

Boström, K. H., K. Simu, A. Hagström & L. Riemann, 2004. Optimization of DNA extraction for quantitative marine bacterioplankton community analysis. Limnology and Oceanography Methods 2: 365–373.

Bradford, M. M., 1976. A rapid and sensitive method for the quantification of microgram quantities of protein utilizing the principle of protein-dye binding. Analytical Biochemistry 72: 248–254.

Carlson, C. A., S. J. Giovannoni, D. A. Hansell, S. J. Goldberg, R. Parsons, M. P. Otero, K. Vergin & B. R. Wheeler, 2002. Effect of nutrient amendments on bacteriplankton production, community structure and DOC utilization in the northwestern Sargasso Sea. Aquatic Microbial Ecology 30: 19–36.

Caron, D. A., P. G. Davis, L. P. Madin & J. Mc N. Sieburth, 1982. Heterotrophic bacteria and bacterivorous protozoa in oceanic macroaggregates. Science 218: 795–797.

Caron, D. A., E. L. Lim, R. W. Sanders, M. R. Dennett & U. G. Berninger, 2000. Responses of bacterioplankton and phytoplankton to organic carbon and inorganic nutrient additions in contrasting oceanic ecosystems. Aquatic Microbial Ecology 22: 175–184.

Cherrier, J. & J. E. Bauer, 2004. Bacterial utilization of transient plankton-derived dissolved organic carbon and nitrogen inputs in surface ocean waters. Aquatic Microbial Ecology 35: 229–241.

Clarke, A., L. J. Holmes & D. J. Gore, 1992. Proximate and elemental composition of gelatinous zooplankton from the Southern Ocean. Journal of Experimental Marine Biology and Ecology 155: 55–68.

Don, R. H., P. T. Cox, B. J. Wainwright, K. Baker & J. S. Mattic, 1991. 'Touchdown' PCR to circumvent spurious priming during gene amplification. Nucleic Acids Research 19: 4008.

Giovannoni, S. J., H. J. Tripp, S. Givan, et al., 2005. Genome streamlining in a cosmopolitan oceanic bacterium. Science 309: 1242–1245.

Grasshoff, K., M. Ehrhardt & K. Kremling, 1983. Methods of Seawater Analysis. Verlag Chemie, Weinheim.

Hamner, W. M. & M. N. Dawson, 2009. A review and synthesis on the systematic and evolution of jellyfish blooms: advantageous aggregations and adaptive assemblages. Hydrobiologia 616: 161–191.

Hansson, L. J. & B. Norrman, 1995. Release of dissolved organic carbon (DOC) by scyphozoan jellyfish *Aurelia aurita* and its potential influence on the production of planktonic bacteria. Marine Biology 121: 527–532.

Hollibaugh, J. T. & F. Azam, 1983. Microbial degradation of dissolved proteins in seawater. Limnology and Oceanography 28: 1104–1116.

Hoppe, H. G., S. J. Kim & K. Gocke, 1988. Microbial decomposition in aquatic environments: combined process of extracellular enzyme activity and substrate uptake. Applied and Environmental Microbiology 54: 784–790.

Houghton, J. D. R., T. K. Doyle, M. W. Wilson, J. Davenport & G. C. Hays, 2006. Jellyfish aggregations and leatherback turtle foraging patterns in a temperate environment. Ecology 87: 1967–1972.

Keil, R. G. & D. L. Kirchman, 1993. Dissolved combined amino acids: chemical form and utilization by marine bacteria. Limnology and Oceanography 38: 1256–1270.

Kirchman, D. L., 2008. Introduction and overview. In Kirchman, D. L. (ed.), Microbial Ecology of the Ocean, 2nd edn. Wiley-Blackwell, New Jersey.

Kirchman, D. L., E. K'Nees & R. Hodson, 1985. Leucin incorporation and its potential as a measure of protein synthesis by bacteria in natural aquatic systems. Applied and Environmental Microbiology 49: 599–607.

Kremer, P., 1975. Excretion and body composition of the ctenophore *Mnemiopsis leidyi* (A. Agassiz): comparison and consequences. In 10th European Symposium on Marine Biology, Ostend, Belgium: 351–362.

Kremer, P., 1977. Respiration and excretion by the ctenophore *Mnemiopsis leidyi*. Marine Biology 71: 149–156.

Larson, R. J., 1986. Water content, organic content, and carbon and nitrogen composition of medusa from the northeast Pacific. Journal of Experimental Marine Biology and Ecology 99: 107–120.

Lee, S. H. & J. A. Fuhrman, 1987. Relationships between biovolume and biomass of naturally derived bacterioplankton. Applied and Environmental Microbiology 53: 1298–1303.

Lo, W. T. & I. L. Chen, 2008. Population succession and feeding of scyphomedusae, *Aurelia aurita*, in a eutrophic tropical lagoon in Taiwan. Estuarine Coastal and Shelf Science 76: 227–238.

Lucas, C. H., 2001. Reproduction and life history strategies of the common jellyfish *Aurelia aurita*, in relation to its ambient environment. Hydrobiologia 451: 229–246.

Malej, A., V. Turk, D. Lučić & A. Benović, 2007. Direct and indirect trophic interactions of *Aurelia* sp. (Scyphozoa) in a stratified marine environment (Mljet Lakes, Adriatic Sea). Marine Biology 151: 827–841.

Martinez, L., D. C. Smith, G. F. Steward & F. Azam, 1996. Variability in the ectohydrolytic enzyme activities of pelagic marine bacteria and its significance for substrate processing in the sea. Aquatic Microbial Ecology 10: 223–230.

Mills, C. E., 1995. Medusae, siphonophores, and ctenophores as planktovorous predators in changing global ecosystem. ICES Journal of Marine Science 52: 575–581.

Muyzer, G. & K. Smalla, 1998. Application of denaturing gradient gel electrophoresis (DGGE) and temperature gradient gel electrophoresis (TGGE) in microbial ecology. Antonie Van Leeuwenhoek 73: 127–141.

Muyzer, G., E. D. Waal & A. G. Uitterlinden, 1993. Profiling of complex microbial populations by denaturing gradient gel electrophoresis analysis of polymerase chain reaction amplified genes coding for 16S rRNA. Applied and Environmental Microbiology 59: 695–700.

Passow, U. & A. L. Alldredge, 1994. Distribution, size, and bacterial colonization of transparent exopolymer particles (TEP) in the ocean. Marine Ecology Progress Series 113: 185–198.

Passow, U., A. L. Alldredge & B. E. Logan, 1994. The role of particulate carbohydrate exudates in the flocculation of diatom blooms. Deep Sea Research 41: 335–357.

Pinhassi, J., F. Azam, J. Hemphala, R. A. Long, J. Martinez, U. L. Zweifel & A. Hagstrom, 1999. Coupling between bacterioplankton species composition, population dynamics, and organic matter degradation. Aquatic Microbial Ecology 17: 13–26.

Pitt, K. A., D. T. Welsh & R. H. Condon, 2009. Influence of jellyfish blooms on carbon, nitrogen and phosphorous cycling and plankton production. Hydrobiologia 616: 133–149.

Porter, K. G. & Y. S. Feig, 1980. The use of DAPI for identifying and counting aquatic microflora. Limnology and Oceanography 25: 943–948.

Purcell, J. E. & M. V. Sturdevant, 2001. Prey selection and dietary overlap among zooplanktivorous jellyfish and juvenile fishes in Prince William Sound, Alaska. Marine Ecology Progress Series 210: 67–83.

Purcell, J. E., A. Malej & A. Benović, 1999. Potential links of jellyfish to eutrophication and fisheries. In Malone, T. C., A. Malej, L. W. Harding, N. Smodlaka & R. E. Turner (eds), Ecosystems at the Land-Sea Margin: Drainage Basin to Coastal Sea (Coastal and Estuarine Studies, 55). American Geophysical Union, Washington, DC: 241–263.

Riemann, L., J. Titelman & U. Båmstedt, 2006. Links between jellyfish and microbes in a jellyfish dominated fjord. Marine Ecology Progress Series 325: 29–42.

Schafer, H., L. Bernard, C. Courties, et al., 2001. Microbial community dynamics in Mediterranean nutrient-enriched seawater mesocosms: changes in genetic diversity of bacterial populations. FEMS Microbiology Ecology 34: 243–253.

Schneider, G., 1989. The common jellyfish *Aurelia aurita*: standing stock, excretion and nutrient generation in the Kiel Bight, western Baltic. Marine Biology 100: 507–514.

Simon, M. & F. Azam, 1989. Protein content and protein synthesis rates of planktonic marine bacteria. Marine Ecology Progress Series 51: 201–213.

Simon, M., A. L. Alldredge & F. Azam, 1990. Bacterial carbon dynamics on marine snow. Marine Ecology Progress Series 65: 205–211.

Smith, D. C. & F. Azam, 1992. A simple, economical method for measuring bacterial protein synthesis rates in seawater using ^{3}H-leucin. Marine Microbial Food Webs 6: 107–114.

Stoecker, D. K., A. E. Michaels & L. H. Davis, 1987. Grazing by the jellyfish, *Aurelia aurita*, on microplankton. Journal of Plankton Research 9: 901–915.

Sullivan, B. K., J. R. Garcia & G. Klein-MacPhee, 1994. Prey selection by the scyphomedusan predator *Aurelia aurita*. Marine Biology 121: 335–341.

Thingstad, T. F., Å. Hagström & F. Rassoulzdadegan, 1997. Accumulation of degradable DOC in surface waters: it is caused by malfunctioning microbial loop? Limnology and Oceanography 42: 398–404.

Titelman, J., L. Riemann, T. A. Sørnes, T. Nilsen, B. Griekspoor & U. Båmstedt, 2006. Turnover of dead jellyfish: stimulation and retardation on microbial activity. Marine Ecology Progress Series 325: 43–58.

Turk, V., D. Lučić, V. Flander-Putrle & A. Malej, 2008. Feeding of *Aurelia* sp. (Scyphozoa) and links to the microbial food web. Marine Ecology 29: 495–505.

West, E. J., D. T. Welsh & K. A. Pitt, 2009. Influence of decomposing jellyfish on the sediment oxygen demand and nutrient dynamics. Hydrobiologia 616: 151–160.

Yamamoto, J., M. Hirose, T. Ohtani, K. Sugimoto, K. Hirase, N. Shimamoto, T. Shimura, N. Honda, Y. Fujimori & T. Mukai, 2008. Transportation of organic matter to the sea floor by carrion falls of the giant jellyfish *Nemopilema nomurai* in the Sea of Japan. Marine Biology 153: 311–317.

Hydrobiologia (2010) 645:193–201
DOI 10.1007/s10750-010-0212-0

Identification of jellyfish from Continuous Plankton Recorder samples

E. J. Baxter · A. W. Walne · J. E. Purcell · R. McAllen · T. K. Doyle

Published online: 21 March 2010

Abstract Previous analysis of the Continuous Plankton Recorder (CPR) long-term data set for the presence of 'coelenterate' tissue revealed changes in the frequency of jellyfish occurrence in the North Atlantic Ocean and North Sea; however, the identities of the jellyfish were unknown, causing uncertainty in interpreting these findings. To improve the utility of the 'coelenterate' data from the CPR, 62 CPR samples were selected for re-analysis from an area where the widespread occurrence of the holoplanktonic scyphomedusan, *Pelagia noctiluca* (Forsskål, 1775), was previously documented from net tows to test if the CPR sampled *P. noctiluca*. Examination of the samples revealed smears (<1 mm) of golden gelatinous tissue on the sampling mesh. Subsequent microscopic examination and measurement of nematocysts identified most of the smears as *P. noctiluca*. Indeed, *P. noctiluca* was identified on 53% of the CPR samples from the area that best coincided with the documented bloom area. Although hydrozoans were also identified in the samples, the characteristic golden colour of *P. noctiluca* and its distinct suite of nematocysts allowed identification to species level. The identification of other jellyfish may be more complicated due to the labour-intensive examination of samples and unclear delineation of nematocyst types when multiple similar species are present. Nevertheless, re-analysis of CPR samples holds great promise for identifying the underlying reason for increases in jellyfish occurrence in the CPR records.

Keywords *Pelagia noctiluca* · Nematocysts · Medusae · North Atlantic Ocean

Guest editors: J. E. Purcell & Dror Angel / Jellyfish Blooms: New Problems and Solutions

E. J. Baxter (✉) · T. K. Doyle
Coastal and Marine Resources Centre, Environmental Research Institute, University College Cork, Naval Base, Haulbowline, Cork, Ireland
e-mail: e.baxter@ucc.ie

E. J. Baxter · R. McAllen
Department of Zoology, Ecology and Plant Science, University College Cork, Distillery Fields, North Mall, Cork, Ireland

A. W. Walne
Sir Alister Hardy Foundation for Ocean Science, Citadel Hill, Plymouth PL1 2PB, UK

J. E. Purcell
Shannon Point Marine Center, Western Washington University, 1900 Shannon Point Road, Anacortes, WA 98221, USA

Introduction

The Continuous Plankton Recorder (CPR) survey is the most extensive plankton monitoring programme in existence (Richardson et al., 2006). The abundance and distribution of some 437 phyto- and zooplankton taxa from near surface waters ($\sim$7 m) have been recorded throughout the North Atlantic Ocean and North Sea since 1946 (Warner & Hays, 1994). This unique monitoring tool provides the best long term measure of the state of the plankton in these areas, providing a historical window to assess the effects of climatic change on phytoplankton, zooplankton and entire marine ecosystems (Beaugrand et al., 2002; Edwards & Richardson, 2004; Richardson et al., 2006). More recently, the CPR data have been used to identify trends in jellyfish occurrences in the North Atlantic Ocean and North Sea (Attrill et al., 2007; Gibbons & Richardson, 2009). Recognition of such trends is important considering the many negative socio-economic impacts of jellyfish blooms (Purcell & Arai, 2001; Purcell et al., 2007).

Before discussing the analysis of data, a number of taxonomic and methodological issues regarding the CPR require some explanation. The CPR is a robust, high-speed plankton sampler, voluntarily towed behind merchant ships on their normal routes of passage (Warner & Hays, 1994). The CPR was originally designed to sample fish larvae from the plankton (Hardy, 1935); therefore, it also quantitatively samples large copepods and other robust mesozooplankton that are captured on the 270-μm mesh. Smaller taxa, such as diatoms and dinoflagellates, are not retained quantitatively on the meshes (Hays, 1994). Jellyfish and other soft bodied animals may be badly damaged during capture by the CPR because they are extremely delicate (Richardson et al., 2006) and may even be forced through the silk mesh. Furthermore, because the aperture of the CPR is small (1.62 cm^2), large gelatinous taxa such as scyphomedusae are unlikely to be sampled as whole individuals. Moreover, when acellular gelatinous material or nematocysts are found in the samples, they are recorded collectively as 'coelenterates' in the CPR data set (Richardson et al., 2006). Thus, this group may include tissues belonging to ctenophores, siphonophores, hydromedusae and scyphomedusae. A subcategory for siphonophores refers solely to the relatively firm calycophoran nectophores, which are rarely found in the samples (T. Jonas pers. comm.); however, the siphonophore nematocysts would be counted in the general 'coelenterate' category (Haddock, 2008; Gibbons & Richardson, 2009).

There are now more than 30,000 positive occurrences of coelenterates entered into the CPR database, making it an extremely valuable resource (Haddock, 2008; Gibbons & Richardson, 2009). Attrill et al. (2007) found an increase in the frequency of occurrence of jellyfish ('coelenterates') in the CPR record in the North Sea from 1958 to 2000 and that it was positively associated with the North Atlantic Oscillation (NAO) and the Atlantic inflow into the North Sea. In contrast, and using the same CPR data set (albeit combining the data from the 'coelenterate' and 'siphonophore' categories), Gibbons & Richardson (2009) found marked cycles of peaks and declines, but no general increase, in the occurrence of jellyfish during 1946–2005 for a much larger area of the North Atlantic Ocean. Gibbons & Richardson (2009) stated that the principle weakness of jellyfish occurrence in the CPR data is the lack of species identifications for the 'coelenterate' signature. Previous authors have speculated that the gelatinous tissue most likely to be sampled by the CPR is from the holoplanktonic hydrozoans (i.e. hydromedusae and siphonophores) because they are smaller and more abundant than scyphomedusae (Attrill et al., 2007; Haddock, 2008; Gibbons & Richardson, 2009).

In October 2007, a CPR was towed through an area where the widespread occurrence of the holoplanktonic scyphozoan, *Pelagia noctiluca*, was concurrently documented (Doyle et al., 2008). The bloom was observed throughout 1,815 km cruise track with large surface aggregations of very small medusa (mean diameter = 14.0 mm ± 7.1 SD, n = 1,867) visible during the day and night (Doyle et al., 2008). The great extent of the bloom presented an opportunity to test whether or not the CPR had sampled *P. noctiluca* and to elucidate how the CPR samples gelatinous zooplankton. Therefore, CPR samples collected from this area were re-examined in order to determine the species sampled by the CPR. It is important to stress that, unlike previous authors (Attrill et al., 2007; Gibbons & Richardson, 2009) who used the existing CPR data set, the CPR samples were re-analysed for cnidarian tissue only (using the presence of nematocysts) and will be referred to as such in the description of this work hereafter.

Materials and methods

Sections of CPR silk (each 104.0 cm^2 of 270-μm filtering mesh) were analysed in November 2008 in the laboratories at the Sir Alister Hardy Foundation for Ocean Science (SAHFOS), Plymouth, UK. Each section represents the filtration of approximately 3 m^3 of water and a towing distance of ~18.5 km. All sections were taken from one CPR route in the North Atlantic (W route) for the months of September, October and November 2007; however, only alternating sections were analysed as is standard CPR protocol (Fig. 1). In addition, to these alternating sections, all sections for a 185-km segment of the October 2007 route also were analysed. This segment corresponded to the cruise track of the aptly named RV Pelagia, which had documented the widespread occurrence of *Pelagia noctiluca* (Fig. 1). In total, 62 sections of silk/samples was analysed on a specialised, mobile glass stage under a Micro Instruments compound microscope (Richardson et al., 2006) for the occurrence of cnidarian tissue. This re-analysis of the samples used visual cues, including tissue colour and texture, as well as nematocyst identification and measurements. The nematocyst measurements were compared to nematocyst samples from cnidarians collected during the RV Pelagia cruise. All cnidarian tissues were recorded when encountered and identified as *P. noctiluca* or other taxa.

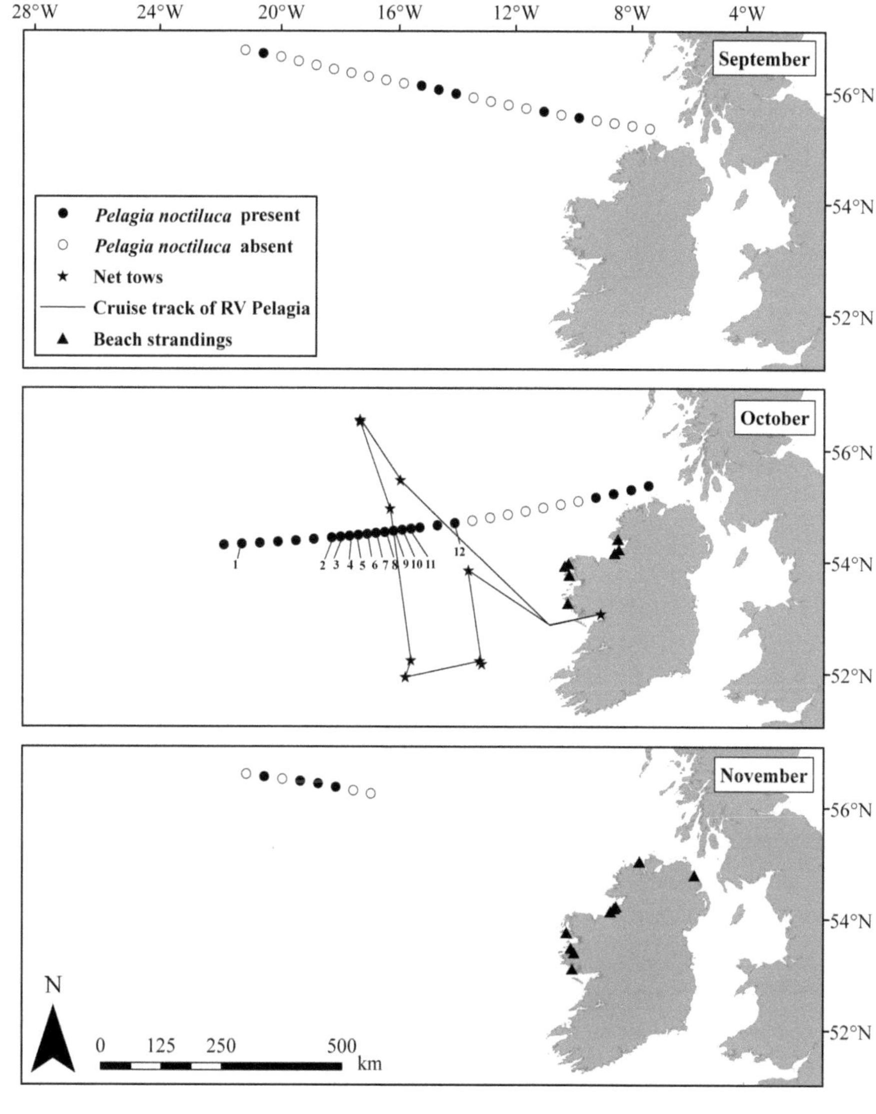

Fig. 1 Transects of the CPR on the W route over 3 months in 2007: 10–11 September; 22–23 October; and 13 November, showing the presence and absence of *Pelagia noctiluca* per CPR sample analysed. Numbered samples in October relate to samples analysed for percentage cover of *P. noctiluca* tissue displayed in Table 1. Net tows were undertaken from 2 to 17 October 2007 by Doyle et al. (2008) on the cruise of the RV Pelagia. *P. noctiluca* was also observed in coastal waters and as reported beach strandings during October and November 2007 (Doyle et al., 2008)

Some samples underwent detailed analyses, specifically, the previously mentioned 185-km segment that traversed the cruise track as well as two other samples for which *P. noctiluca* was present. We analysed the samples at 62.5× magnification with a stepped traverse, in which the width of the sample is examined in a series of steps moving down the silk. This replicated the standard CPR protocol detailed in Richardson et al. (2006), except we had a larger field of view (FOV, 2.97 mm in diameter). The number of tissue patches identified as *P. noctiluca* by the distinctive golden-brown colour and nematocysts encountered throughout the traverse were recorded in the following categories defined by tissue patch size: (a) <1 mesh square (7.3×10^{-4} cm^2); (b) between 1 and 2 mesh squares ($>7.3 \times 10^{-4}$ to 1.5×10^{-3} cm^2) and (c) 2–4 mesh squares ($>1.5 \times 10^{-3} \leq 2.9 \times 10^{-3}$ cm^2). This method traversed 1/34 of the sample, which is an area larger than the 1/50 accepted for standard CPR analysis (Richardson et al., 2006) due to the size of the FOV. A subsequent qualitative estimation of abundance was made using the number of mesh squares covered by *P. noctiluca* tissue within the traversed area multiplied by 34 to equal the area of the sample (104.0 cm^2) and then converted into percentage cover of the sample (Table 1).

The types and sizes of discharged nematocysts from tissues on the CPR samples, preserved samples of medusae, and published data were compared to determine if the golden tissues in the samples were from *P. noctiluca* medusae (Tables 2, 3). The capsule

Table 1 Qualitative estimates of the abundance of *Pelagia noctiluca* in terms of percentage cover of CPR samples

Sample no.	Date	Time (GMT)	Longitude (°W)	Latitude (°N)	% Mesh covered
1	22	15:57	20.49	53.98	26
2	22	21:39	17.71	54.34	44
3	22	21:13	17.43	54.37	30
4	22	22:47	17.15	54.41	28
5	22	23:21	16.87	54.44	26
6	22	23:55	16.59	54.48	19
7	22	00:30	16.31	54.51	27
8	23	01:04	16.03	54.55	27
9	23	01:38	15.75	54.58	23
10	23	02:12	15.47	54.62	27
11	23	02:46	15.19	54.65	32
12	23	05:32	13.78	54.82	12

Analysis of samples 2–11 covers a continuous 185-km segment of the October 2007 (W route) transect across the eastern North Atlantic Ocean (sample numbers indicated on Fig. 1)

Table 2 Comparison of *Pelagia noctiluca* nematocyst capsule lengths

Type		Heterotrichous microbasic euryteles			Holotrichous O-isorhizas		
Samples		CPR	RV Pelagia	Avian et al. (1991)	CPR	RV Pelagia	Avian et al. (1991)
Length (μm)	Min	8.00	10.00	10.90	16.00	16.00	14.77
	Max	14.00	14.00	15.11	22.00	24.00	24.97
	Mean	9.99	11.49	12.88	19.93	20.60	20.82
	Median	10.00	12.00	–	20.00	22.00	–
	SD	1.21	1.21	0.90	1.29	2.15	1.94
	n	384	160	148	116	67	196

Measurements presented are from discharged nematocysts; heterotrichous microbasic euryteles and holotrichous O-isorhizas (images in Fig. 2) from October 2007 CPR samples, samples collected on RV Pelagia cruise (30 September to 22 October 2007) (both from fixed samples) and data of Avian et al. (1991) for individually isolated, unfixed nematocysts

Table 3 Length ranges (μm) of undischarged nematocysts of pelagic scyphozoans, hydrozoans and anthozoans common in the North Atlantic Ocean

Nematocyst type	Scyphozoa				Hydrozoa		Anthozoa
	Aurelia aurita[a]	*Chrysaora hysoscella*[b]	*Cyanea capillata*[c]	*Pelagia noctiluca*[d]	*Aglantha digitale*[e]	*Muggiaea atlantica*[f]	Anthozoan larvae[g]
Stenoteles	–	–	–	–	7.0–10.0[h]	–	–
Euryteles	6.9–11.9	10.0–13.0	8.5–17.0	10.1–16.9	7.0–10.0[h]	–	–
a-isorhizas	4.0–6.5	7.0–9.0 (S) 11.0–14.0 (L)	5.0–10.5	4.4–7.9	–	–	–
Isorhizas	–	15.0–17.0	–	12.2–21.7	–	–	–
O-isorhizas	–	–	10.0–18.0	15.20–30.1	–	–	–
A-isorhizas	–	–	7.0–12.0 (S) 11.0–19.5 (M) 20.0–28.0 (L) 17.0–32.0 (sp.)	–	–	–	–
Microbasic birhopaloids	–	–	8.0–10.0 (S) 10.5–16.5 (M) 17.0–17.5 (L)	–	–	–	–
Desmonemes	–	–	–	–	–	7.0–10.0	
Microbasic p-mastigophores	–	–	–	–	–	36.0–43.0	–
Anisorhizas	–	–	–	–	–	13.0–18.0	–
Spirocysts	–	–	–	–	–	–	30.8–32.6

All measurements refer to unfixed material except that of the anthozoan larvae for which the measurements are from unfixed material. Size classes of small, medium and large refer to those defined by the original author

S small, *M* medium, *L* large, *sp.* spherical

[a] Calder (1977); [b] Russell (1970); [c] Östman & Hydman (1997); [d] Avian et al. (1991); [e] Russell (1940); [f] Russell (1938); [g] This study

[h] Nematocyst type evolutionary between two forms

lengths of 50 discharged nematocysts from golden tissues on the same 10 samples were measured to the nearest 2 μm with the aid of an ocular micrometer. Two out of four easily identifiable nematocyst types (heterotrichous microbasic euryteles and holotrichous O-isorhizas) were measured. Nematocysts from *P. noctiluca* medusae collected from net tows during the RV Pelagia cruise (October 2007) in the same area as the CPR samples (Fig. 1) were also re-analysed and measured. These preserved nematocysts were prepared for examination by taking a small piece of tentacle, oral arm, and umbrella from five individuals ranging from ephyrae to small medusae and squashing each segment between a microscope slide and cover slip. The mean lengths for each nematocyst type were compared among the sample sets using a Mann–Whitney *U* test. A comparison was also made to previously published data on the lengths of unfixed *P. noctiluca* nematocysts (Avian et al., 1991) (Table 2).

Results

Initial analysis of the samples revealed smears of gelatinous tissue, rather than whole jellyfish, identified as cnidarian tissues by the presence of nematocysts. In fact, cnidarian tissues were observed on 94% of the CPR samples analysed from September to November 2007 on the W route. Fifty-three per cent of samples contained cnidarian tissue with a golden colour and nematocyst types that were consistent with preserved *Pelagia noctiluca* medusae. Fifty-two per cent of the CPR samples contained other cnidarian tissues of indistinct colour and often hydrozoan-specific nematocysts (e.g. an intact physonect siphonophore nematocyst battery) that were not classified as *P. noctiluca*.

Nematocysts found in the CPR samples were compared to nematocysts of hydrozoans, scyphozoans and anthozoans, and our data then was compared to published data. Table 3 summarises the nematocyst types found in common epipelagic cnidarians of the North Atlantic Ocean; additional classification of nematocyst types on the basis of spine and shaft types were removed for simplicity [e.g. hetertrichous = two or more kinds of spines; microbasic = shaft <1.5-times capsule length (Östman, 2000)].

In the tissues identified as *P. noctiluca* according to tissue colour and nematocyst signature, the two predominant nematocyst types were heterotrichous microbasic euryteles and holotrichous O-isorhizas (Fig. 2). These were consistent between samples from the CPR and RV Pelagia cruise, and were comparable to the data in Avian et al. (1991). The lengths (means and ranges) of both types of discharged nematocysts were very similar in fixed *P. noctiluca* samples from both the CPR and the RV Pelagia cruise. The frequency distributions of nematocyst lengths of both nematocyst types were within the same range, but the mean lengths differed significantly between CPR and RV Pelagia samples (heterotrichous microbasic euryteles: Mann–Whitney $U = 13{,}651$, $P > 0.05$; holotrichous O-isorhizas: $U = 2{,}879$, $P > 0.05$). The median values for each nematocyst type were consistently 2 μm smaller in the CPR samples than in the RV Pelagia samples. The nematocyst length measurements of unfixed *P. noctiluca* nematocysts from the Adriatic Sea (Avian et al., 1991) were also within a similar range to those compared from the CPR and RV Pelagia samples (Table 2).

Our analysis of the CPR samples indicated that *P. noctiluca* was present in CPR samples from September to November on the W route (Fig. 1). In October, the occurrence of *P. noctiluca* was recorded at intervals for ~900 km across the eastern North Atlantic Ocean and occurred continuously (in samples analysed) for a maximum distance of 480 km between 22 and 23 October 2007. There was an increase in the frequency of occurrence of *P. noctiluca* in the CPR samples from 25% in September to 72% of equivalent October samples analysed. Although the November CPR route was shortened due to technical reasons, with only the more westerly half producing viable samples, *P. noctiluca* was recorded in 50% of the samples analysed.

In an attempt to quantify the amount of *P. noctiluca* tissue that was present on the CPR samples examined, some of the samples were analysed in detail. The percentage of the mesh covered by

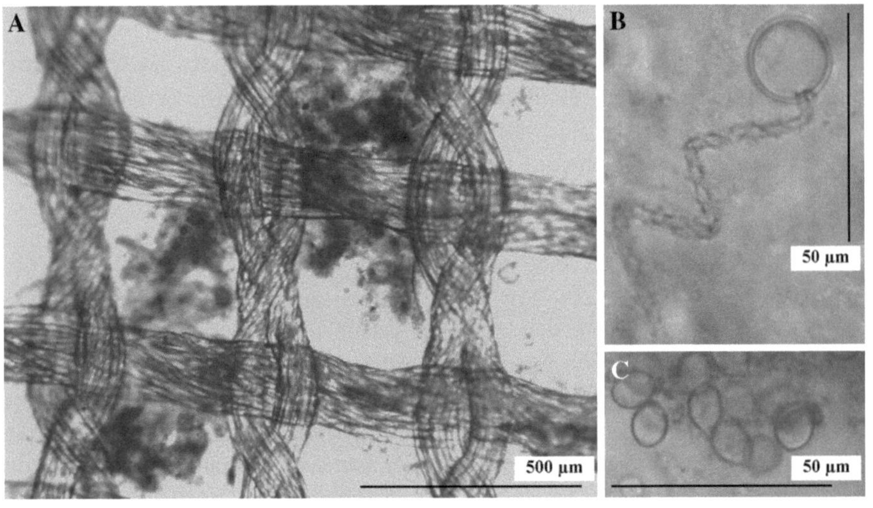

Fig. 2 **A** Light micrograph of *Pelagia noctiluca* tissue on the CPR sample mesh. **B**, **C** High magnification light micrographs of a discharged *P. noctiluca* nematocysts: **B** holotrichous O-isorhiza, showing shaft and spines and **C** heterotrichous microbasic euryteles without visible shafts

P. noctiluca tissue per sample ranged from 12 to 44% (Table 1) with a mean of 27% over the area of October samples that were analysed by this method.

Discussion

The prediction of 'a more gelatinous future for the North Sea' (Attrill et al., 2007) triggered a debate on the identity of jellyfish sampled by the CPR. By re-analysing the CPR samples, our study provides important insights into how the CPR sampled a single jellyfish species (*Pelagia noctiluca*) that was known to be widespread in the area traversed by the selected CPR route (W route, Fig. 1). Initial analysis revealed that rather than individual jellyfish being visible on the CPR samples, only smears of cnidarian tissue were present. Nonetheless, it was apparent that a large proportion (53%) of the samples contained cnidarian tissue with a golden colour characteristic of *P. noctiluca*.

Our investigation into the nematocysts found in the cnidarian tissue indicated that the mean length and length ranges of capsules from the CPR samples and the RV Pelagia cruise were very similar (Table 2) and the median values for each nematocyst type differed systematically by 2 μm between the two sample sets. We believe this to be an artefact of measuring nematocysts from globular tissue masses in the CPR samples versus thin sections of tissues analysed on microscope slides for the RV Pelagia samples. The difference between the mean lengths of heterotrichous microbasic euryteles from the two sample sets may be further explained by the ellipsoid shape of the nematocyst capsule, which could have been measured end-on if examined within a mass of tissue, and therefore measuring a smaller length. Despite this, the disparity between the values is less than interspecific differences of the same nematocyst type. Thus, the nematocyst signature of the golden smears matched only that of *P. noctiluca* (Table 3).

A principle weakness of the 'coelenterate' signature from the CPR database is the lack of species identification (Gibbons & Richardson, 2009); the findings here are the first to show conclusively that approximately 50% of the samples we examined contained tissue attributable to *P. noctiluca*. None of the previous studies (Attrill et al., 2007; Attrill & Edwards, 2008; Haddock, 2008; Gibbons & Richardson, 2009) suggested *P. noctiluca* as a likely contributor to the cnidarians in the 'coelenterate' tissue collected by the CPR.

Doyle et al. (2008) found *P. noctiluca* to be present in all 13 net tows randomly deployed throughout 1,815 km cruise track off the west coast of Ireland (Fig. 1). In addition, several observations of large surface aggregations of *P. noctiluca* (>1,000 jellyfish seen during a 5-min visual survey) were documented by Doyle et al. (2008) during both day and night. At other times, only 1–10 individual jellyfish were recorded per 5-min survey interval. Despite such observed patchiness, all CPR samples in October 2007 that crossed the cruise track of the RV Pelagia (Fig. 1) contained *P. noctiluca* tissue. Moreover, CPR samples analysed within and to the west of the cruise track had very high percentage coverage of *P. noctiluca* tissue (Table 1). Little information exists on the broad-scale distribution of jellyfish (Doyle et al., 2007; Purcell, 2009). The CPR data presented here suggest that *P. noctiluca* was present over a linear distance of 480 km. These findings extend the widespread occurrence of *P. noctiluca* 350 km westward of that described by Doyle et al. (2008).

There was an apparent absence of *P. noctiluca* in the CPR data for 220 km to the east of the RV Pelagia cruise track before it was recorded again along the northern Irish coastline (Fig. 1). The samples spanning this area of apparent absence contained both day- and night-time samples; therefore, it is unlikely that diel vertical migration (DVM) was responsible for this absence. Indeed, because *P. noctiluca* was present during both day- and night-time sampling of the CPR throughout the October transect and was observed during day and night by Doyle et al. (2008), we believe that other factors were responsible for the observed spatio-temporal variation in the distribution of this species. However, analysis of a longer time series of samples would be required before conclusions could be drawn.

Pelagia noctiluca was present on 25% of the September samples analysed compared to 72% of equivalent, more southerly, samples analysed from October 2007. Due to problems with the CPR sampling in November 2007 and the absence of the easterly samples from this month, it was not possible to accurately extrapolate these data to compare the distribution of *P. noctiluca* across this area with that of the September and October distributions. It is likely that the *P. noctiluca* distribution was very widespread, as a result of the presence of this species in the western samples and from beach stranding data for Ireland throughout November (Doyle et al., 2008). Analysis of the CPR samples can provide essential information on the occurrence and distribution of *P. noctiluca* over wide areas, and when this is combined with ship-board observations and discrete net tow data, the resolution of the distribution can be increased.

Jellyfish blooms can have serious negative socio-economic affects on tourism (Macrokanis et al., 2004; Ariza et al., 2008), aquaculture and fisheries (Purcell & Arai, 2001; Purcell et al., 2007). Sabatés et al. (this volume) showed that the ephyrae of *P. noctiluca* on the Catalan coast positively select for fish larvae (at night) as part of their diet, including commercially important species, including anchovy. The bloom documented here and in Doyle et al. (2008) proceeded to cause a substantial fish kill in November 2007, killing 250,000 harvest-sized salmon, thus destroying Northern Ireland's only salmon farm. Hence, a combined approach to investigate the occurrences of jellyfish blooms could provide useful information for resource managers.

Although the present study specifically set out to determine whether the CPR had adequately sampled the widespread bloom of *P. noctiluca* in the North Atlantic, other cnidarian taxa were also clearly identified. Both *P. noctiluca* and other cnidarians were recorded in the same sample on some occasions (15% of samples) and it was possible to differentiate patches of *P. noctiluca* tissues from those of other cnidarians on the basis of colour and nematocyst types. This highlights the fact that both scyphozoan and hydrozoan cnidarians are sampled by the CPR.

Confirmation that the *P. noctiluca* bloom was sampled by the CPR demonstrates that, with some effort, CPR samples can be re-analysed for the presence of jellyfish taxa sometimes to species level. Subsequent, targeted re-analysis of samples for a specific area (e.g. the northern coast of Ireland) over a much longer time period could be conducted, with the objective of producing a time series of jellyfish within the CPR samples; however, there are several important considerations. The analysis presented here of 62 samples required approximately 45 h of microscope time to examine the tissues of relatively easily identified species (due to colour). Similar analyses for other taxa may therefore be more complicated, especially when nematocyst signatures may not be easily distinguished. Although the use of molecular techniques will enable identification of jellyfish species from CPR samples, as has been demonstrated for non-gelatinous plankton (Kirby & Lindley, 2005), our study highlights the potential for cost-effective, visual re-analysis of CPR samples for the presence of *P. noctiluca* or other cnidarians. Future re-analysis will need to consider the sample size required to construct robust time series given the inherent variability in plankton abundance (Hays et al., 1993) and the loss of colour in *P. noctiluca* specimens that may occur over time.

Finally, with respect to previous analyses of the CPR data set (Attrill et al., 2007; Attrill & Edwards, 2008; Gibbons & Richardson, 2009), the confirmation of *P. noctiluca* in CPR samples has wide implications for the North Atlantic Ocean and the North Sea. Late autumn peaks in the frequency of jellyfish occurrences were documented in the North Sea by Attrill et al. (2007), which agrees with the observed seasonal advection of jellyfish from oceanic areas onto the shelf by Witt et al. (2007) and Gibbons & Richardson (2009). Specifically, *P. noctiluca* is also known to enter the North Sea when the timing of its arrival in the eastern North Atlantic coincides with periods of strong Atlantic inflow (Fraser, 1955; Reid et al., 2003). Although other cosmopolitan, epipelagic species may contribute to this signal, considering the abundance of *P. noctiluca* in the eastern North Atlantic Ocean during October 2007 (and in 2008 and 2009, pers. obs.) and historically (Doyle et al., 2008), it is possible that these small, oceanic scyphomedusae may, at times, represent a large fraction of the gelatinous material recorded by the CPR in the North Sea.

Acknowledgements This project [Grant-Aid Agreement No PBA/AF/08/002(01)] was carried out under the Sea Change strategy with the support of the Marine Institute and the Marine Research Sub-Programme of the National Development Plan 2007–2013, co-financed under the European Regional Development Fund. EJB was supported by the above grant aid. TKD was supported by an Irish Research Council for Science Engineering and Technology fellowship. The authors would like to acknowledge all past and present members and supporters of the CPR survey, especially the ships that voluntarily tow CPRs on regular routes and the Sir Alister Hardy Foundation for Ocean Science (SAHFOS) for facilitating the use of the CPR samples and the SAHFOS laboratories. We thank G. Hays and two other anonymous reviewers for their comments on the article.

References

Ariza, E., J. Jiménez & R. Sardá, 2008. A critical assessment of beach management on the Catalan coast. Ocean and Coastal Management 51: 141–160.

Attrill, M. J. & M. Edwards, 2008. Reply to Haddock, S.H.D. Reconsidering evidence for potential climate-related increases in jellyfish. Limnology and Oceanography 53: 2763–2766.

Attrill, M. J., J. Wright & M. Edwards, 2007. Climate-related increases in jellyfish frequency suggest a more gelatinous future for the North Sea. Limnology and Oceanography 52: 480–485.

Avian, M., P. Negro & L. Sandrini, 1991. A comparative analysis of nematocysts in *Pelagia noctiluca* and *Rhizostoma pulmo* from the North Adriatic Sea. Hydrobiologia 52: 480–485.

Beaugrand, G., F. Ibanez, J. A. Lindley & P. C. Reid, 2002. Diversity of calanoid copepods in the North Atlantic and adjacent seas: species associations and biogeography. Marine Ecology Progress Series 232: 179–195.

Calder, D., 1977. Nematocysts of the ephyra stages of *Aurelia*, *Chrysaora*, *Cyanea*, and *Rhopilema* (Cnidaria, Scyphozoa). Transactions of the American Microscopical Society 96: 13–19.

Doyle, T. K., J. D. R. Houghton, S. M. Buckley, G. C. Hays & J. Davenport, 2007. The broad-scale distribution of five jellyfish species across a temperate coastal environment. Hydrobiologia 579: 29–39.

Doyle, T. K., H. De Haas, D. Cotton, B. Dorschel, V. Cummins, J. D. R. Houghton, J. Davenport & G. C. Hays, 2008. Widespread occurrence of the jellyfish *Pelagia noctiluca* in Irish coastal and shelf waters. Journal of Plankton Research 30: 963–968.

Edwards, M. & A. J. Richardson, 2004. Impact of climate change on marine pelagic phenology and trophic mismatch. Nature 430: 881–884.

Fraser, J. H., 1955. The plankton of the waters approaching the British Isles in 1953. Marine Research 1: 1–12.

Gibbons, M. J. & A. J. Richardson, 2009. Patterns of jellyfish abundance in the North Atlantic. Hydrobiologia 616: 51–65.

Haddock, S. H. D., 2008. Reconsidering evidence for potential climate-related increases in jellyfish. Limnology and Oceanography 53: 2759–2762.

Hardy, A., 1935. The continuous plankton recorder: a new method of survey. Rapports et Proces-Verbaux des Reunions. Conseil International pour l'Exploration de la Mer 95: 36–47.

Hays, G. C., 1994. Mesh selection and filtration efficiency of the Continuous Plankton Recorder. Journal of Plankton Research 16: 403–412.

Hays, G. C., M. R. Carr & A. H. Taylor, 1993. The relationship between Gulf Stream position and copepod abundance derived from the Continuous Plankton Recorder Survey: separating biological signal from sampling noise. Journal of Plankton Research 15: 1359–1373.

Kirby, R. & J. Lindley, 2005. Molecular analysis of Continuous Plankton Recorder samples, an examination of echinoderm larvae in the North Sea. Journal of the Marine Biological Association of the UK 85: 451–459.

Macrokanis, C., N. Hall & J. Mein, 2004. Irukandji syndrome in northern Western Australia: an emerging health problem. Medical Journal of Australia 181: 699–702.

Östman, C., 2000. A guideline to nematocyst nomenclature and classification, and some notes on the systematic value of nematocysts. Scientia Marina 64: 31–46.

Östman, C. & J. Hydman, 1997. Nematocyst analysis of *Cyanea capillata* and *Cyanea lamarckii* (Scyphozoa, Cnidaria). Scientia Marina 61: 313–344.

Purcell, J. E., 2009. Extension of methods for jellyfish and ctenophore trophic ecology to large-scale research. Hydrobiologia 616: 23–50.

Purcell, J. E. & M. N. Arai, 2001. Interactions of pelagic cnidarians and ctenophores with fish: a review. Hydrobiologia 451: 27–44.

Purcell, J. E., S. Uye & W. T. Lo, 2007. Anthropogenic causes of jellyfish blooms and their direct consequences for humans: a review. Marine Ecology Progress Series 350: 153–174.

Reid, P. C., M. Edwards, G. Beaugrand, M. Skogen & D. Stevens, 2003. Periodic changes in the zooplankton of the North Sea during the twentieth century linked to oceanic inflow. Fisheries Oceanography 12: 260–269.

Richardson, A. J., A. W. Walne, A. G. W. John, T. D. Jonas, J. A. Lindley, D. W. Sims, D. Stevens & M. Witt, 2006. Using continuous plankton recorder data. Progress in Oceanography 68: 27–74.

Russell, F. S., 1938. On the nematocysts of hydromedusae. Journal of the Marine Biological Association of the United Kingdom 23: 145–165.

Russell, F. S., 1940. On the nematocysts of hydromedusae. III. Journal of the Marine Biological Association of the United Kingdom 24: 515–523.

Russell, F. S., 1970. The Medusae of the British Isles: II. Cambridge University Press, Cambridge.

Sabatés, A., F. Pagès, D. Atienza, V. L. Fuentes, J. E. Purcell & J.-M. Gili. Planktonic cnidarian distribution and feeding of *Pelagia noctiluca* from near shore to open sea in the NW Mediterranean Sea. Hydrobiologia. doi:10.1007/s10750-010-0221-z.

Warner, A. J. & G. C. Hays, 1994. Sampling by the Continuous Plankton Recorder Survey. Progress in Oceanography 34: 237–256.

Witt, M. J., A. C. Broderick, D. J. Johns, C. Martin, R. Penrose, M. S. Hoogmoed & B. J. Godley, 2007. Prey landscapes help identify potential foraging habitats for leatherback turtles in the NE Atlantic. Marine Ecology Progress Series 337: 231–243.

Hydrobiologia (2010) 645:203–212
DOI 10.1007/s10750-010-0227-6

Separation and analysis of different types of nematocysts from *Cyanea capillata* (L.) medusae

Annika Wiebring · Heike Helmholz · Stephan Lassen · Andreas Prange · Gerhard Jarms

Published online: 21 March 2010

Abstract Medusae play an important role in marine ecosystems, as competitors of many invertebrate and fish species. Additionally, jellyfish stings can cause severe pain, inflammation of the affected skin, and allergic reactions in human. Climate and environmental changes are likely to affect the medusae, but it is not yet clear whether these will affect their distribution, physiology, and their toxicity. Very little is known about the effect of biotic and abiotic factors on the proliferation and the distribution of medusan nematocysts. In this study, we compared three types of nematocysts (euryteles and A- and O-isorhizas) and venoms of *Cyanea capillata* medusae (Scyphozoa) obtained from the North Sea and the Baltic Sea, which have different salinity and temperature ranges. Different types of nematocysts were separated by laser microdissection and pressure catapulting (LMPC), and the proteinaceous contents of the nematocysts were analyzed by matrix-assisted laser desorption ionization time-of-flight mass spectrometry (MALDI-TOF MS). Medusae from the brackish Baltic Sea possessed more euryteles than those from the North Sea. The O-isorhizas and A-isorhizas were smaller in the Baltic Sea sample compared to the North Sea samples and the length-to-width ratios were larger in the Baltic Sea sample. Moreover, the pattern of proteins (potential toxins) obtained from the separated nematocysts showed differences among samples and nematocyst types, but no clear pattern was observable. This study displays the novel LMPC/MALDI-TOF MS approach as a useful tool to investigate the function and venom of cnidarian nematocysts types.

Keywords Jellyfish · Sting · Protein pattern · LMPC · Mass spectrometry · Cnidom · Venom

Guest editors: J. E. Purcell & Dror Angel / Jellyfish Blooms: New Problems and Solutions

A. Wiebring (✉) · H. Helmholz · S. Lassen · A. Prange
Department for Marine Bioanalytical Chemistry, GKSS Research Centre, Institute for Coastal Research, Max-Planck-Street 1, 21502 Geesthacht, Germany
e-mail: annika.wiebring@gkss.de

A. Wiebring · G. Jarms
Department of Biology, Biocenter Grindel and Zoological Museum, University of Hamburg, 20147 Hamburg, Germany

Introduction

In recent years, human problems with jellyfish blooms, and their effects on marine ecosystems appear to be increasing on a worldwide basis (reviewed in Purcell et al., 2007). Many species of jellyfish are known to cause injuries or even fatalities by stinging (Burnett et al., 1988; Fenner, 1998; Segura-Puertas et al., 2002; Tibballs, 2006; Šuput, 2009). Despite the wide-spread concern about the damage caused by jellyfish stings to the tourism and aquaculture industries (Doyle et al., 2008), surprisingly little is known about the effect of various environmental factors on the toxins of the medusae. Differences in the stinging potencies of the same species from different habitats have been documented (Burnett et al., 1988; Radwan et al., 2001; Segura-Puertas et al., 2002).

Medusae compete with fish over food resources and consume large numbers of zooplankton, fish eggs and larvae, thereby affecting marine food webs (Purcell & Arai, 2001; Brodeur et al., 2002; Purcell, 2003; Lynam et al., 2005). Several factors affect the occurrence of medusae, including abiotic factors, such as temperature, salinity, solar irradiance, and currents and biotic factors, such as food availability (Lynam et al., 2004; Purcell, 2005; Purcell & Decker, 2005; Hay, 2006, Purcell et al., 2009).

Medusae, like all cnidarians, are equipped with nematocyst-bearing stinging cells. The cells may contain a several different types of nematocysts, which and are assumed to have different functions, e.g., entangling or paralyzing prey, defence, or locomotion (Rifkin & Endean, 1983; Purcell, 1984; Heeger & Möller, 1987; Purcell & Mills, 1988; Carrette et al., 2002; Jarms et al., 2002; Kintner et al., 2005; Colin & Costello, 2007). The proportions of the various types of nematocysts may vary among individuals within the same species. Carrette et al. (2002) described different ratios of distinctive nematocyst types in *Chironex fleckeri* Southcott and *Chiropsalmus* sp. and suggested that different diets during the growth of the medusae may be responsible for the different frequencies of nematocyst types. Furthermore, Burnett et al. (1988), Radwan et al. (2001), and Segura-Puertas et al. (2002) suggested that the variation in stinging potency and toxic potential in the medusae they studied was related to their different habitat conditions.

The objective of this study was to compare the cnidom and the capsule contents of separated nematocysts of *C. capillata* (L.) medusae collected from the North Sea and from the Baltic Sea. Frequencies and sizes of three nematocyst types were determined microscopically. Other aims of this study were to assess the utility of the LMPC/MALDI-TOF MS approach and the mass spectrometric characterisation of the proteinaceous contents of separated capsules.

Materials and methods

Organisms and preparation

Three different groups of *C. capillata* medusae were used to compare the proportions of distinctive nematocyst types and their potential toxic contents in different size classes and in animals from different habitats. Group NS1 consisted of medusae from a research cruise to the Orkney Islands (Scotland) in June 2005. Seawater salinity (S) and temperature was 35 and 12°C, respectively. The group consisted of seven animals with an average umbrella diameter of 24 cm (range: 16–31 cm). Medusae in group NS2 were collected in June 2007 at the same site, where water salinity and temperature were the same. This group was composed of 48 medusae with an average umbrella diameter of 13 cm (range: 11–20 cm). The third sample (BS) was collected in the harbour of Eckernförde, Baltic Sea in September 2008, where the salinity was 20 and temperature 16°C. The BS group consisted of 72 medusae with an average umbrella diameter of 21 cm (range: 11–28 cm). In each of the sampling expeditions, medusan fishing tentacles were sliced off immediately after collection and combined in one container. The fishing-tentacle mixture was macerated by gently stirring the material in distilled water on ice. After washing the samples two to three times with sterile filtered seawater in order to remove unwanted material, samples were stored on board at −80°C (Helmholz et al., 2007).

At the laboratory, the frozen fishing tentacles mixture was thawed on ice and a fraction of this material was macerated again with cold distilled water (4°C) using a roller mixer (SRT9, Stuart, Staffordshire, Great Britain). The resulting suspension was centrifuged at 4°C at 11.5 g for 3 min. After discarding the supernatant, the pellet was washed with filtered seawater. These steps were repeated three to four times (Helmholz et al., 2007), and the resulting samples were stored at −80°C.

Microscopic analysis of the nematocyst suspension

In order to compare the cnidom of the medusae samples, three different types of capsules were analysed (euryteles, A- and O-isorhizas) because their relatively large sizes were necessary for the individual separation and protein analysis below. These three nematocyst types were classified according to Östman & Hydman (1997) based on their morphology. The work of Östman & Hydman (1997) is based on the classification of Weill (1934) and Mariscal (1974) modified by Calder (1974) and Östman et al. (1995). The other two nematocyst types classified by Östman & Hydman (1997), birhopaloids and a-isorhizas, were not included in this analysis. Furthermore, no distinction was made between large spherical and large ovate A-isorhizas.

The percentages of the three nematocyst types in the suspension was examined microscopically (Olympus BX 51/M, Hamburg, Germany) in a counting chamber with a depth of 0.1 mm (Neubauer Improved, LO-Laboroptik, Friedrichshof, Germany). Prior to microscopy, the frozen samples were thawed on ice, centrifuged at 4°C for 3 min at 11.5 g, and the counting chamber was loaded with 30 µl of the nematocyst suspension. Four large corner squares with a total surface area of 4 mm^2 were counted. The counting procedure was repeated with 10 aliquots for each nematocyst suspension. Comparison among the three groups of medusae was performed by calculating relative frequencies (mean $\pm$ SD) of the three nematocyst types (sum of frequencies of analysed nematocyst types in a sample = 100%). Relative rather than absolute frequencies were calculated because absolute frequencies of nematocyst types may exhibit a large measurement uncertainty due to partial discharge of capsules and other types of losses during the preparation procedure.

Capsule lengths and widths were determined from photographs taken with a digital camera (Color View, Soft Imaging System, Olympus, Hamburg, Germany) mounted on the Olympus microscope and the software analySIS (Olympus, Hamburg, Germany). In all three suspensions, 15 randomly chosen nematocysts were measured per capsule type in each of the three samples (NS1, NS2, and BS).

Statistical analysis

Statistical analysis of all data was performed by means of winSTAT for Excel (version 2005.1, Bad Krozingen, Germany); the Tukey test was used for the analysis of variances between multiple groups.

LMPC and protein contents of different nematocyst types

In order to examine the protein contents of the discrete nematocyst types, they first were separated by the LMPC method described in Wiebring et al. (2010). The separation of different types of nematocysts from the nematocyst suspension was conducted with a PALM® MicroBeam (P.A.L.M. Microlaser Technologies GmbH, subsidiary company of Carl Zeiss MicroImaging GmbH, Bernried, Germany) as follows: 1.2 µl of the nematocyst suspension was pipetted onto a covered microscope slide (PALM® MembraneSlide) with a UV light-absorbing membrane, and 0.4 µl of an octyl-β-D-glucopyranoside solution (1%, Merck, Darmstadt, Germany) was added in order to lower the surface tension of the sample. Nematocysts were catapulted vertically upward with a single laser shot and collected in a drop of distilled water, which had been placed in the lid of an Eppendorf tube (Wiebring et al., 2010). This method was most effective for the large nematocysts chosen.

Potential toxin components contained in the nematocysts was investigated by matrix-assisted laser desorption ionization time-of-flight mass spectrometry (MALDI-TOF MS). The nematocysts were mixed with freshly prepared matrix solution (sinapinic acid, SA (Bruker Daltonic, Bremen, Germany), $c = 1$ mg ml^{-1} acetonitrile (Merck, Darmstadt, Germany)/0.1% trifluoroacetic acid (Fluka, Selze, Germany) at a ratio of 90:10 (v/v)), This was achieved by adding 3 µl of SA solution to the collection Eppendorf tube, closing the cap, and centrifuging (30 s at 2500 rpm). The resulting nematocyst/matrix mixture was treated in a sonicator (Branson Sonifier 450, G. Heinemann Ultraschall- und Labortechnik, Schwäbisch Gmünd, Germany) and was subsequently transferred onto the MALDI target (AnchorChip™ 800/384 T F). The sample spot was washed with 2 µl washing solution (10 mM monobasic ammonium phosphate (Merck, Darmstadt, Germany)) and recrystallised with 2 µl of recrystallisation solution (ethanol (Merck, Darmstadt, Germany), acetone (LGC Promochem, Wesel, Germany), and trifluoroacetic acid (Fluka, Selze, Germany) at a ratio of 6:3:1 (v/v/v)) according to Bruker Daltonik (2004). Due to this preparation method, it can be assumed that the soluble proteinaceous content of the nematocysts was ionised rather than the components of the capsule wall or the thread. Protein standard I (Bruker Daltonic, Bremen, Germany) was used for MS calibration. The analysis of the soluble protein content was performed with an Ultraflex II (Bruker, Bremen, Germany) mass spectrometer in the linear positive mode. The resulting spectra were annotated with the FlexControl 2.4 software (Bruker, Bremen, Germany).

The number of mass spectrometrically analysed nematocysts varied as a result of their variable relative abundances in the samples; 70–190 A-isorhizas, 100–220 O-isorhizas, and 120–180 euryteles were examined. Due to the small number of euryteles in sample NS2, which were from small medusae in the North Sea, it was not possible to isolate a sufficient number of euryteles. Therefore, only A- and O-isorhizas were analysed for the NS2 sample.

Results

Frequencies of different nematocyst types

The relative frequencies of A-isorhizas, O-isorhizas, and euryteles (Fig. 1) in nematocyst suspensions from the North Sea samples taken in 2005 and 2007 (NS1, NS2) were rather similar, but very different from frequencies found in the Baltic Sea (BS) sample (Fig. 2). The mean numbers of counted nematocysts varied from 66 to 155; therefore, the nematocyst counts were converted to percentages to enable them to be compared.

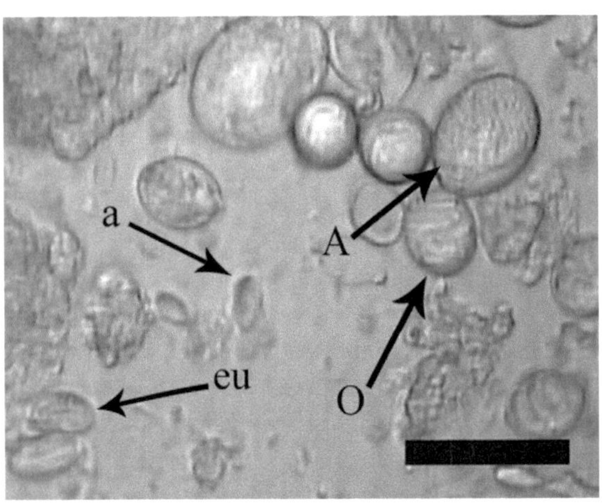

Fig. 1 Nematocyst suspension from *Cyanea capillata* medusae collected in the North Sea (NS1). *A* A-isorhizas, *a* a-isorhizas, *eu* euryteles, *O* O-isorhizas. Scale bar = 30 μm

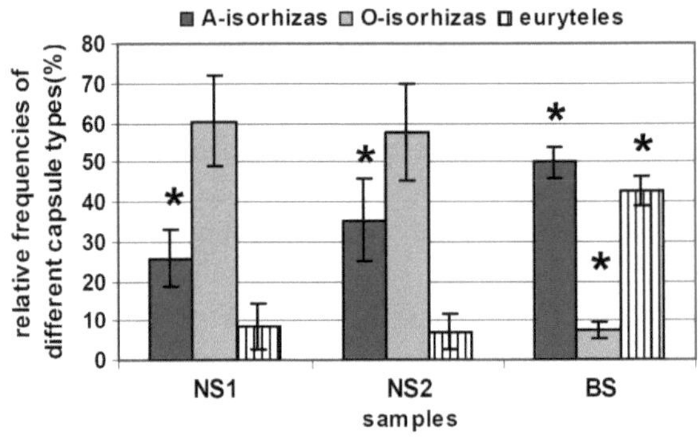

Fig. 2 Relative frequencies (mean ± SD; %) of A-isorhizas, O-isorhizas, and euryteles in the nematocyst suspensions from *Cyanea capillata* medusae from the North Sea (NS1 and NS2) and from the Baltic Sea (BS). The nematocysts of ten aliquots from each sample (NS1, NS2, and BS) were counted ($n = 10$). * Statistical difference, $P > 0.05$

The percentages of the three nematocysts types (mean ± standard deviation) varied significantly among the samples. In NS1 and NS2, O-isorhizas were the prevalent nematocysts (~60%), A-isorhizas averaged 26 and 35%, and euryteles averaged 7 and 9%. The BS sample contained higher percentages of A-isorhizas (50%) and euryteles (43%) than the North Sea samples. The percentages of the A-isorhizas differed significantly among all samples ($P < 0.05$). Percentages of O-isorhizas and percentages of euryteles did not differ significantly between the two North Sea samples (NS1 and NS2); however, the Baltic Sea sample had significantly fewer O-isorhizas and more euryteles than did NS1 and NS2 samples.

Nematocyst sizes

Sizes of the three nematocyst types were not significantly different in the two NS samples (Fig. 3); however, the mean lengths and widths of the A- and O-isorhizas in the BS sample differed significantly from those in the North Sea samples (Fig. 3a, b). None of the measured parameters, length, width, or the length-to-width ratio of the euryteles differed significantly among samples (Fig. 3c, d). The length-to-width ratio showed a significant difference only for A-isorhizas, which were relatively more elongated in the BS samples than in the NS samples (Fig. 3d).

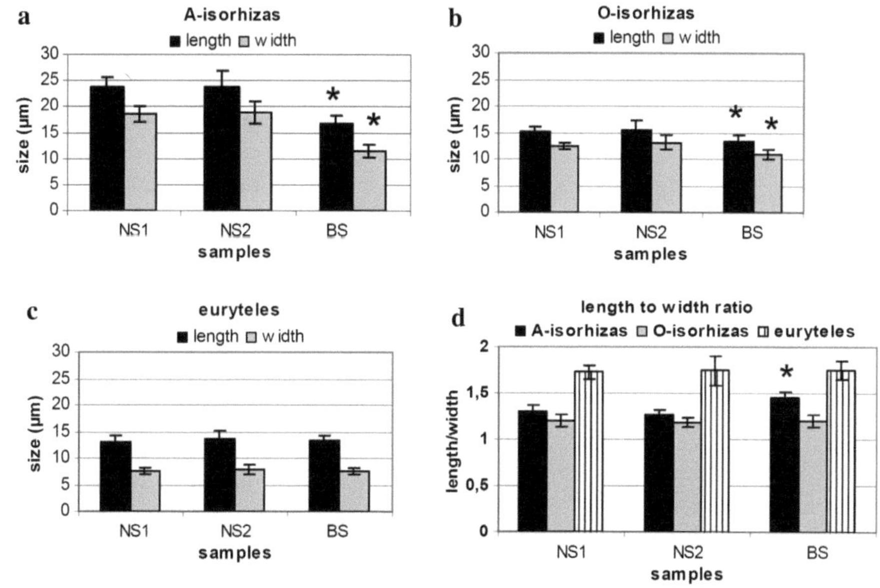

Fig. 3 Lengths (*black bars*) and widths (*grey bars*) (mean ± SD; μm) of the three nematocyst types of *Cyanea capillata* medusae from the North Sea (NS1, NS2) and the Baltic Sea (BS): **a** A-isorhizas, **b** O-isorhizas, **c** euryteles, **d** shows the length-to-width ratio of the three nematocysts (*black bars*: A-isorhizas; *grey bars*: O-isorhizas; *striped bars*: euryteles). * Statistical differences, $P > 0.05$

Analysis of nematocyst contents

The protein contents of different nematocyst types from the three medusan samples were assessed by MALDI-TOF-mass spectrometry (Figs. 4 and 5; Table 1). There were no spectrometry data for euryteles from the NS2 sample because the procedure did not work. The spectrometry yielded a variety of peaks, but only peaks that exceeded 10% of the height of the main component (largest peak) were considered in this analysis (Table 1). Comparison of the peaks showed that certain proteins in the same mass (Daltons, Da) range recurred in the different nematocyst types and in the different samples analyzed, including 3544, 4127, 4266, 5089, 5672, and 7744 Da. Despite the pronounced protein patterns, it was not possible to make a clear distinction between the BS and the two NS medusa groups. It is noteworthy that O-isorhizas in all three samples contained the same main component (NS1: 5672.0 Da; NS2: 5673.7 Da; BS: 5673.1 Da), whereas the main components of the A-isorhizas (NS1: 5796.2 Da; NS2: 4266.2 Da; BS: 5090.7 Da) and of the euryteles (NS1: 8214.0 Da; BS: 5089.7 Da) differed between locations.

Discussion

In this study, nematocysts from three different samples of *C. capillata* were analysed for size distribution and protein content. The O-isorhizas and A-isorhizas were smaller in the Baltic Sea (BS) sample as compared to those in the North Sea (NS) samples, and the length-to-width ratios of the A-isorhizas were larger in the BS sample. The lengths and widths of the nematocysts agreed within 10–20% with the measurements of Östman & Hydman (1997) for *C. capillata* from the North Sea with one exception. The A-isorhizas of the BS sample were 57% shorter and 76% narrower than the nematocysts analysed by Östman & Hydman (1997). Moreover, the capsule suspensions showed different relative frequencies of certain nematocyst types. The Baltic Sea sample contained significantly lower percentage of O-isorhizas and a significantly higher percentage of euryteles than the samples from the North Sea. To the best of our knowledge, similar investigations of Scyphozoa, in general, and *C. capillata*, in particular, have not yet been reported.

With respect to the measurement and enumeration of nematocysts, it is important to note that nematocysts may discharge or get lost during the preparation procedures, thereby affecting the cnidom results. Therefore, future studies should include microscopic investigation of the cnidom, both in the nematocyst suspensions and also in intact cnidarian tissues.

Several studies have attempted to assign specific functions to the various types nematocysts (Rifkin & Endean, 1983; Purcell, 1984; Heeger & Möller, 1987; Östman & Hydman, 1997; Jarms et al., 2002; Colin &

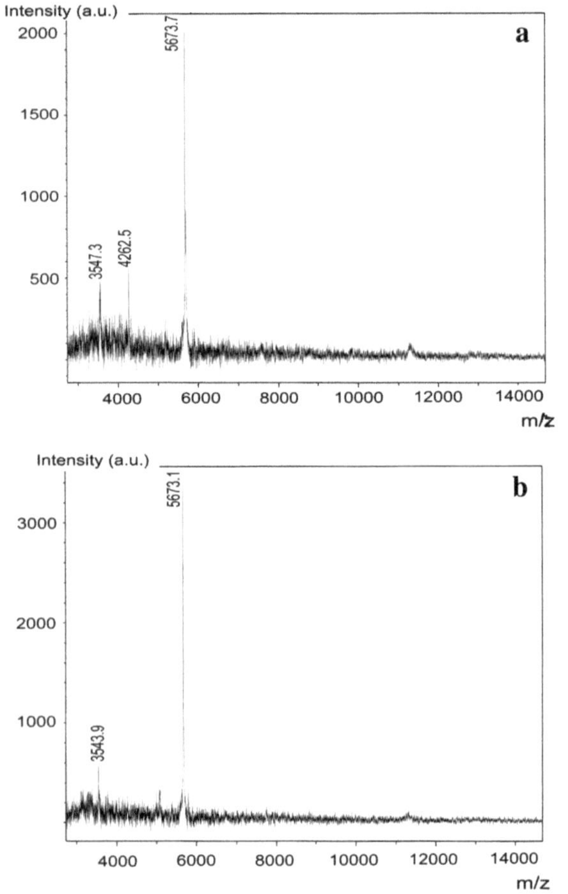

Fig. 4 Mass spectra (m/z = mass to charge ratio) of polypeptides extracted from O-isorhizas isolated from *Cyanea capillata* collected in: **a** the North Sea (NS2) and **b** the Baltic Sea (BS). The intensity of the peaks are presented in arbitrary units (a.u.) and the numbers in bold at the top of these peaks indicate the mass in Daltons (Da) of the polypeptide molecules

Costello, 2007). Most authors conclude that the euryteles are penetrating nematocysts, which inject toxins into the prey. The isorhizas either penetrate partly through the skin of the prey or adhere to the surface structures and entangle the prey.

It is likely that the relative frequencies of nematocysts found in a species are correlated with the available prey types. In this regard, changes of the cnidom during ontogenesis of two species of cubomedusae (box jellyfish) were described in association with the changing diet of the medusae (Carette et al., 2002). Purcell (1984) and Purcell & Mills (1988) showed that the types of nematocysts (entangling vs. penetrating) differed among different groups of hydrozoans and those differences were reflected in the diets (crustacean vs. soft-bodied); the predators of soft-bodied prey possessed soft-tissue-penetrating nematocyst types (isorhizas), but crustacean predators contained adhesive/entangling and exoskeleton-penetrating nematocyst types (euryteles). *C. capillata* is known to be an important predator of both soft-bodied and crustacean prey (e.g., Purcell, 2003). Although the turn-over time of nematocysts in the tentacles is unknown, it seems reasonable that as prey population composition changes, for example from abundant gelatinous prey to predominantly crustacean prey, the proportions of the various nematocyst types might change. Thus, in the present study, the differing cnidoms of the medusae from the Baltic and North seas may have reflected different prey types available.

Mass spectrometric analysis of the contents of the A-isorhizas, euryteles, and O-isorhizas showed specific protein patterns. The main protein component in the O-isorhizas was present in all three medusan samples, whereas the main components in the A-isorhizas and in the euryteles were different in the spectra from the different samples. On the basis of our data, it is not clear whether there are distinct differences in the capsule proteins of the North Sea and the Baltic Sea medusae.

The toxins of the nematocysts are thought to be proteinaceous (Hessinger, 1988), which assist by paralysing the prey. Lassen et al. (2010, this issue) detected a group of peptides in *C. capillata* venom that also was found in the separated A-isorhizas in the present study (Fig. 5b). Certain fractions of the venom affected the cell membranes of neurocytes. Before now, it was not possible to study the functions or toxicity of individual nematocyst types. It is likely that the contents of the various nematocyst types differ depending on their roles in prey capture.

The toxic components and toxic potencies of cnidarians differ between medusae of the same species from different habitats (Helmholz et al., 2010, this issue). Kintner et al. (2005) found changes in the venom in association with diet changes during the ontogenesis of two cubozoan species. Variation of venom composition in *Aurelia aurita* (L.) was observed by Burnett et al. (1988) and Segura-Puertas et al. (2002). Both studies demonstrated a higher stinging potency of *A. aurita* in Mexican waters;

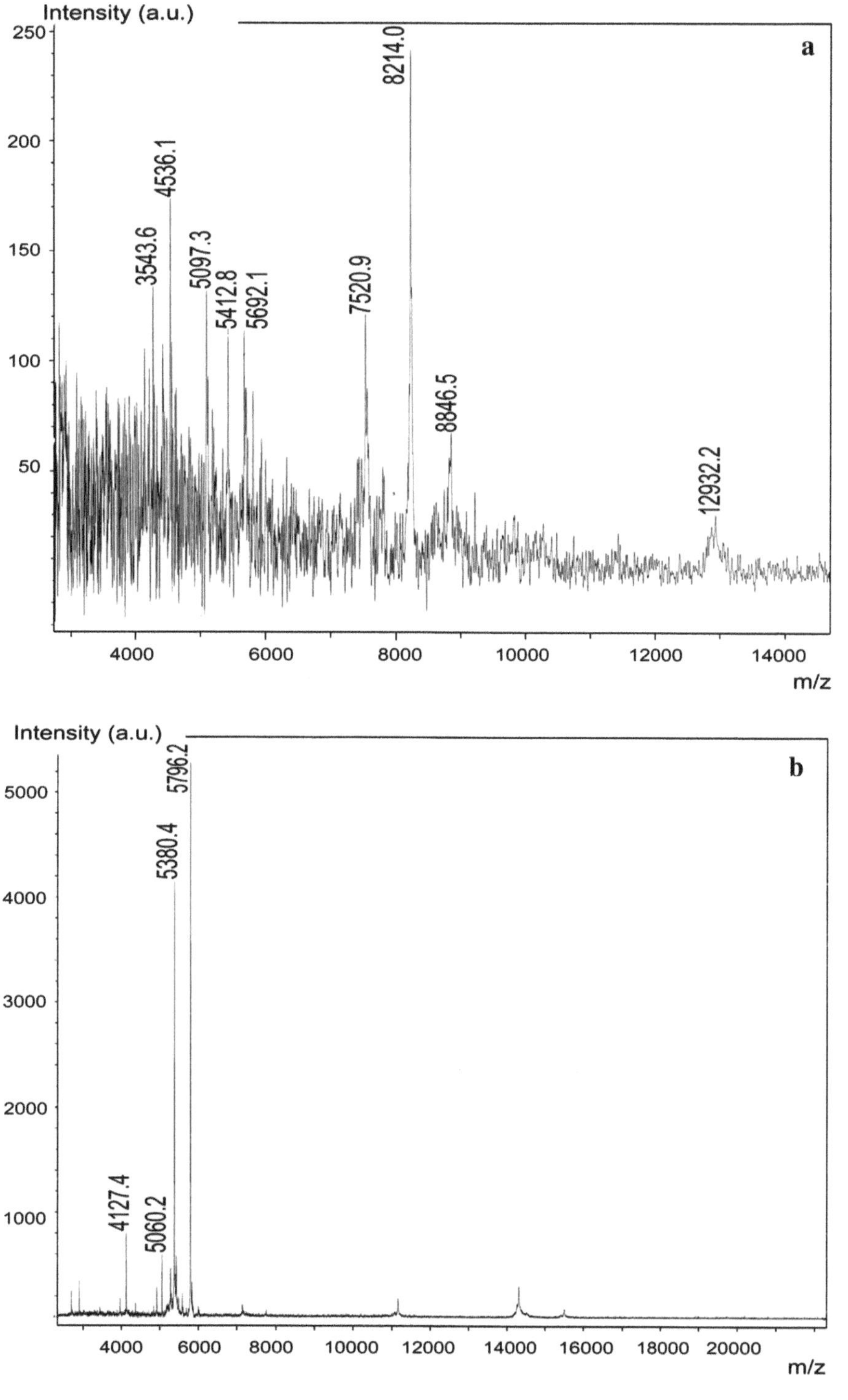

Fig. 5 Mass spectra (m/z = mass to charge ratio) of polypeptides extracted from: **a** euryteles isolated from *Cyanea capillata* medusae collected in the North Sea (NS1) and **b** A-isorhizas from *C. capillata* collected in the North Sea (NS1). The intensity of the peaks are presented in arbitrary units (a.u.) and the numbers in bold at the top of these peaks indicate the mass in Daltons (Da) of the polypeptide molecules

however, no possible explanations were given for his phenomenon. The crude venom of *A. aurita* had different patterns in the gel electrophoresis from two different habitats and the toxin of medusae from the Red Sea was more potent than from the Chesapeake Bay (Radwan et al., 2001).

The separation of nematocysts via the micromanipulative LMPC method and the mass spectrometric

Table 1 Molecular mass distribution in Daltons (Da) in capsules types from the three capsule types from three different samples *Cyanea capillata* medusae collected in the North and Baltic seas (NS1, NS2, and BS). The mass peaks that exceeded 10% of the height of mass peak of the main component (presented in bold font) are included

Nematocyst type	Sample		
	NS1 (Da)	NS2 (Da)	BS (Da)
A-isorhizas	–	4035.4	–
	4127.4	–	–
	–	**4266.2**	–
	–	4883.4	–
	5060.2	–	–
	–	5091.5	**5090.7**
	5375.3	–	–
	5380.4	–	–
	–	5674.3	–
	–	5699.3	–
	–	5734.8	–
		–	–
	5796.2	5795.6	–
	–	7583.5	–
	–	7744.4	7741.6
	–	7848.3	–
O-isorhizas	3544.4	3547.3	3543.9
	4125.7	–	–
		4262.5	–
	4818.4	–	–
	5089.3	–	–
	5106.6	–	–
	5178.5	–	–
	5380.9	–	–
	5672.0	**5673.7**	**5673.1**
Euryteles	3543.6	No results	3544.0
	–		–
	4536.1		–
	–		5067.7
	–		**5089.7**
	5097.3		–
	–		5213.6
	5412.8		
	–		5672.9
	5692.1		–
	7520.9		–
	–		7740.4
	8214.0		–
	8846.5		–
	12932.9		–

investigation of the proteinaceous contents of the separated nematocysts are part of a novel approach toward the study of the scyphomedusae. This research complements another related study of *C. capillata* (Lassen et al., 2010), which focused on potential neuroactive polypeptides in the nematocyst venom. This group of polypeptides had the same m/z values that were observed in the A-isorhizas in this study (Fig. 5b; Table 1).

Conclusions and outlook

We developed a new approach for isolating nematocyst capsules utilizing the micromanipulative LMPC method and subsequent mass spectrometric characterisation of the proteins in specific nematocyst types. This method allows investigation of the properties of individual nematocyst types for the first time. The LMPC method yielded intact capsules that could be directly prepared on a MALDI (matrix-assisted laser desorption ionization) target. The TOF MS (time-of-flight mass spectrometry) analyses provided distinct protein patterns from the nematocyst contents of A-isorhizas, O-isorhizas, and euryteles from groups of medusae from the North Sea and the Baltic Sea. This approach should enable further characterisation of the venom in individual nematocyst types of jellyfish. For an amino acid sequence analysis of single proteinaceous components, either the number of capsules harvested by LMPC or the sensitivity of the mass spectrometry analysis would need to be increased.

In addition to the development of the LMPC/TOF MS approach, the medusan cnidom was investigated. It appears that the sizes and relative frequencies of A-isorhizas, O-isorhizas, and euryteles in *C. capillata* collected from the North Sea are not the same as in medusae sampled from the Baltic Sea. One of our conclusions from the cnidom work is that due to potential procedural artifacts, future studies should include visual (microscopic) inspection of both capsule suspensions and intact medusan tissues in order to determine if the preparation procedures may cause partial discharge or loss of nematocysts.

Acknowledgements We are grateful to Dr. Jürgen Gandraß (GKSS Research Centre Geesthacht, Institute for Costal

Research, Department for Marine Bioanalytical Chemistry) for critically reading the manuscript.

References

Brodeur, R. D., H. Sugisaki & G. L. Hunt, 2002. Increases in jellyfish biomass in the Bering Sea: implications for the ecosystem. Marine Ecology Progress Series 233: 89–103.

Bruker Daltonik GmbH, 2004. AnchorChip^{TM} technology, revision 2.3. Preparation for ultra-sensitive automated MALDI-TOF MS. Bruker Daltonik, Bremen: 17.

Burnett, J. W., G. J. Calton & J. B. Larsen, 1988. Significant envenomation by *Aurelia aurita*, the moon jellyfish. Toxicon 26: 215–217.

Calder, D. R., 1974. Nematocysts of the coronate scyphomedusa, *Linuche unguiculata*, with a brief re-examination of scyphozoan nematocyst classification. Chesapeake Science 15: 170–173.

Carrette, T., P. Alderslade & J. Seymour, 2002. Nematocyst ratio and prey in two Australian cubomedusans, *Chironex fleckeri* and *Chiropsalmus* sp. Toxicon 40: 1547–1551.

Colin, S. P. & J. H. Costello, 2007. Functional characteristics of nematocysts found on the scyphomedusa *Cyanea capillata*. Journal of Experimental Marine Biology and Ecology 351: 114–120.

Doyle, T. K., H. de Hass, D. Cotton, B. Dorshel, V. Cummins, J. D. R. Houghton, J. Davenport & G. C. Hays, 2008. Widespread occurrence of the jellyfish *Pelagia noctiluca* in Irish coastal and shelf waters. Journal of Plankton Research 30: 963–968.

Fenner, P. J., 1998. Dangers in the ocean: The traveler and the marine envenomation. I. Jellyfish. Journal of Travel Medicine 5: 135–141.

Hay, S., 2006. Marine ecology: gelatinous bells may ring change in marine ecosystems. Current Biology 16: R679–R682.

Heeger, T. & H. Möller, 1987. Ultrastructural observations on prey capture and digestion in the scyphomedusa *Aurelia aurita*. Marine Biology 96: 391–400.

Helmholz, H., C. Ruhnau, C. Schütt & A. Prange, 2007. Comparative study on the cell toxicity and enzymatic activity of two northern species *Cyanea capillata* (L.) and *Cyanea lamarckii* (Péron & Léslieur). Toxicon 50: 53–64.

Helmholz, H., B. D. Johnston, C. Ruhnau & A. Prange, 2010. Gill cell toxicity of northern boreal scyphomedusae *Cyanea capillata* and *Aurelia aurita* measured by an in vitro cell assay. Hydrobiologia. doi:10.1007/s10750-010-0216-9.

Hessinger, D. A., 1988. Nematocyst venoms and toxins. In Hessinger, D. A. & H. N. Lenhoff (eds), The Biology of Nematocysts. Academic Press, San Diego: 333–367.

Jarms, G., H. Tiemann & A. Båmstedt, 2002. Development and biology of *Periphylla periphylla* (Scyphozoa: Coronatae) in a Norwegian fjord. Marine Biology 141: 647–657.

Kintner, A. H., J. E. Seymour & S. L. Edwards, 2005. Variation in lethality and effects of two Australian chirodropid jellyfish venoms in fish. Toxicon 46: 699–708.

Lassen, S., H. Helmholz, C. Ruhnau & A. Prange, 2010. Characterisation of neurotoxic polypeptides from *Cyanea capillata* medusae (Scyphozoa). Hydrobiologia. doi: 10.1007/s10750-010-0215-x.

Lynam, C., S. Hay & A. Brierley, 2004. Interannual variability in abundance of North Sea jellyfish and links to the North Atlantic Oscillation. Limnology and Oceanography 49: 637–643.

Lynam, C. P., M. R. Heath, S. J. Hay & A. S. Brierley, 2005. Evidence for impacts by jellyfish on North Sea herring recruitment. Marine Ecology Progress Series 298: 157–167.

Mariscal, R. N., 1974. Nematocysts. In Muscatine, L. & H. M. Lenhoff (eds), Coelenterate Biology. Academic Press, New York: 129–178.

Östman, C. & J. Hydman, 1997. Nematocyst analysis of *Cyanea capillata* and *Cyanea lamarckii* (Scyphozoa, Cnidaria). Scientia Marina 61: 313–344.

Östman, C., A. Aquirre, M. Myrdal, P. Nyvall, J. Lindström & M. Björklund, 1995. Nematocysts in *Tubularia larynx* (Cnidaria, Hydrozoa) from Scandinavia and the northern coast of Spain. Scientia Marina 59: 165–179.

Purcell, J. E., 1984. The functions of nematocysts in prey capture by epipelagic siphonophores (Coelenterata, Hydrozoa). Biological Bulletin 166: 310–327.

Purcell, J. E., 2003. Predation on zooplankton by large jellyfish, *Aurelia labiata*, *Cyanea capillata* and *Aequorea aequorea*, in Prince William Sound, Alaska. Marine Ecology Progress Series 246: 137–152.

Purcell, J. E., 2005. Climate effects on formation of jellyfish and ctenophore blooms: a review. Journal of the Marine Biological Association of the UK 85: 461–476.

Purcell, J. E. & M. N. Arai, 2001. Interactions of pelagic cnidarians and ctenophores with fish: a review. Hydrobiologia 451: 27–44.

Purcell, J. E. & M. B. Decker, 2005. Effects of climate on relative predation by scyphomedusae and ctenophores on copepods in Chesapeake Bay during 1987–2000. Limnology and Oceanography 50: 376–387.

Purcell, J. E. & C. E. Mills, 1988. The correlation between nematocyst types and diets in pelagic Hydrozoa. In Hessinger, D. A. & H. M. Lenhoff (eds), The Biology of Nematocysts. Academic Press, San Diego: 463–485.

Purcell, J. E., S.-I. Uye & W.-T. Lo, 2007. Anthropogenic causes of jellyfish blooms and direct consequences for humans: a review. Marine Ecology Progress Series 350: 153–174.

Purcell, J. E., R. A. Hoover & N. T. Schwarck, 2009. Interannual variation of strobilation by the scyphozoan *Aurelia labiata* in relation to polyp density, temperature, salinity, and light conditions in situ. Marine Ecology Progress Series 375: 139–149.

Radwan, F. F. Y., J. W. Burnett, D. A. Bloom, T. Coliano, M. E. Eldefrawi, H. Erderly, L. Aurelian, M. Torres & E. P. Heimer-de la Cotera, 2001. A comparison of the toxinological characteristics of two *Cassiopea* and *Aurelia* species. Toxicon 39: 245–257.

Rifkin, J. & R. Endean, 1983. The structure and function of the nematocysts of *Chironex fleckeri* Southcott, 1956. Cell and Tissue Research 233: 563–577.

Segura-Puertas, L., G. Avila-Soria, J. Sanchez-Rodgriguez, M. Ramos-Aguilar & J. W. Burnett, 2002. Some toxinological aspects of *Aurelia aurita* (Linne) from the Mexican Caribbean. Journal of Venomous Animals and Toxins 8: 113–128.

Šuput, D., 2009. In vivo effects of cnidarian toxins and venoms. Toxicon 54: 1190–1200.

Tibballs, J., 2006. Australian venomous jellyfish, envenomation syndromes, toxins and therapy. Toxicon 48: 830–859.

Weill, R., 1934. Contribution à ľ étude des cnidaires et de leurs nématocystes. Travaux de la Station Zoologique de Wimereux 10(11): 1–701.

Wiebring, A., H. Helmholz, I. Sötje, S. Lassen, A. Prange & H. Tiemann, 2010. A new method for the separation of different types of nematocysts from Scyphozoa and investigation of proteinaceous toxins utilizing laser catapulting and subsequent mass spectrometry. Marine Biotechnology. doi:10.1007/s10126-010-9261-7.

Hydrobiologia (2010) 645:213–221
DOI 10.1007/s10750-010-0215-x

Characterisation of neurotoxic polypeptides from *Cyanea capillata* medusae (Scyphozoa)

Stephan Lassen · Heike Helmholz ·
Christiane Ruhnau · Andreas Prange

Published online: 21 March 2010

Abstract Cnidarian venoms include neurotoxins, which are able to paralyse prey organisms immediately. Important targets for neurotoxins are voltage-gated ion channels in membranes of excitable cells. By blocking specific receptor sites, neurotoxic components disturb the physiological ion channel functions. Here, we describe the isolation and characterisation of potential neurotoxic polypeptides from the crude tentacle venom of the boreal scyphomedusan *Cyanea capillata*. Partially purified venom fractions were obtained by size-exclusion and subsequent reversed-phase chromatography. To assess the blocking activity of the venom on voltage-gated sodium channels, we modified a mouse neuroblastoma (MNB) cell assay. Venom fractions containing channel-blocking activity were analysed by matrix-assisted laser desorption ionisation time-of-flight mass spectrometry (MALDI-TOF MS). The resulting mass spectra revealed a cluster of singly charged peptides within a mass range from 3,900 to 7,000 Da. A group of three potentially neurotoxic peptides with molecular masses of 3983.4, 5795.4 and 6962.1 Da could be tracked throughout the purification process. This investigation of the crude venom is part of a multidimensional assay-guided approach for the isolation and structural characterisation of toxic polypeptides in northern Scyphozoa.

Keywords Jellyfish · Voltage-gated sodium channel · Mass spectrometry · Venom · Toxin

Guest editors: J. E. Purcell & Dror Angel / Jellyfish Blooms: New Problems and Solutions

S. Lassen (✉) · H. Helmholz · C. Ruhnau · A. Prange
Department for Marine Bioanalytical Chemistry, Institute for Coastal Research, GKSS Research Centre, Max-Planck-Street 1, 21502 Geesthacht, Germany
e-mail: stephan.lassen@gkss.de

Introduction

The scyphozoan *Cyanea capillata* (L.) is widespread in northern temperate waters and is well-known to humans due to their stinging effects. In addition to skin irritations, Walker (1977) described cardiotoxic effects on mammalian hearts caused by venomous tentacle material from *C. capillata*. Like many pelagic cnidarians, *C. capillata* medusae prey on crustaceans and vertebrates such as small fish and fish larvae. They use a wide array of toxic polypeptides and proteins, which can be divided into three main categories: high molecular mass cytolytic pore-forming proteins, phospholipases and neurotoxic polypeptides (Zhang et al., 2003). Neurotoxic polypeptides usually interact with ion channels of excitable cell membranes. In general, neurotoxins are fast-acting compounds that influence the neuronal excitability and the neuromuscular transmission of prey organisms (Wu & Narahashi, 1988). A disruption of the normal ion channel function leads to rapid paralysis (Messerli & Greenberg, 2006).

Voltage-gated sodium (Na_v) channels play an essential role in the initiation and propagation of action potentials in neurons and other electrically excitable cells (Yu & Catterall, 2003). The Na_v channel is considered to have at least six neurotoxin receptor sites that recognise the following classes of toxins (Catterall, 1995; Cestèle & Catterall, 2000; Nicholson, 2007): (1) the guanidinium alkaloids tetrodotoxin (TTX) and saxitoxin (STX) as well as the Tx1 peptide toxin from the spider *Phoneutria nigriventer* (Keyserling) (Nicholson, 2007), which block sodium conductance by physically occluding the extracellular entry of the channel; (2) lipophilic compounds such as batrachotoxin and veratridine, which stabilise the open form of the channel; (3) α-scorpion toxins and sea anemone polypeptides that retard channel inactivation; (4) β-scorpion toxins that enhance channel activation: (5) ciguatoxin and brevetoxin, which cause persistent activation, and (6) δ-conotoxin, TxVIa, which effectuates prolongation of action potentials due to the inhibition of sodium current inactivation.

Whereas many terrestrial venomous animals possess a variety of neurotoxic polypeptides that target Na_v channels, there have been very few descriptions of neurotoxins from marine species that specifically block Na_v channels at receptor site 1. Several sea anemones possess polypeptides that bind to receptor site 3 in Na_v channels, e.g. ATX II, from *Anemonia sulcata* (Pennant) or Anthopleurin A and B, from *Anthopleura xanthogrammica* Durchassaing de Fonbressin & Michelotti (Honma & Shiomi, 2006; Messerli & Greenberg, 2006). The μ-conotoxin GIIIA, a 22-amino-acid polypeptide, isolated from the venom of the snail, *Conus geographus* L., acts on skeletal muscle Na_v channels in a similar manner to TTX and STX (Dudley et al., 1995; Blumenthal & Seibert, 2003; Terlau & Olivera, 2004). μ-Conotoxin PIIIA from the purple cone snail (*Conus purpurascens* Sowerby) specifically blocks neuronal sodium channels at receptor site 1 (Safo et al., 2000).

We investigated the venom of *Cyanea capillata* medusae with respect to polypeptides blocking receptor site 1 of vertebrate neuronal Na_v channels. The analytical approach consisted of a bioassay-guided isolation and chemical characterisation of the neurotoxic peptides in the jellyfish venom. The bioassay was chosen following established in vitro protocols for the detection and characterisation of algal neurotoxins (Manger et al., 1993; Gallacher et al., 1997; Kirchner et al., 2001).

Materials and methods

Preparation of cnidocysts

Medusae were collected at the Orkney Islands, Scotland, during a research cruise with the RV "Heincke" in 2006. Fishing tentacles of six *Cyanea capillata* medusae with umbrella diameters larger than 30 cm were removed immediately after the animals were brought aboard. The tentacles were pooled and further prepared for cnidocyst extraction on board following Helmholz et al. (2007), as described below. The tentacle material was mixed with four volumes of ice-cold distilled water (volume ratio 1:5) and stirred for 10 h on ice. The mixture was filtered through a nylon sieve (mesh size 500 μm) and the filtrate was centrifuged at 4°C and 1100 × *g* for 5 min. The supernatant was discarded and residues were washed three times with sterile filtered sea water. The resulting cnidocyst suspensions were stored at −80°C.

Venom preparation and purification

Thawed cnidocyst suspensions were diluted ten-fold in ice-cold 10 mM ammonium acetate (Merck, Germany) buffer, pH 5.5. The cnidocysts were discharged by sonication in a Branson sonifier 450 (Heinemann Ultaschall- und Labortechnik, Germany). A subsequent centrifugation step at 4°C and 14400 × *g* for 5 min yielded the crude tentacle venom. The venom was fractionated by size-exclusion chromatography (SEC) using a High Load 16/60 Superdex 75 prep column connected to an Äkta FPLC system (Amersham Biosciences, Germany) with 10 mM ammonium acetate buffer (pH 5.5) at a flow-rate of 1 ml min^{-1}. The integrated UV detector was set at 280 nm and 5 ml of the crude venom was loaded onto the SEC column for each chromatographic run. Corresponding venom fractions from several SEC runs were combined and concentrated by lyophilisation and consecutively ultra-filtrated with Vivaspin 2 cartridges (MWCO 2 kDa; Satorius Stedim Biotech, Germany) at 4°C and 8000 × *g*. The desalted fractions were stored at −80°C.

The SEC-fractions were further purified by reversed-phase (RP) chromatography on a HPLC system (Agilent, Germany) equipped with a Synergi Hydro-RP (150 × 3.00 mm, 4 μm, 80 Å) analytical column (Phenomenex, Germany) at 30°C. The flow rate was set to 0.5 ml min^{-1} and 50 μl of the venom fractions were injected into the liquid chromatography (LC) system. The compounds were eluted isocratically under completely aqueous conditions with 100% mobile phase A (A: 0.09% formic acid (Merck)/0.01% trifluoroacetic acid (TFA; Fluka, Germany) in ultrapure water) for 8 min, followed by a linear gradient to 25% mobile phase B (B: 80% acetonitrile (Merck) with 0.09% formic acid/0.01% TFA) for 22 min and a second isocratic step for 5 min at 25% B. Chromatograms were recorded at 280 nm with a diode array detector (Agilent). The collected RP-fractions of three LC runs were pooled, dried in a vacuum centrifuge (Thermo Electron, Germany), redissolved in ultrapure water and stored at −20°C for further analysis.

Tetrazolium-based mouse neuroblastoma (MNB) cell assay

A mouse neuroblastoma cell line Neuro 2A (CCL 131) purchased from LGC Standards GmbH (Germany) was cultivated in Roswell Park Memorial Institute (RPMI) 1640 cell culture medium (Invitrogen, Germany) with 10% foetal calf serum (FCS; PAA Laboratories, Germany) and 100 IU penicillin ml^{-1} (Sigma, Germany) and 100 μg streptomycin ml^{-1} (Sigma) at 37°C in a humidified 5% carbon dioxide (CO_2) atmosphere. Cells from a continuous culture were seeded into 96-well microtiter plates at a density of 2×10^5 cells ml^{-1} in 250 μl RPMI medium per well with penicillin/streptomycin but without FCS and incubated at 37°C/5% CO_2 for 24 h. Subsequently, culture wells were inoculated with 10 μl of thawed medusa venom fractions at various concentrations, ranging from 0.01 to 4.5 μg protein (see below) and 10 μl additions of 10 mM ouabain (Sigma) in purified water and 1 mM veratridine (Sigma) in 0.01 M HCl, pH 2. Each venom sample and concentration was tested in eight replicates. The venom fractions were added to replicate wells in parallel in the presence and in the absence of ouabain/veratridine. Eight wells/plate were prepared as untreated negative controls (without ouabain/veratridine and without sample), eight wells were processed as only ouabain/veratridine-treated positive controls, and eight wells received culture medium without cells. The inoculated culture plates were incubated for 48 h. After removing the overlaying medium, 60 μl of 3-[4,5-dimethylthiazol-2-yl]-2,5-diphenyltetrazolium bromide (MTT; Sigma) solution (2 mM in phosphate-buffered saline (1×, pH 7.2; Invitrogen) was added to each well. The well plates were then incubated for 30–40 min at 37°C. For cell lysis, 200 μl of dimethyl sulfoxide (DMSO; Sigma) was added to each well and mixed carefully. The plates were immediately measured on a multiwell scanning photometer (Victor 3, Perkin Elmer, Germany) at 550 nm.

In the presence of ouabain, veratridine enhances the sodium influx in mouse neuroblastoma cells leading to altered cell morphology and subsequent death. Toxins blocking receptor site 1 of Na_v channels antagonise this effect and the cells survive. Vital cells maintain their metabolic activity and reduce MTT to a coloured formazan derivative. The percentage of viable cells reflecting the neurotoxic activity (NA; % viable cells) is photometrically determined after lysis with DMSO (Manger et al., 1993). The neurotoxic activity was calculated as the difference between the normalised sample values (S) and the normalised control values (C): $\mathrm{NA}\ (\%) = S - C = (A_{S_{\mathrm{vi+v/o}}} \times A_{S_{\mathrm{vi}}}^{-1} - A_{C_{\mathrm{v/o}}} \times A_C^{-1}) \times 100$ (A: median ($n = 8$) of the absorbance measured at 550 nm; $S_{\mathrm{vi\,+\,v/o}}$: sample treated with venom and ouabain/veratridine; S_{vi}: sample treated with venom; $C_{\mathrm{v/o}}$: positive control treated with ouabain/veratridine; C: negative control). Venom samples with calculated neurotoxic activity values <5% were considered neurotoxically inactive.

Protein determination

The protein concentration, as a standard for the venom concentration in all chromatographic fractions, was determined using the Bradford assay, which is based on complexing the proteins with Brilliant Blue G. Venom fractions were mixed with Bradford reagent (Sigma) in 96-well plates following the manufacturer's instructions and then read out at 595 nm after 10-min incubation at room temperature.

Mass spectrometry

The matrix for a dried droplet preparation was prepared at ambient temperature as a saturated solution of 3,5-dimethoxy-4-hydroxycinnamic acid (sinapinic acid; Bruker, Germany) in 50% acetonitrile containing 0.1% trifluoroacetic acid (TFA). Samples and matrix were premixed at a ratio of 1:1 and subsequently 1 µl of the mixture was spotted on a polished steel target (Bruker). After crystallisation of the droplets at room temperature, an on-target wash step with 2 µl 10 mM monobasic ammonium phosphate in 0.1% TFA was carried out. The steel target was introduced into the ion source of an Ultraflex II matrix-assisted laser desorption ionisation time-of-flight mass spectrometer (MALDI-TOF MS; Bruker) and the samples were analysed in the linear positive mode. The Ultraflex II was controlled by the FlexControl 3.1 software (Bruker) and externally calibrated with a protein standard (Bruker) containing insulin, ubiquitin, cytochrome C and myoglobin. Accumulated mass spectra were processed and the peptide masses (*m*/*z*) were annotated using the FlexAnalysis 3.0 software.

Results

Establishment of the MNB cell assay and calculation of the neurotoxic activity

The mouse neuroblastoma (MNB) cell assay was established utilising the known Na_v channel blocker saxitoxin (STX; Institute of Marine Biosciences, Canada). The calculated neurotoxic activity was plotted against the logarithm of the STX concentration and revealed a typical sigmoid dose–response curve (Fig. 1). A median effective concentration (EC_{50}) of 0.9 ng $well^{-1}$, corresponding to 3.2 ng ml^{-1} or 11 pmol ml^{-1}, was deduced from the curve. Cells treated with oubain/veratridine and active venom samples were more viable than cells treated only with oubain/veratridine (controls). Increased viability results in higher formazan concentrations and absorbance values, respectively. Therefore, the difference between the absorbance of a normalised sample and the normalised control was used as an indicator for neurotoxic activity. The calculation with normalised values considered the test-to-test variance of the cell activity, reflected by the normalised control in each experiment, and in the case of the normalised samples, possible cytotoxic effects of the purified venom samples on the cells.

Venom purification and detection of neurotoxic activity

The crude tentacle venom was separated into different fractions by SEC. Two chromatographic run series, series A and B consisting of six and ten runs, respectively, were performed and the corresponding fractions were combined. Four consecutive SEC-fractions from series A (F1–F4; Fig. 2) within the calibrated mass range 25 kDa $> m <$ 1.5 kDa were concentrated and examined with regard to their neurotoxic activity. Each fraction was diluted three times resulting in four samples, which were measured with the MNB assay. SEC-fraction F3A exhibited a dose-dependent Na_v channel blocking activity (Fig. 3). Fraction F2A, tested in a comparable concentration range of 1 to 4.5 µg protein $well^{-1}$, corresponding to 3.6–16.2 µg protein ml^{-1}, did not have a clear neurotoxic effect. Fractions F1A and F4A also showed no activity (Fig. 3), possibly due to their low concentration range. To validate the inherent neurotoxicity of fraction 3, a second fraction from series B (F3B) was tested within the same concentration range and it indicated channel blocking activity as well (Fig. 3).

The active fraction F3B was further purified by reversed-phase liquid chromatography (RP-LC) into three sub-fractions, yielding the chromatogram in Fig. 4. One group of components eluting between 1

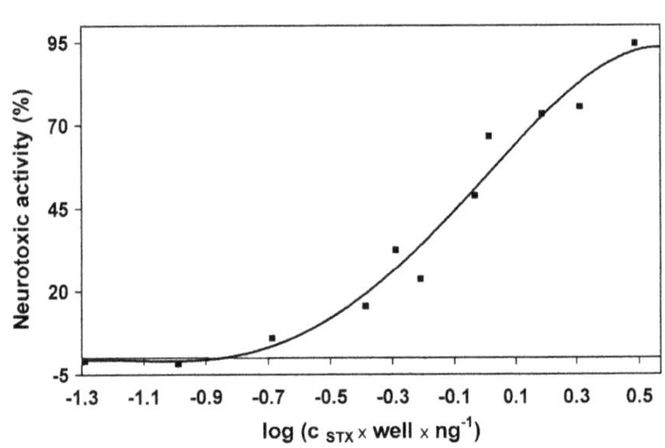

Fig. 1 Normalised dose–response curve of saxitoxin neurotoxic acivity (%) in the mouse neuroblastoma cell assay (c_{STX}: concentration of saxitoxin in ng × $well^{-1}$)

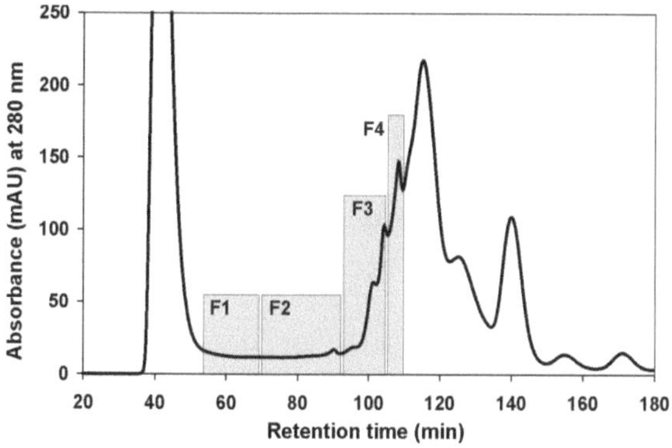

Fig. 2 Size-exclusion chromatogram of the crude tentacle venom of *Cyanea capillata* medusae. Indicated fractions F1–F4 were investigated with the mouse neuroblastoma cell assay

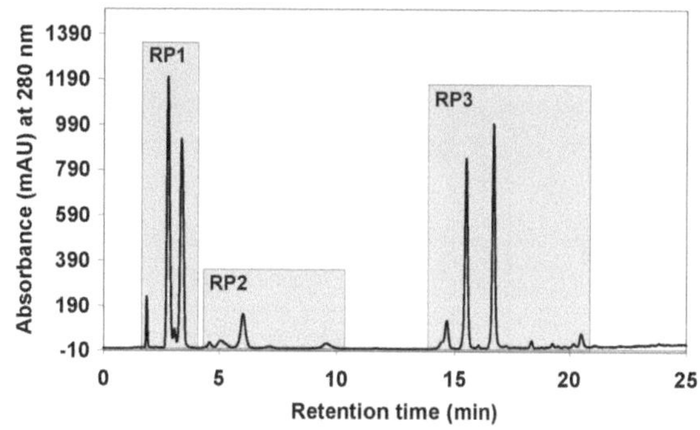

Fig. 4 Reversed-phase chromatogram of the active fraction F3B of the tentacle venom of *Cyanea capillata* medusae. Indicated sub-fractions RP1–RP3 were investigated with the mouse neuroblastoma cell assay

and 4 min in 100% aqueous conditions yielded the fraction RP1. A second group of compounds, eluting between 14 and 21 min was combined to form fraction RP3. All other compounds eluting between 4 and 14 min constituted RP2. Each fraction was diluted in the same manner as the SEC fractions and measured with the MNB assay. Protein concentrations ranged from 0.01 to 0.07 μg protein well^{-1}, corresponding to 0.036–0.25 μg ml^{-1}. In contrast to RP2 and RP3, sub-fraction RP1 showed Na_V channel-blocking activity. Moreover, it appears that the purification of fraction F3B to sub-fraction RP1 caused a 50-fold increase in neurotoxic activity, because 0.06 μg well^{-1} of RP1 yielded a neurotoxic activity of 20%, which was comparable to the activity exerted by 3.0 μg well^{-1} of F3B. The assay results for the subfractions RP1–RP3 are summarised in Fig. 3.

Mass fingerprints of the active fractions

The fractions with channel blocking activity were investigated by matrix-assisted laser desorption ionisation time-of-flight mass spectrometry to get information about the number and the molecular mass distribution of potential neurotoxic polypeptides. The mass spectrum of fraction F3B exhibited 17 components in a mass range from 2,000 to 8,000 Da (Fig. 5, annotated as 1–17 in Table 1). Compared to the peaks registered for F3B, the mass spectrum of fraction RP1

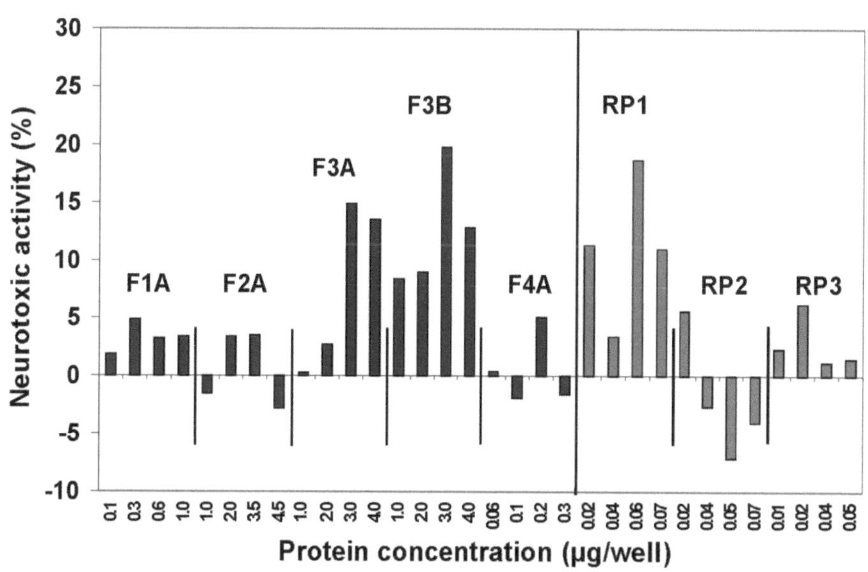

Fig. 3 Neurotoxic activity of purified fractions of the tentacle venom of *Cyanea capillata* medusae. Protein concentrations were calculated from measured protein concentrations in undiluted fractions and subsequent dilution steps (F1A–F4A: fractions from size-exclusion chromatography, series A; F3B: fraction from size-exclusion chromatography, series B; RP1–RP3: fractions from reversed-phase chromatography, sub-fractions from F3B)

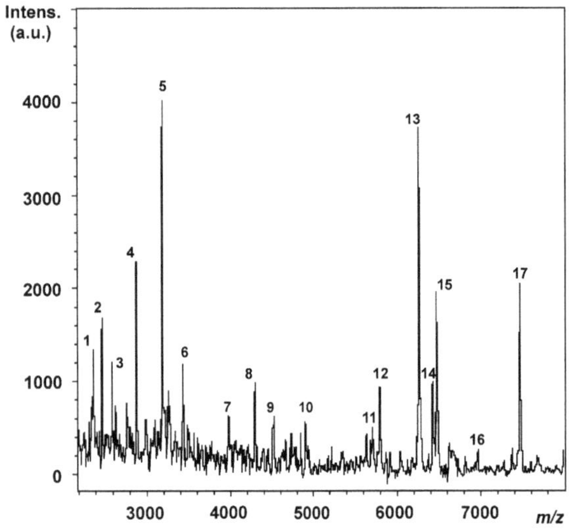

Fig. 5 Mass fingerprint of the active fraction F3B of the tentacle venom of *Cyanea capillata* medusae. (*m/z*: mass-to-charge ratio; a. u.: arbitary units). Charged peptides $[M + H]^+$ with a signal-to-noise ratio >2 are annotated (1–17)

indicated nine compounds in the mass range between 3,900 and 7,000 Da (Fig. 6; Table 1). The success of the purification process was obvious from the increased neurotoxic activity per protein equivalent and also in the reduced complexity of the venom samples. Three singly positive charged peptides ($[M + H]^+$) with masses of 3981.5–3983.4, 5793.0–5795.4 and 6962.1–6967.0 Da occurred in both fractions (values in bold in Table 1). Thus, preparation of the samples with saturated sinapinic acid on a polished steel target was appropriate for sensitive mass spectrometric detection of pico-molar amounts of potential neurotoxic polypeptides in the partially purified venom.

Discussion

This is the first investigation of neurotoxic polypeptides in the venom of the scyphozoan *Cyanea capillata*. Generally, neurotoxins from jellyfish are not well characterised. Sanchez-Rodriguez et al. (2006) isolated a 120-kDa protein from the cubomedusa *Carybdea marsupialis* (L.) with a strong neurotoxic activity on marine crabs (*Ocypode quadrata* (Fabricius)). Large venom proteins could be aggregates of smaller molecules that agglomerate as a result of non-covalent interactions during purification steps. This effect was described by Long & Burnett (1989) for a purified haemolytic peptide from *Chrysaora quinquecirrha* (Desor), which tended to aggregate after size-exclusion chromatography. Therefore, the neurotoxicity described by Sanchez-Rodriguez et al. (2006) may have resulted from smaller polypeptides.

Molecular masses of neurotoxic polypeptides of sea anemones (Cnidaria) are distributed between 3,000 and 5,000 Da (Norton, 1991), which is in the same size range as the polypeptides detected in our study (sub fraction RP1). Moreover, smaller sea anemone polypeptides interact with specific receptor sites of Na_v channels (Wu & Narahashi, 1988; Messerli & Greenberg, 2006), as did the venom polypeptides of *C. capillata.*

A MTT-based mouse neuroblastoma (MNB) cell assay was established as a tool for a bioassay-guided isolation of neurotoxins from the crude tentacle venom of *C. capillata*. Although the MNB assay has been applied for neurotoxins from different marine sources, e.g. marine bacteria (Gallacher & Birkbeck, 1992) and dinoflagelates (Gallacher et al., 1997; Kirchner et al., 2001), this is its first use for the detection and isolation of jellyfish neurotoxins. Moreover, this assay was suitable for the detection of Na_v channel-blocking compounds in partially purified venom.

Site 1 blocking activity was detected in one SEC fraction and in a sub-fraction obtained from the SEC material by RP-LC. Because of the design of this assay, in which veratridine serves as a Na_v channel activator and ouabain as inhibitor of Na^+/K^+-ATPase, our results suggest that *C. capillata* neurotoxins may exert a site 1-blocking activity that consequently paralyses prey organisms. The ability of these polypeptides to block receptor site 1 depends on the presence of strong basic guanidinium groups at arginine residues (Shon et al., 1998). For tetrodotoxin, saxitoxin and μ-conotoxins, guanidinium groups interact electrostatically with the extracellular binding site 1, located at the outer entrance of Na_v channels (French et al., 1996; Jones & Bulaj, 2000; Hui et al., 2002; Arias, 2006).

In order to discover the substances that cause this effect, the bioactivity-guided two-dimensional (2D) chromatographic purification process combined with mass spectrometry was a successful isolation strategy. Sub-fraction RP1 from the 2D purification

Table 1 Main components in the active fraction F3B and sub-fraction RP1 of the tentacle venom of *Cyanea capillata* medusae

Fraction F3B		Sub-fraction RP1	
Mass peak number	Mass of the charged peptide (Da)	Mass peak number	Mass of the charged peptide (Da)
1	2373.7		
2	2475.6		
3	2588.0		
4	2875.6		
5	3185.0		
6	3428.4		
		1	3921.8
7	**3981.5**	2	**3983.4**
		3	4127.5
8	4291.4		
9	4515.2	4	4531.7
10	4895.1	5	4822.0
		6	5236.7
		7	5379.5
11	5710.0		
12	**5793.0**	8	**5795.4**
13	6255.8		
14	6420.3		
15	6470.0		
16	**6967.0**	9	**6962.1**
17	7461.4		

Values in bold peptides that were present in both fractions

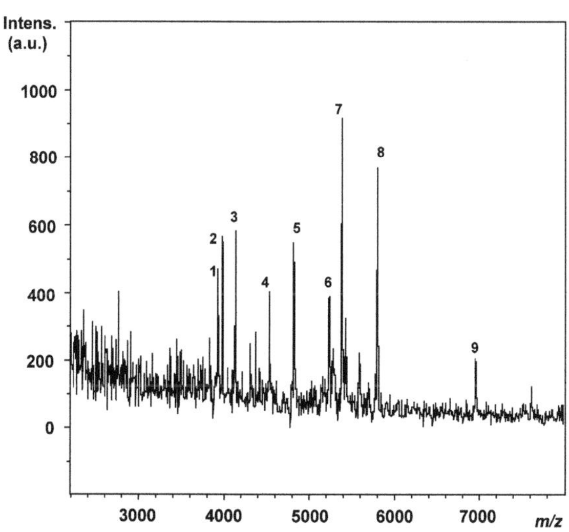

Fig. 6 Mass fingerprint of the active sub-fraction RP1 of the tentacle venom of *Cyanea capillata* medusae. (*m/z*: mass-to-charge ratio; a. u.: arbitrary units). Charged peptides $[M + H]^+$ with a signal-to-noise ratio >2 are annotated (1–9)

highlighted a group of hydrophilic, potentially neurotoxic peptides, which exhibited clear signals in the molecular mass range 3,900–7,000 Da. The hydrophilic property is probably due to a number of basic amino acids, e.g. lysine, histidine or arginine.

Wiebring et al. (2010, this issue) analysed isolated cnidocysts (euryteles, O-isorizhas, and A-isorizhas) from *C. capillata* by matrix-assisted laser desorption ionisation time-of-flight mass spectrometry. Three potential neurotoxic polypeptides with *m/z*-values of 4127.5, 5379.5 and 5795.4 from sub-fraction RP1 were shown to be main components in isolated A-isorizhas, but not in the other two cnidocyst types. This possible link between A-isorizhas and paralysing neurotoxins is notable and to our knowledge has not been reported for *C. capillata* yet. Investigations on cystonect siphonophores like *Physalia physalis* (L.), which eat only soft-bodied prey and have only isorizhas, support that these cnidocyst types can penetrate soft-bodied prey, like fish larvae (Purcell, 1984; Purcell & Mills, 1988). Östman & Hydman (1997) hypothesised that A-isorizhas in *C. capillata* can serve also to entangle prey.

Conclusion and outlook

An MNB cell assay was found to be suitable for the investigation of neurotoxins in jellyfish venom. This assay is sensitive and requires small amounts (<1 μg) of sample material. We demonstrated the presence of potentially neurotoxic polypeptides in the partially purified tentacle venom of *C. capillata*. The venom purification process achieved a 50-fold increase in neurotoxic activity and the neurotoxic activity was attributable to a group of low molecular weight polypeptides (3,900–7,000 Da). Further assay-guided purification steps will enable identification of the active components; sequence information obtained by tandem mass spectrometry will lead to structural characterisation of novel neurotoxins.

Acknowledgments The authors would like to thank colleagues from the Alfred-Wegener-Institute Helgoland and the crew of the RV "Heincke" for their substantial help and support and Dr. J. Gandrass (Institute for Costal Research, Department for Marine Bioanalytical Chemistry) for critically reading the manuscript.

References

Arias, H. R., 2006. Marine toxins targeting ion channels. Marine Drugs 4: 37–69.

Blumenthal, K. M. & A. L. Seibert, 2003. Voltage-gated sodium channel toxins – poisons, probes, and future promise. Cell Biochemistry and Biophysics 38: 215–237.

Catterall, W. A., 1995. Structure and function of voltage-gated ion channels. Annual Review of Biochemistry 64: 493–531.

Cestèle, S. & W. A. Catterall, 2000. Molecular mechanisms of neurotoxin action on voltage-gated sodium channels. Biochimie 82: 883–892.

Dudley, S. C., H. Todt, G. Lipkind & H. A. Fozzard, 1995. A μ-conotoxin-insensitive Na + channel mutant: possible localization of a binding site at the outer vestibule. Biophysical Journal 69: 1657–1665.

French, R. J., E. Prusak-Sochaczewski, G. W. Zamponi, S. Becker, A. S. Kularatna & R. Horn, 1996. Interactions between a pore-blocking peptide and the voltage sensor of the sodium channel: an electrostatic approach to channel geometry. Neuron 16: 407–413.

Gallacher, S. & T. H. Birkbeck, 1992. A tissue-culture assay for direct detection of sodium-channel blocking toxins in bacterial culture supernates. FEMS Microbiology Letters 92: 101–108.

Gallacher, S., K. J. Flynn, J. M. Franco, E. E. Brueggemann & H. B. Hines, 1997. Evidence for production of paralytic shellfish toxins by bacteria associated with *Alexandrium* spp. (Dinophyta) in culture. Applied and Environmental Microbiology 63: 239–245.

Helmholz, H., C. Ruhnau, C. Schuett & A. Prange, 2007. Comparative study on the cell toxicity and enzymatic activity of two northern scyphozoan species *Cyanea capillata* (L.) and *Cyanea lamarckii* (Péron & Léslieur). Toxicon 50: 53–64.

Honma, T. & K. Shiomi, 2006. Peptide toxins in sea anemones: structural and functional aspects. Marine Biotechnology 8: 1–10.

Hui, K. Y., G. Lipkind, H. A. Fozzard & R. J. French, 2002. Electrostatic and steric contributions to block of the skeletal muscle sodium channel by μ-conotoxin. Journal of General Physiology 119: 45–54.

Jones, R. M. & G. Bulaj, 2000. Conotoxins – new vistas for peptide therapeutics. Current Pharmaceutical Design 6: 1249–1285.

Kirchner, M., A. Wichels, A. Seibold, G. Sahling & C. Schuett, 2001. New and potentially toxic isolates from *Noctiluca scintillans* (Dinoflagellata). In Hallegraeff, G. M., S. I. Blackburn, J. S. Bolch & R. J. Lewis (eds), Proceedings of the IX International Conference on Harmful Algal Blooms. Intergovernmental Oceanographic Commission, Hobart, Australia: 379–382.

Long, K. O. & J. W. Burnett, 1989. Isolation, characterization, and comparison of hemolytic peptides in nematocyst venoms of 2 species of jellyfish (*Chrysaora quinquecirrha* and *Cyanea capillata*). Comparative Biochemistry and Physiology Part B: Biochemistry & Molecular Biology 94: 641–646.

Manger, R. L., L. S. Leja, S. Y. Lee, J. M. Hungerford & M. M. Wekell, 1993. Tetrazolium-based cell bioassay for neurotoxins active on voltage-sensitive sodium-channels – semiautomated assay for saxitoxins, brevetoxins, and ciguatoxins. Analytical Biochemistry 214: 190–194.

Messerli, S. M. & R. M. Greenberg, 2006. Cnidarian toxins acting on voltage-gated ion channels. Marine Drugs 4: 70–81.

Nicholson, G. M., 2007. Insect-selective spider toxins targeting voltage-gated sodium-channels. Toxicon 49: 490–512.

Norton, R. S., 1991. Structure and structure-function-relationships of sea anemone proteins that interact with the sodium-channel. Toxicon 29: 1051–1084.

Östman, C. & J. Hydman, 1997. Nematocyst analysis of *Cyanea capillata* and *Cyanea lamarckii* (Scyphozoa, Cnidaria). Scientia Marina 61: 313–344.

Purcell, J. E., 1984. The functions of nematocysts in prey capture by epipelagic siphonophores (Coelenterata, Hydrozoa). Biological Bulletin 166: 310–327.

Purcell, J. E. & C. E. Mills, 1988. The correlation between nematocyst types to diets in pelagic Hydrozoa. In Hessinger, D. A. & H. M. Lenhoff (eds), The Biology of Nematocysts. Academic Press, San Diego: 463–485.

Safo, P., T. Rosenbaum, A. Shcherbatko, D. Y. Choi, E. Han, J. J. Toledo-Aral, B. M. Olivera, P. Brehm & G. Mandel, 2000. Distinction among neuronal subtypes of voltage-activated sodium-channels by μ-conotoxin PIIIA. Journal of Neuroscience 20: 76–80.

Sanchez-Rodriguez, J., E. Torrens & L. Segura-Puertas, 2006. Partial purification and characterization of a novel neurotoxin and three cytolysins from box jellyfish (*Carybdea marsupialis*) nematocyst venom. Archives of Toxicology 80: 163–168.

Shon, K. J., B. M. Olivera, M. Watkins, R. B. Jacobsen, W. R. Gray, C. Z. Floresca, L. J. Cruz, D. R. Hillyard, A. Brink & H. Terlau, 1998. μ-Conotoxin PIIIA, a new peptide for discriminating among tetrodotoxin-sensitive Na channel subtypes. Journal of Neuroscience 18: 4473–4481.

Terlau, H. & B. M. Olivera, 2004. *Conus* venoms: a rich source of novel ion channel-targeted peptides. Physiological Reviews 84: 41–68.

Walker, A. J. A., 1977. The cardiac actions of a toxin-containing material from the jellyfish, *Cyanea capillata*. Toxicon 15: 15–27.

Wiebring, A., H. Helmholz, S. Lassen, A. Prange & G. Jarms, 2010. Separation and analysis of different types of nematocysts from *Cyanea capillata* (L.) medusae. Hydrobiologia. doi:10.1007/s10750-010-0227-6.

Wu, C. H. & T. Narahashi, 1988. Mechanism of action of novel marine neurotoxins on ion channels. Annual Review of Pharmacology and Toxicology 28: 141–161.

Yu, F. H. & W. A. Catterall, 2003. Overview of the voltage-gated sodium channel family. Genome Biology 4(207): 1–7.

Zhang, M. L., Y. Fishman, D. Sher & E. Zlotkin, 2003. Hydralysin, a novel animal group-selective paralytic and cytolytic protein from a noncnidocystic origin in hydra. Biochemistry 42: 8939–8944.

Hydrobiologia (2010) 645:223–234
DOI 10.1007/s10750-010-0216-9

Gill cell toxicity of northern boreal scyphomedusae *Cyanea capillata* and *Aurelia aurita* measured by an in vitro cell assay

Heike Helmholz · Blair D. Johnston · Christiane Ruhnau · Andreas Prange

Published online: 21 March 2010

Abstract Scyphozoan medusae are very successful foragers which occasionally occur in high abundances in boreal waters and may impact many different groups in the marine ecosystem by means of a variety of toxins. A rainbow trout gill cell line, RTgill-W1, was tested for its suitability as quantitative indicator of the cytotoxicity of *Cyanea capillata* and *Aurelia aurita*; the major scyphozoan species in the North and Baltic seas. Cultures of rainbow trout gill cells were exposed to whole venoms extracted from fishing tentacles and oral arms at increasing protein concentrations. The venom caused detachment, clumping and lysis of cells, as well as a drop in vitality, in a dose-dependent manner. Morphological changes in the cells were evident within 1 h after venom addition. The damage to gill cells was quantified by measuring the metabolic activity of the cells by means of the fluorescence of resorufin derived from the nonfluorescent substrate, resazurin. In general, a decrease in the metabolic activity of the cells was detected at a venom (protein) concentration above 2.0 μg ml^{-1} (corresponding to 0.2 μg 10^4 $cells^{-1}$), and a total loss of activity was observed above 40.0 μg ml^{-1} (corresponding to 4.0 μg 10^4 $cells^{-1}$). *C. capillata* venoms had increased cytotoxic activity as compared to *A. aurita* venoms at the same concentration. Cnidocyst extracts from oral arms of *A. aurita* induced an 85% loss of gill cell viability at concentrations of 0.2 μg 10^4 $cells^{-1}$, whereas crude venoms from fishing tentacles reduced cell viability by 18% at the same concentration. Gel electrophoresis of the venoms indicated that these consist of a large number of proteins in a fairly wide size range, from 6 to 200 kDa, including some that are the same size as those found in cubomedusae. It also appears that larger (i.e., older) medusae have more complex venoms and, in some cases, more potent venoms than smaller animals.

Keywords Jellyfish · RTgill-W1 · Venom · Toxin

Guest editors: J. E. Purcell & Dror Angel / Jellyfish Blooms: New Problems and Solutions

H. Helmholz (✉) · C. Ruhnau · A. Prange
Department for Marine Bioanalytical Chemistry, Institute for Coastal Research, GKSS Research Centre, Max-Planck-Street 1, 21502 Geesthacht, Germany
e-mail: heike.helmholz@gkss.de

B. D. Johnston
Ecotoxicology and Aquatic Biology Research Group, School of Biosciences, Hatherly Laboratories, University of Exeter, Prince of Wales Road, Exeter EX4 4PS, UK

Introduction

The investigated scyphozoan species *Aurelia aurita* (L.) and *Cyanea capillata* (L.) are among the most abundant jellyfish species in the North Sea region (Gröndahl, 1988; Båmstedt et al., 1994). As a member of the phylum Cnidaria, their main characteristic is the proliferation of specialized cells (cnidocytes) harbouring cnidocysts as cell organelles that contain a complex mixture of highly active and structurally diverse toxins. The cnidocyst capsules are discharged as a response to adequate chemical and mechanical stimuli elicited by prey organisms (Kass-Simon & Acappaticci, 2002). The toxin mixture released as whole venom induces different effects on prey, e.g. paralysis. The development and distribution of certain types of cnidocytes and the injection of bioactive compounds are the crucial factors for prey capture and digestion by cnidarians (Kintner et al., 2005; Regula et al., 2009).

The ecological impact of gelatinous predators on zooplankton species and fish fry has been getting a lot of attention in recent years due to an increasing frequency of jellyfish outbreaks and abundance (Hay, 2006). Factors such as over-fishing, increased availability of substrates for polyp settlement and climate change may play an important role in causing such outbreaks (Purcell et al., 2007). There are numerous records of possible negative effects of high abundances of gelatinous zooplankton on fish recruitment (Båmstedt, 1990; Båmstedt et al., 1994; Behrends & Schneider, 1995; Omori et al., 1995; Lynam et al., 2005, 2006; Barz & Hirche, 2007; Malej et al., 2007; Møller & Riisgård, 2007; Titelman et al., 2007).

In order to assess the ecological role of jellyfish in the food web, a number of in vivo experiments with different types of potential prey organisms were performed (Bailey & Batty, 1984; Båmstedt & Martinussen, 2000; Martinussen & Båmstedt, 2001; Hansson, 2006) to determine feeding rates, feeding behaviour and prey selection among jellyfish (Sullivan et al., 1997; Suchman & Sullivan, 2000). In addition, in vitro assays may be used for detecting cellular and molecular effects on potential prey organisms, to characterize the stinging capacity of jellyfish and to isolate and identify toxins relevant for envenomation responses of humans.

There are only a few studies concerning the toxic potential of *A. aurita*, and no detailed biochemical information about the composition of the venom is available (Burnett et al., 1988; Radwan et al., 2001; Segura-Puertas et al., 2002).

Contrary to *A. aurita*, which is relatively harmless to humans, *C. capillata* causes severe dermal irritation in humans, and a few in vitro and in vivo studies have investigated the toxic potential of *C. capillata* venoms (Rice & Powell, 1972; Walker, 1977a, b; Walker et al., 1977; Fenner & Fitzpatrick, 1986; Long & Burnett, 1989; Heeger, 1998; Helmholz et al., 2007). *C. capillata* has a strong haemolytic and cytotoxic as well as cardiotoxic activity. The analysis of the proteinaceous components of the venom showed a spectrum of proteins in a broad molecular size range, but no structure–effect relationships have been found (Burnett & Calton, 1987).

Since cell-damaging in vitro assays can serve as an indicator for the potential toxic effects, cell-based approaches are useful tools for a comparative and sensitive toxicological venom analysis. The haemolysis of erythrocytes from different animal species or damaging effects on cell lines, e.g. hepatocytes, are often used as responsive elements to test the venom (Helmholz et al., 2007). Due to the fact that gills are very sensitive to the action of cnidarian venoms, an in vitro assay utilizing a gill cell line as the target (responsive element) was developed. An existing rainbow trout gill cell line, RTgill-W1, was used in this assay because it is a difficult and laborious process to establish a novel stable cell line. This cell line was selected because it has proven to be a valuable tool in marine and freshwater ecotoxicological studies investigating the effects of polycyclic aromatic hydrocarbons (PAHs), metals and waste water (Schirmer et al., 1998; Dayeh et al., 2002; Dayeh et al., 2005; Lee et al., 2008). To the best of our knowledge, this gill cell line has not been used for the detection and assessment of toxic effects of natural marine substances. In this publication, the sensitivity of gill cells to crude scyphozoan venoms is demonstrated utilizing the developed in vitro assay. In addition to the quantitative in vitro determination of gill cell toxicity, this assay enables the clarification of cellular responses and modes of action.

Materials and methods

Leibovitz L15 Medium (L15) (Lonza group, Wuppertal, Germany), penicillin/streptomycin solution and phosphate buffered saline (PBS) (10×) were purchased from Invitrogen (Karlsruhe, Germany). L-glutamin, Bradford reagent, bovine serum albumin (BSA) fraction V were obtained from Sigma (Munich, Germany) and foetal calf serum (FCS) from PAA Laboratories (Cölbe, Germany).

Organisms

Aurelia aurita (L.) were collected during research cruises to the Scottish Orkney Islands and Western Isles in June 2006 and 2008, and *C. capillata* (L.) were collected from the waters near the Scottish Orkney Islands and along the coast of the German Baltic Sea in October 2006 and 2008. The animals were separated into groups (see Tables 1, 2) based on the average umbrella diameter. The groups were chosen according to the equal distribution of individuals of the different size classes and following Long & Burnett (1989). Oral arms and fishing tentacles were removed immediately after sample collection and used for subsequent processing. Although medusae were collected in different years, sample handling and preparation were identical (see Table 1).

Preparation of cnidocysts

Fishing tentacles and oral arms were used for the preparation of intact cnidocysts following the methods described in Endean (1987), Burnett et al. (1992), Avian et al. (1995) and Bloom et al. (1998). The methodology was modified to accommodate large scale preparation for biochemical and analytical purposes and to suit available laboratory facilities on board the research vessel. Detailed description of the modifications can be found in Helmholz et al. (2007). The preparation of intact cnidocysts was performed within 24 h of medusae collection in a temperature-controlled laboratory. The medusae tissues were stirred gently (*A. aurita* tissue material was not stirred) in distilled water on ice. The ratio of organic tissue to distilled water was approximately 1:5. After 10 h, the suspension was filtered through a nylon sieve (500 μm mesh size) to discard the mesoglea residue. The filtrate was centrifuged at 4°C for 5 min at 3,000 rpm. After discarding the supernatant, the residues were washed two to three times with sterile-filtered seawater. The content, purity and integrity of cnidocysts were controlled microscopically, and the cnidocyst concentrate was stored in sterile seawater at −80°C until further use.

Cnidocyst lysis and protein extraction

Prior to lysis, the isolated cnidocysts were suspended in ice cold 10 mM acetate buffer, pH 5.5, and discharged in a cooled sonicator (Branson Sonifier 450, G. Heinemann Ultraschall- und Labortechnik, Schwäbisch Gmünd, Germany). At least 2/3 of the cnidocysts needed to be discharged at this stage, otherwise the protein concentration would have been too low for testing, thereby necessitating a repeated process of discharging. The suspension was centrifuged at 11,000 rpm for 5 min at 4°C. The supernatant was carefully removed, sterile filtered and an aliquot was taken for protein determination. The protein content was measured by means of the Bradford protein microassay using BSA as standard protein (Bradford, 1976). The protein extracts were immediately used for bioactivity assays or stored at −80°C until further use. In the following, all mention of "venom concentration" refers to protein concentration expressed in units of $\mu g\ ml^{-1}$.

Cell toxicity assay

A cell viability assay (CellTiter-Blue®, Promega, Mannheim, Germany) was used to detect the acute toxicity of the extracts. A rainbow trout cell line RTgill-W1 ATCC No: CRL-2523 (LCG Promochem,

Table 1 Details of the aggregated samples of *Aurelia aurita* (A. aur 1–6) including sample size (*n*), umbrella diameter expressed as mean ± standard deviation and size range [minimum and maximum diameters] and the region and year medusae were collected

Sample	Mean umbrella diameter (cm)	Region/year
A. aur 1	8.2 ± 0.9 (n = 66) [6–10]	Lewis/West Coast Western Isles/2007
A. aur 2	13.1 ± 1.9 (n = 888) [11–20]	Lewis/West Coast Western Isles/2007
A. aur 3	23.4 ± 3.5 (n = 651) [12–29]	Lewis/West Coast Western Isles/2008
A. aur 4	32.1 ± 1.9 (n = 82) [30–38]	Lewis/West Coast Western Isles/2008
A. aur 5	9.1 ± 0.9 (n = 81) [7–10]	Mainland/ Orkney Islands/2008
A. aur 6	12.0 ± 1.2 (n = 90) [11–16]	Mainland/ Orkney Islands/2008

Table 2 Details of the aggregated samples of *Cyanea capillata* (C. cap 1–5) including sample size (n), umbrella diameter expressed as mean ± standard deviation and size range [minimum and maximum diameters] and the region and year medusae were collected

Sample	Mean umbrella diameter (cm)	Region/year	Cell viability OA venom (%)
C. cap 1	21.5 ± 6.0 (n = 23) [11–30]	Rousay/Orkney Islands/2006	7.42 (±4.06)
C. cap 2	33.0 ± 0.9 (n = 6) [32–34]	Rousay/Orkney Islands/2006	16.32 (±7.54)
C. cap 3	20.7 ± 5.5 (n = 32) [11–30]	Priwall/Baltic Sea/2006	36.40 (±10.87)
C. cap 4	20.9 ± 4.5 (n = 72) [11–28]	Eckernfoerde/Baltic Sea/2008	29.13 (±9.75)
C. cap 5	33.0 ± 2.8 (n = 2) [31–35]	Eckernfoerde/Baltic Sea/2008	32.05 (±10.32)

Gill cell viability (% ± standard deviation; $n > 10$) after exposure to 2 μg ml^{-1} oral arm (OA) venom

Wesel, Germany) was cultivated in Leibovitz L15 cell culture medium with 10% FCS, 2 mM L-Glutamine, 100 IU penicillin and 100 μg streptomycin per ml medium at 20°C modified after Bols et al. (1994). Cells of a continuous culture were sowed into a black microtiter plate at a density of 10^4 cells per well in 75 μl L15 medium without FCS and allowed to settle and reattach for 24 h. After this adaptation step, 25 μl of protein extracts was applied to each well in concentrations of 0.1–4.0 μg protein per well (corresponding to 1–40.0 μg protein ml^{-1}). Extracts were diluted with L15 medium. The following controls were used for the detection of the relative toxicity of the extracts: a positive control with cells growing in 100 μl L15 medium, a negative control with same volume of medium but without cells and a buffer control containing 25 μl of 10 mM acetate buffer. After an incubation of 48 h at 20°C, 20 μl of CellTiter-Blue® reagent (resazurin) was added, and the fluorescence intensity of the metabolized resorufin was recorded after 4 h incubation at 20°C at 560 excitation/590 emission (Victor 3, 1420 Multilabel Counter, Perkin Elmer, Rodgau-Jügesheim, Germany). This assay involved the enzymatic transformation of the fluorescent dye by dehydrogenases whereby the intensity of the fluorescence signal depends on cell vitality, cell proliferation, and on the dye incubation time. Values of cell vitality above 100% occurred due to slight variations in handling and cell growth. Therefore, the assay was performed with eight replicates of each extract concentration and controls. The experiments were repeated independently at least three times. The percentage of vital cells was calculated by defining the fluorescence of positive controls minus the values of no-cell control as "100%." Mean and standard deviation values were taken for statistical analysis, using the WinStat (statistics add-in for MS Excel) software. The nonparametric U test (Mann–Whitney test for comparison of two data sets) was also used because it does not require data to be normally distributed (significance is indicated as $P < 0.05$).

Gel electrophoresis

Tricine-sodium dodecyl sulphate-gel electrophoresis was performed with self-casted polyacrylamide gels according to Schaegger & von Jagow (1987). A 10% separating gel was combined with a 4% stacking gel. Dimensions of the gel were 20 × 20 cm. A SERVA unstained SDS Page protein marker 6.5–200 kDa (Serva electrophoresis GmbH, Heidelberg, Germany) was used for the estimation of molecular weights. This size range was chosen because previous results and literature values show that the majority of venom proteins are smaller than 200 kDa. The one-dimensional electrophoresis was performed with the Protein II electrophoresis system (BioRad, Munich, Germany). After silver staining (Blum et al., 1987), the gels were analysed with the Bio Imaging System Genegenius (Syngene, Cambridge, UK). Molecular weight determination was based on at least four independent SDS Page experiments.

Results

Development of the gill cell toxicity assay

Cells of the continuous culture of RTgill-W1 reproducibly developed a dense monolayer of cells with a polygonal, epithelial-like shape and oriented aggregation of the cells (Fig. 1), characteristic for this cell line.

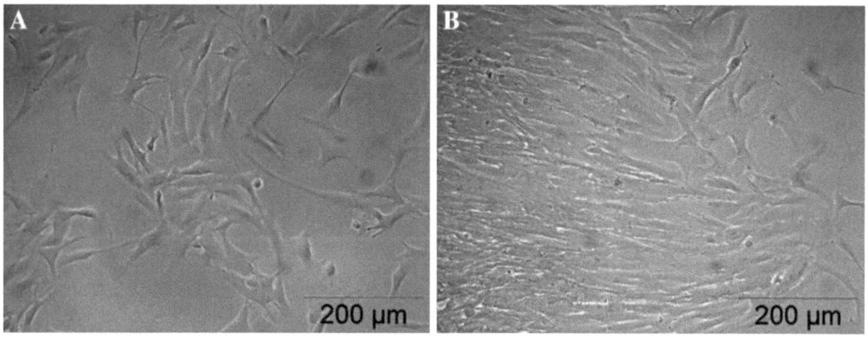

Fig. 1 Micrograph showing a monolayer of RTgill-W1 cells growing in continuous culture on a cell culture flask surface; **A** 8-day old culture, **B** 14-day old culture

After exposure to whole jellyfish venom at different protein concentrations, cells responded with changes in morphology and growth. Morphological changes were observed 1 h after treatment and became more obvious within the 48 h of incubation. The cells lost the flask surface contact, cell conjunctions and extension. Detachment (Fig. 2B), clumping (Fig. 2C) and cell lysis (Fig. 2D) occurred in a dose-dependent manner. Even though cell damage was obvious during microscopic examination, cell viability was measured fluorometrically to quantify the gill cell toxicity of the venom.

We demonstrated the suitability of the CellTiter-Blue® assay for estimating the number of viable fish gill cells by measuring their metabolic capacity. As the commercial cell viability test was developed and applied for mammalian cells, the assay conditions had to be adapted to the culture conditions that were optimal for fish cells. As a result of this optimization procedure, the recommended cell number was

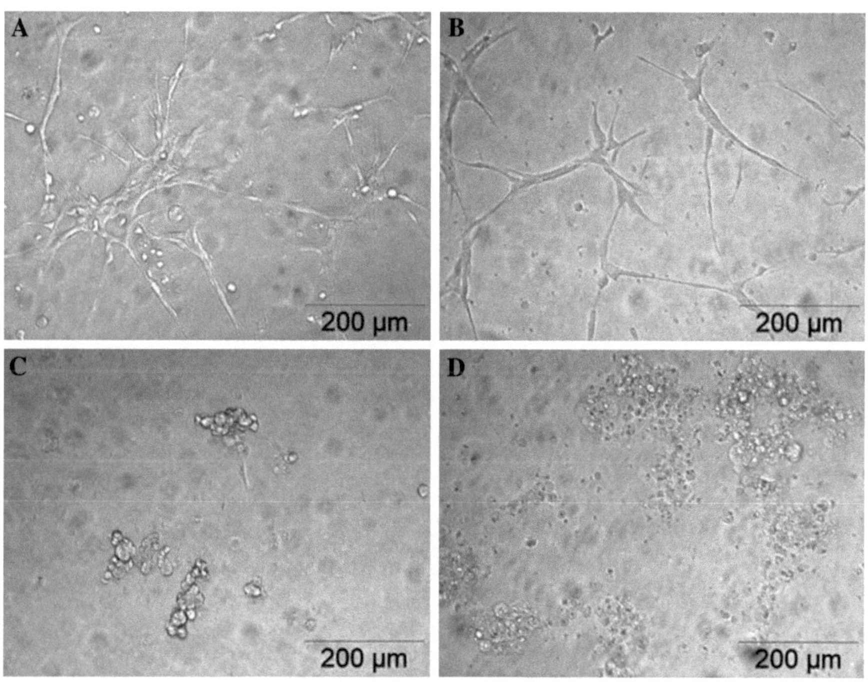

Fig. 2 Influence of increasing doses of *Aurelia aurita* (A. aur 2; umbrella diameter > 10 cm; Western Isles) whole animal venom on RTgill-W1 cell growth: **A** positive control (no venom added); **B** 5 µg ml^{-1} (detachment); **C** 20 µg ml^{-1} (clumping) and **D** 50 µg ml^{-1} (lysis)

reduced to 10^4 cells per well. This initial cell number was both sufficient to form interacting cell aggregates during the exposure period and did not cause growth inhibition that might result from limited nutrients and space. In addition, it was high enough for fluorescence signal detection showing clear differences between controls and disturbed cells.

A reliable performance of the assay was ensured by using the fluorescence signal of the positive cell control as an indicator for RTgill cell vitality over a continuous cultivation period. After a phase of 3 months with relatively homogenous metabolic activity, a slight but consistent decrease of the fluorescence signal was observed. Within this stable period, a variation coefficient of 8.5% ($n = 341$) was recorded. Therefore, a loss of cell viability of 10% due to the impact of jellyfish venom was judged as a true toxic effect.

Gill cell toxicity of *Aurelia aurita* venoms

Whole venoms of fishing tentacles and oral arms obtained from an enriched suspension of all types of cnidocysts induced a concentration-dependent loss of RTgill cell viability. Extracts of oral arms showed stronger cytotoxic effects, at an effective concentration range of 1.0–20.0 μg ml^{-1}, as compared to extracts of fishing tentacles in the concentration range of 2.0–40.0 μg ml^{-1}. *A. aurita* venom showed dose-dependent toxic effects (Fig. 3; Table 3).

The stronger cytotoxic effect of venoms obtained from oral arms was most obvious at a concentration of 10 μg protein ml^{-1} corresponding to 1 μg $(10^4 \text{ cells})^{-1}$ where a loss of more than 90% cell viability in each sample could be observed, whereas up to 80% of the gill cells remained vital after exposure to venoms from fishing tentacles.

The average umbrella diameter of individuals from the Orkney Islands in 2008 was smaller compared to those from Western Islands. Also a reduced toxic effect can be seen for both size groups from the Orkney Islands by comparing the gill cell toxicity of 10 μg ml^{-1} fishing tentacle venom (Fig. 4) and for 5 μg ml^{-1} oral arm venom (Table 3), respectively.

A significant dependency of gill toxicity on umbrella size could be observed at a concentration of 5.0 μg ml^{-1} gastric filament venom and 10.0 μg ml^{-1} fishing tentacle venom. As the fishing tentacle venoms of the larger animals were more active than from the animals <10 cm bell diameter (Fig. 4), the opposite trend was observed for the crude venom of oral arms.

Gill cell toxicity of *Cyanea capillata* venoms

The application of *C. capillata* venoms obtained by the preparation of cnidocysts from oral arms and fishing tentacles induced similar morphological changes in gill cells as observed after exposure to *A. aurita* venom. Toxic effects were observed microscopically and measured via the cell viability fluorescence assay at a dose >1.0 μg ml^{-1} crude venom of oral arms and >5.0 μg ml^{-1} crude venom of fishing tentacles. These indicated an increased gill cell toxicity of oral arms as compared to fishing tentacles.

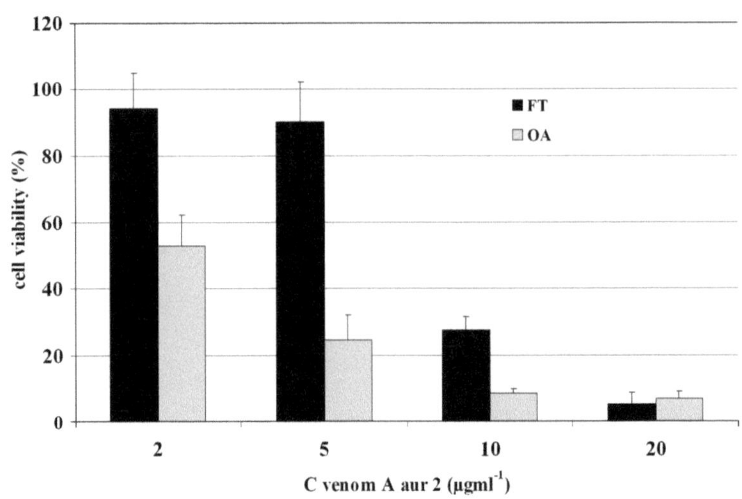

Fig. 3 Dose-dependent (2–20 μg ml^{-1}) effect of *Aurelia aurita* (sample A. aur 2 umbrella diameter > 10 cm; Western Isles) venom from oral arms (OA) and fishing tentacles (FT) (mean ± standard deviation; $n > 9$) on cell viability (%)

Table 3 Dose-dependent (protein concentration in µg ml^{-1}) cytotoxic effect on gill cell viability (%) of whole venom of *Aurelia aurita* obtained from oral arms (OA) and fishing tentacles (FT) (mean; $n > 9$)

Sample	C_{venom} 2 µg ml^{-1}		C_{venom} 5 µg ml^{-1}		C_{venom} 10 µg ml^{-1}		C_{venom} 20 µg ml^{-1}	
	OA	FT	OA	FT	OA	FT	OA	FT
A. aur 1	49.2	101.7	15.5	82.1	6.2	43.5	7.4	8.3
A. aur 2	53.0	94.1	22.4	90.9	6.5	27.6	6.0	6.3
A. aur 3	53.9	n.d.	26.8	87.8	4.9	62.5	3.1	9.8
A. aur 4	52.4	n.d.	48.8	90.0	9.3	39.1	1.9	10.1
A. aur 5	66.9	n.d.	34.7	102.3	4.8	82.6	2.9	27.8
A. aur 6	59.0	n.d.	53.1	99.5	6.0	59.4	3.2	11.3

n.d. not determined

Samples from the coast of Rousay/Orkney Islands and two sampling sites in the Baltic Sea produced different gill cell toxicities (Table 2). The viability of RT gill cells was reduced to 2.7% at a fishing tentacle venom concentration of 10.0 µg ml^{-1} from *C. capillata* with a mean umbrella diameter of 21.5 cm sampled at the Orkney Islands. This was significantly stronger toxicity than that observed for similar sized specimens collected along the Baltic Sea coast (39.9% cell viability C. cap 3 and 12.6% cell viability C. cap 4). Whole venoms of oral arms obtained from *C. capillata* sampled at the Orkney Islands induced an average loss of cell viability of 85%, whereas samples from the Baltic Sea induced a cell viability loss of about 70% (protein concentration 2.0 µg ml^{-1}) (Table 2).

A comparison of the toxicity (against gill cells) of the two species indicated that the venom of large *C. capillata* (>30 cm) fishing tentacles had threefold greater toxic activity and 20-fold greater toxic activity (*C. capillata* > 20 cm), respectively, than *A. aurita* of equivalent size classes at a concentration of 10 µg ml^{-1}. The venom of the oral arms of *C. capillata* was fivefold more toxic than that of *A. aurita* for both size classes at a concentration of 5.0 µg ml^{-1}.

Comparative analytical gel electrophoresis

A comparative overview of the protein content in the crude venoms was obtained by one-dimensional gel electrophoresis (Fig. 5A, B). Proteins of a broad size range from 6 to 170 kDa were detected. Venoms extracted from *A. aurita* had more proteins than did venoms of *C. capillata*.

An increased number of protein bands were observed in venoms from larger *A. aurita* (A. aur 4, A. aur 3; Table 1) reflecting a higher degree of venom complexity. A number of small proteins (6.5, 13, 17 and 23 kDa) seemed to be highly conserved, appearing in most of the venom extracts. The clusters at 30–37, 41–47 and 51–59 kDa were remarkable since a pattern of four protein bands with a regular difference of 2,000 Da was detected. A comparison of the venoms from *A. aurita* oral arms and fishing tentacles revealed a cluster of four proteins at 95, 100, 116 and 140 kDa that occurred only in the extract obtained from the oral arms. Regional

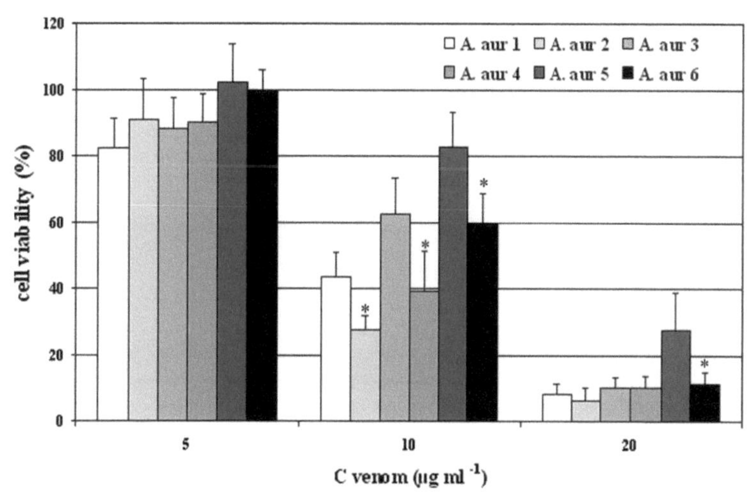

Fig. 4 The effect of increasing doses (5–20 µg ml^{-1}) of *Aurelia aurita* fishing tentacle venom on mean cell viability (%). *Error bars* indicated standard deviation ($n > 10$). * $P > 0.05$; Mann–Whitney test

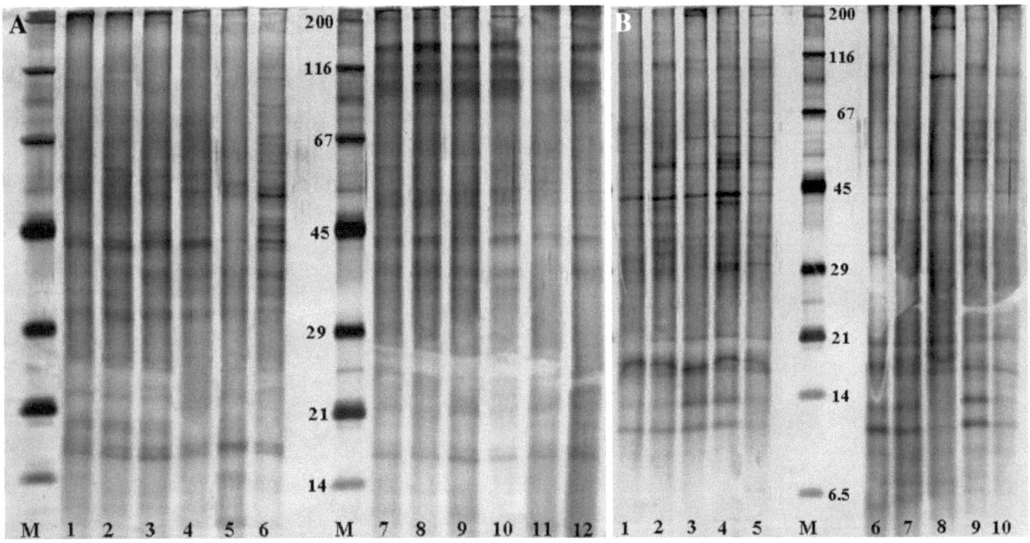

Fig. 5 **A** Gel electrophoresis (SDS Page) of crude venoms (protein equivalents 2 μg ml^{-1}) of fishing tentacles and oral arms of *Aurelia aurita*; M—marker (*numbers* indicate the molecular weight in kDa). *Lanes 1–6* fishing tentacles of A. aur 1–6, *lanes 7–12* oral arms of A. aur 1–6. **B** Gel electrophoresis (SDS Page) of crude venoms (protein equivalents 2 μg ml^{-1}) of fishing tentacles and oral arms of *Cyanea capillata*; M—marker (*numbers* indicate the molecular weight in kDa). *Lanes 1–6* fishing tentacles of C. cap 1–5, *lanes 6–10* oral arms of C. cap 1–5

variations can be detected in the venoms of *C. capillata* with higher complexity in the samples collected from the Baltic Sea. Venoms extracted from *C. capillata* fishing tentacles showed a greater number of protein bands as compared to extracts from oral arms, including proteins in a molecular weight range of approximately 70 kDa exclusively found in fishing tentacle venom.

Discussion

As the observed and predicted frequency of gelatinous zooplankton blooms increases, the investigation of the toxic potential and toxins of dominant scyphozoan species becomes more and more important (Hay, 2006). Classical laboratory and field experiments for the analysis of gut contents, prey selection, feeding behaviour and feeding rates have been applied to investigate trophic interactions of Scyphozoa (Bailey, 1984; Purcell, 1997). In addition, in vitro cell-based approaches have been utilized to compare the toxic potential of different medusa species and to understand their effect on different prey organisms. In general, whole venoms are tested in haemolysis and cell assays with mammalian cells (Allavena et al., 1998; Radwan et al., 2000; Bloom et al., 2001; Sun et al., 2002; Nagai, 2003; Bailey et al., 2005; Brinkman & Burnell, 2007). The novel contribution of this study is the use of fish gill cells as a major responsive element to test the toxicity of jellyfish. One of the major groups impacted by jellyfish is fish (Möller, 1980; Hay et al., 1990; Doyle et al., 2008) which are often impaired by stinging and toxic effects that are manifested in their gills. The RTgill W1 cell line is a cell line derived from a freshwater fish species which fulfils the prerequisites of an established and available cell line with the characteristics of gill cells. Although the feasibility of such a fish cell line for ecotoxicological studies of pollutants has been shown (Lee et al., 2008), the effects of natural marine toxins on these gill cells have not been tested. Therefore, the detected toxic potential of the northern boreal scyphozoan species *A. aurita* and *C. capillata* on gill cells can only be compared to haemolysis and mammalian cell assays.

Aurelia aurita is ubiquitously distributed in coastal areas, and its ecological impact on pelagic organisms is especially noticeable when it occurs in high abundances (Möller, 1980; Lynam et al., 2005). Biochemical investigations regarding the composition of *A. aurita* venom and its toxicology are rare. In vivo experiments whereby proteinaceous extracts from this medusa were administered to rabbits showed an induction of paralysis and glomerulonephritis (Wiersbitzky et al., 1973). A protein extract from whole *A. aurita* specimens caused muscle depolarization, indicating changes in sodium channel activity (Kihara et al., 1988). However, these two studies did not deal with causative toxins that may have been involved in the physiological observations. A general lytic effect was observed when red-tide flagellate species were treated with *A. aurita* autolysate (Hiromi et al., 1997). Segura-Puertas et al. (2002) found that dermonecrotic activity, haemolysis and neurotoxicity were caused by exposure to whole venoms of tentacular margins.

The traditional methods to test cytotoxic effects have involved the use of haemolysis and mammalian cell assays. Lytic effects on human erythrocytes have been shown at venom concentrations above 400 $\mu g\ ml^{-1}$. The advantage of the RTgill cell assay is that it is more than an order of magnitude more sensitive (employed at venom concentrations of 1–40 $\mu g\ ml^{-1}$) than the traditional methods employed thus far, and it uses a gill cell line derived from fish, which in many ways is a more realistic test of medusa toxicity.

Size exclusion chromatography and gel electrophoresis of the bioactive venom were performed by Segura-Puertas et al. (2002) and indicated that the toxic proteins were 45 and 66 kDa which matches the detected values in the present investigation. The broad molecular weight range (Fig. 5) reflects the complexity of whole venoms as also shown by Radwan et al. (2001). In that study, major protein bands at 50, 66 and 100 kDa have been shown, but peptides in the lower molecular weight range, as shown in this study (Fig. 5), were not detected.

Due to the high morphological and genetic variance detected among *A. aurita* species, it is likely that this species will exhibit considerable variety in physiological and toxinological characteristics (Dawson & Martin, 2001; Dawson, 2003). Radwan et al. (2001) compared medusae from the Red Sea and Chesapeake Bay and found increased haemolytic, dermonecrotic and phospholipase activities in medusae originating from the Red Sea. Regional differences have also been found in this study when comparing medusae collected at the Western Islands, the Scottish Orkney Islands and the Baltic Sea. It is noteworthy that medusae collected in a Western Isles bay with aquaculture facilities were more toxic than animals sampled at the Orkney Islands.

Another interesting finding of this study was the increased toxicity of fishing tentacle venom as a function of animal size. Moreover, both the size of medusae and their geographic origin might be related to the observed differences in venom composition, toxin concentration, or/and structure. The higher protein pattern complexity of venoms from larger (compared to smaller) *A. aurita* individuals (Fig. 5A) may be related to a change in venom composition with the development of the medusae.

In a comparison among oral arms and fishing tentacles, the venom obtained from cnidocysts of oral arms showed stronger activity at the same protein concentration than the venom obtained from fishing tentacles. This may be related to the presence of proteolytic enzymes in the oral arms, which function as digestive enzymes and affect the gill cells non-specifically, whereas the toxins in the fishing tentacles include various neurotoxins to induce rapid paralysis of prey organisms which probably do not affect the gill cells in the same way (see Lassen et al., 2010).

The predation impact of *C. capillata* on planktonic organisms, especially fish larvae, has been shown in field observations and laboratory studies (Fancett & Jenkins, 1988; Brewer, 1989; Hay et al., 1990; Purcell, 2003). As *C. capillata* has a higher envenomation capacity, in vivo and in vitro studies concerning the toxinological potency and biochemical composition of the venom have been performed. Detailed pharmacological studies by Walker et al. (1977) indicate that cardiotoxic activity of the venom is related to a 70 kDa protein. In addition, Long & Burnett (1989) discovered that haemolytic activity is related to proteins smaller than 10 kDa and that haemolytic activity did not occur in venoms extracted from *C. capillata* smaller than 15 cm. As no animals smaller than 20 cm were tested in the present investigation, this size-dependent toxic relationship could not be confirmed. In vitro experiments with hepatocytes used as the venom target showed a cytotoxic effect at protein concentrations >7 $\mu g\ ml^{-1}$ (Helmholz et al., 2007). The fish gill cells utilized in this study showed greater sensitivity (as compared to mammalian cell assays) to *C. capillata* venoms (Table 2).

Rice & Powell (1972) tested the ability of *C. capillata* fishing tentacle venom to induce dermal reactions and found only mild erythema in a few cases. The proteinaceous nature of the venom and the nematocyst origin of the active substance were indicated in their findings. The dermal reaction and the nematocyst discharge were also observed by Heeger et al. (1992) but neither of the above studies provided information on the biochemical composition of the venom.

Since only a few toxins have been isolated from jellyfish thus far, it is not possible to carry out an interspecies comparison among the detected proteins, but it is likely that there are biochemical similarities. Lethal toxins from box jellyfish species *Carybdea alata* and *Carybdea rastoni* with molecular weights of 43, 45 and 46 kDa have been isolated and sequenced (Nagai, 2003; Brinkman & Burnell, 2007). A remarkable cluster of protein bands in the same molecular weight range was detected in the venom of both *A. aurita* and *C. capillata* making this region especially interesting for further structural and biochemical investigations.

Conclusion and outlook

The objective of this study was the development of a novel tool for the determination of the toxic potential of jellyfish against a specific target gill cell. The application of the rainbow trout gill cell line for determination of the toxic potential of whole venoms derived from purified cnidocyst suspensions from *C. capillata* and *A. aurita* showed a strong susceptibility of these cells. Morphological changes in cell growth were observed within 1 h after venom exposure, and a fluorescence-based quantification of the cell viability showed a measurable dose-dependent effect at protein equivalent concentrations above 1 $\mu g\ ml^{-1}$. The comparatively high susceptibility of the gill cells enables the analysis of samples containing very low levels of venom. This approach provides a stable and reproducible in vitro test system for the detection and comparison of the toxicity of different cnidarian species, such as the *C. capillata* and *A. aurita* examined in this study. The toxic potential in these two boreal species is animal-size dependent and venoms of *C. capillata* expressed greater activity as compared to *A. aurita* venoms. Moreover, the venom obtained from cnidocysts of oral arms showed stronger activity at the same protein concentration than the venom obtained from fishing tentacles.

Biochemical analysis of whole venoms by gel electrophoresis showed high venom complexity with a broad size range of proteins. Some of the proteins were similar in molecular weights to those isolated toxins from box jellyfish, and mass spectrometric analysis will be performed to clarify structural homologies. There were clear differences in the composition of the venoms of the two species from different regions and among their oral arms and fishing tentacles, but further structural analyses have to be performed to reveal structure–effect relationships. Further work is planned to examine the quantitative impact of scyphomedusae venom on prey organisms in vivo to evaluate the relevance and applicability of in vitro cell-based assays.

Acknowledgements The authors thank Dr. Christian Schuett and colleagues from the Alfred-Wegener-Institute Station Helgoland and the crew of the RV Heincke for their substantial help and support, and Dr. Juergen Gandrass (Institute for Coastal Research, Department for Marine Bioanalytical Chemistry) for critical reading of this manuscript.

References

Allavena, A., G. L. Mariottini, A. M. Carli, S. Contini & A. Martelli, 1998. In vitro evaluation of the cytotoxic, hemolytic and clastogenic activities of *Rhizostoma pulmo* toxin(s). Toxicon 36: 933–936.

Avian, M., E. Spanier & B. Galil, 1995. Nematocysts of *Rhopilema nomadica* (Scyphozoa; Rhiztostomae): an immigrant jellyfish in the eastern Mediterranean. Journal of Morphology 224: 221–231.

Bailey, K. M., 1984. Comparison of laboratory rates of predation of five species of marine fish larvae by three planktonic invertebrates: effects of larval size on vulnerability. Marine Biology 79: 303–309.

Bailey, K. M. & R. S. Batty, 1984. Laboratory study of predation by *Aurelia aurita* on larvae of cod, flounder, plaice and herring: development and vulnerability to capture. Marine Biology 83: 287–291.

Bailey, P. M., A. J. Bakker, J. E. Seymour & J. A. Wilce, 2005. A functional comparison of the venom of three Australian jellyfish-*Chironex fleckeri*, *Chiropsalmus* sp., and *Carybdea xaymacana* – on cytosolic Ca^{2+}, haemolysis and *Artemia* sp. lethality. Toxicon 45: 233–242.

Båmstedt, U., 1990. Trophodynamics of the scyphomedusae *Aurelia aurita* – predation rates in relation to abundance, size and type of prey organism. Journal of Plankton Research 12: 215–229.

Båmstedt, U. & M. B. Martinussen, 2000. Estimating digestion rate and the problem of individual variability, exemplified by a scyphozoan jellyfish. Journal of Experimental Marine Biology and Ecology 251: 1–15.

Båmstedt, U., M. B. Martinussen & S. Matsakis, 1994. Trophodynamics of the two scyphozoan jellyfishes, *Aurelia aurita* and *Cyanea capillata*, in western Norway. ICES Journal of Marine Science 51: 369–382.

Barz, K. & H.-J. Hirche, 2007. Abundance, distribution and prey composition of scyphomedusae in the southern North Sea. Marine Biology 151: 1021–1033.

Behrends, G. & G. Schneider, 1995. Impact of *Aurelia aurita* medusae (Cnidaria, Scyphozoa) on the standing stock and community composition of mesozooplankton in the Kiel Bight (western Baltic Sea). Marine Ecology Progress Series 127: 39–45.

Bloom, D. A., J. W. Burnett & P. Alderslade, 1998. Partial purification of box jellyfish (*Chironex fleckeri*) nematocyst venom isolated at the beachside. Toxicon 36: 1075–1085.

Bloom, D. A., F. F. Y. Radwan & J. W. Burnett, 2001. Toxinological and immunological studies of capillary electrophoresis fractionated *Chrysaora quinquecirrha* (Desor) fishing tentacle and *Chironex fleckeri* Southcott nematocyst venoms. Comparative Biochemistry and Physiology Part C: Toxicology & Pharmacology 128: 75–90.

Blum, H., H. Beier & H. J. Gross, 1987. Improved silver staining of plant proteins, RNA and DNA in polyacrylamide gels. Electrophoresis 8: 93–99.

Bols, N., A. Barlian, M. Chirino-Trejo, S. J. Caldwell, P. Geogan & L. E. J. Lee, 1994. Development of a cell line from primary cultures of rainbow trout, *Oncorhynchus mykiss* (Walbaum), gills. Journal of Fish Diseases 17: 601–611.

Bradford, M., 1976. A rapid and sensitive method for the quantitation of microgram quantities of protein utilizing the principle of protein-dye binding. Analytical Biochemistry 72: 248–254.

Brewer, R. H., 1989. The annual pattern of feeding, growth, and sexual reproduction in *Cyanea* (Cnidaria: Scyphozoa) in the Niantic River Estuary, Connecticut. Biological Bulletin 176: 272–281.

Brinkman, D. & J. Burnell, 2007. Identification, cloning and sequencing of two major venom proteins from the box jellyfish, *Chironex fleckeri*. Toxicon 50: 850–860.

Burnett, J. W. & G. J. Calton, 1987. Venomous pelagic coelenterates: chemistry, toxicology, immunology and treatment of their stings. Toxicon 25: 581–602.

Burnett, J. W., G. Calton & J. B. Larsen, 1988. Significant envenomation by *Aurelia aurita*, the moon jellyfish. Toxicon 26: 215–217.

Burnett, J. W., K. O. Long & H. M. Rubinstein, 1992. Beachside preparation of jellyfish nematocyst tentacles. Toxicon 30: 794–796.

Dawson, M. N., 2003. Macro-morphological variation among cryptic species of the moon jellyfish, *Aurelia* (Cnidaria: Scyphozoa). Marine Biology 143: 369–379.

Dawson, M. N. & L. E. Martin, 2001. Geographic variation and ecological adaptation in *Aurelia* (Scyphozoa, Semaeostomeae): some implications from molecular phylogenetics. Hydrobiologia 451: 259–273.

Dayeh, V. R., K. Schirmer & N. C. Bols, 2002. Applying whole-water samples directly to fish cell cultures in order to evaluate the toxicity of industrial effluent. Water Research 36: 3727–3738.

Dayeh, V. R., D. H. Lynn & N. C. Bols, 2005. Cytotoxicity of metals common in mining effluent to rainbow trout cell lines and to the ciliated protozoan, *Tetrahymena thermophila*. Toxicology in vitro 19: 399–410.

Doyle, T. K., H. De Haas, D. Cotton, B. Dorschel, V. Cummins, J. D. R. Houghton, J. Davenport & G. C. Hays, 2008. Widespread occurrence of the jellyfish *Pelagia noctiluca* in Irish coastal and shelf waters. Journal of Plankton Research 30: 963–968.

Endean, R., 1987. Separation of two myotoxins from the nematocysts of the box jellyfish (*Chironex fleckeri*). Toxicon 25: 483–492.

Fancett, M. S. & G. P. Jenkins, 1988. Predatory impact of scyphomedusae on ichthyoplankton and other zooplankton in Port Phillip Bay. Journal of Experimental Marine Biology and Ecology 116: 63–77.

Fenner, P. & P. Fitzpatrick, 1986. Experiments with the nematocysts of *Cyanea capillata*. The Medical Journal of Australia 145: 174.

Gröndahl, F., 1988. A comparative ecological study on the scyphozoans *Aurelia aurita, Cyanea capillata* and *Cyanea lamarckii* in the Gullmar Fjord, Western Sweden, 1982 to 1986. Marine Biology 97: 541–550.

Hansson, L. J., 2006. A method for in situ estimation of prey selectivity and predation rate in large plankton exemplified with the jellyfish *Aurelia aurita* (L.). Journal of Experimental Marine Biology and Ecology 328: 113–126.

Hay, S., 2006. Marine ecology: gelatinous bells may ring change in marine ecosystems. Current Biology 16: 679–682.

Hay, S., J. Hislop & A. M. Shanks, 1990. North Sea Scyphomedusae; summer distribution, estimated biomass and significance particularly for 0-group gadoid fish. Netherlands Journal of Sea Research 25: 113–130.

Heeger, T., 1998. Quallen – Gefährliche Schönheiten. Wissenschaftliche Verlagsgesellschaft, Stuttgart.

Heeger, T., H. Möller & U. Mrowietz, 1992. Protection of human skin against jellyfish (*Cyanea capillata*) stings. Marine Biology 113: 669–678.

Helmholz, H., C. Ruhnau, C. Schütt & A. Prange, 2007. Comparative study on the cell toxicity and enzymatic activity of two northern scyphozoan species *Cyanea capillata* (L.) and *Cyanea lamarckii* (Peron & Leslueur). Toxicon 50: 53–64.

Hiromi, J., S. Handa & T. Sekine, 1997. Lethal effect of an autolysate of *Aurelia aurita* on red tide flagellates. Fisheries Sciences 63: 478–479.

Kass-Simon, G. & A. Acappaticci, 2002. The behavioral and developmental physiology of nematocysts. Canadian Journal of Zoology 80: 1772–1794.

Kihara, H., M. Anraku & O. Hirokazu, 1988. Tetrodotoxin-unaffected depolarization of frog muscles induced by the venom of jellyfish (Genus: *Aurelia*). Japanese Journal of Physiology 38: 839–849.

Kintner, A. H., J. E. Seymour & S. L. Edwards, 2005. Variation in lethality and effects of two Australian chirodropid jellyfish venoms in fish. Toxicon 46: 699–708.

Lassen, S., H. Helmholz, C. Ruhnau & A. Prange, 2010. Characterisation of neurotoxic polypeptides from *Cyanea capillata* medusae (Scyphozoa). Hydrobiologia. doi: 10.1007/s10750-010-0215-x.

Lee, L. E. J., K. Schirmer & N. C. Bols, 2008. Applications and potential uses of RTgill-W1 – a cell line derived from gills of rainbow trout. In Vitro Cellular & Developmental Biology – Animal 44: 17.

Long, K. O. & J. W. Burnett, 1989. Isolation, characterization, and comparison of hemolytic peptides in nematocyst venoms of two species of jellyfish (*Chrysaora quinquecirrha* and *Cyanea capillata*). Comparative Biochemistry and Physiology Part B: Biochemistry and Molecular Biology 94: 641–646.

Lynam, C. P., M. R. Heath, S. J. Hay & A. S. Brierley, 2005. Evidence for impacts by jellyfish on North Sea herring recruitment. Marine Ecology Progress Series 298: 157–167.

Lynam, C. P., M. J. Gibbons, B. E. Axelsen, C. A. J. Sparks, J. Coetzee, B. G. Heywood & A. S. Brierley, 2006. Jellyfish overtake fish in a heavily fished ecosystem. Current Biology 16: 1976.

Malej, A., V. Turk, D. Lučić & A. Benović, 2007. Direct and indirect trophic interactions of *Aurelia* sp. (Scyphozoa) in a stratified marine environment (Mljet Lakes, Adriatic Sea). Marine Biology 151: 827–841.

Martinussen, M. B. & U. Båmstedt, 2001. Digestion rate in relation to temperature of two gelatinous planktonic predators. Sarsia 86: 21–35.

Möller, H., 1980. Scyphomedusae as predators and food competitors of larval fish. Meeresforschung 28: 90–100.

Møller, L. F. & H. U. Riisgård, 2007. Population dynamics, growth and predation impact of the common jellyfish *Aurelia aurita* and two hydromedusae, *Sarsia tubulosa* and *Aequorea vitrina*, in Limfjorden (Denmark). Marine Ecology Progress Series 346: 153–165.

Nagai, H., 2003. Recent progress in jellyfish toxin study. Journal of Health Science 49: 337–340.

Omori, M., H. Ishii & A. I. Fujinaga, 1995. Life history strategy of *Aurelia aurita* (Cnidaria, Scyphomedusae) and its impact on the zooplankton community of Tokyo Bay. ICES Journal of Marine Sciences 52: 597–603.

Purcell, J. E., 1997. Pelagic cnidarians and ctenophores as predators: selective predation, feeding rates, and effects on prey populations. Annales de l'Institut Oceanographique 73: 125–137.

Purcell, J. E., 2003. Predation on zooplankton by large jellyfish, *Aurelia labiata*, *Cyanea capillata* and *Aequorea aequorea*, in Prince William Sound, Alaska. Marine Ecology Progress Series 246: 137–152.

Purcell, J. E., S. Uye & W.-T. Lo, 2007. Anthropogenic causes of jellyfish blooms and their direct consequences for humans: a review. Marine Ecology Progress Series 350: 153–174.

Radwan, F. F. Y., L. A. Gershwin & J. W. Burnett, 2000. Toxinological studies on the nematocyst venom of *Chrysaora achlyos*. Toxicon 38: 1581–1591.

Radwan, F. F. Y., J. W. Burnett, D. A. Bloom, T. Coliano, M. E. Eldefrawi, H. Erderly, L. Aurelian, M. Torres & E. P. Heimer-de la Cotera, 2001. A comparison of the toxinological characteristics of two *Cassiopea* and *Aurelia* species. Toxicon 39: 245–257.

Regula, C., S. P. Colin, J. H. Costello & H. Kordula, 2009. Prey selection mechanism of ambush-foraging hydromedusae. Marine Ecology Progress Series 374: 135–144.

Rice, N. & W. Powell, 1972. Observations on three species of jellyfish from Chesapeake Bay with special reference to their toxins. II Cyanea capillata. Biological Bulletin 143: 617–622.

Schaegger, H. & G. von Jagow, 1987. Tricine-sodium dodecyl sulfate-polyacrylamide gel electrophoresis for the separation of proteins in the range from 1 to 100 kDa. Analytical Biochemistry 166: 368–379.

Schirmer, K., D. G. Dixon, B. M. Greenberg, D. G. Dixon & N. C. Bols, 1998. Ability of 16 priority PAHs to be directly cytotoxic to a cell line from the rainbow trout gill. Toxicology 127: 129–141.

Segura-Puertas, L., G. Avila-Soria, J. Sanchez-Rodgriguez, M. E. Ramos-Aguilar & J. W. Burnett, 2002. Some toxinological aspects of *Aurelia aurita* (Linne) from the Mexican Caribbean. Journal of Venomous Animals and Toxins 8: 113–128.

Suchman, C. L. & B. K. Sullivan, 2000. Effect of prey size on vulnerability of copepods to predation by the scyphomedusae *Aurelia aurita* and *Cyanea* sp. Journal of Plankton Research 22: 2289–2306.

Sullivan, B. K., C. L. Suchman & J. H. Costello, 1997. Mechanics of prey selection by ephyrae of the scyphomedusa *Aurelia aurita*. Marine Biology 130: 213–222.

Sun, L.-K., Y. Yoshii, A. Hyodo, H. Tsurushima, A. Saito, T. Harakuni, Y.-P. Li, M. Nozaki & N. Morine, 2002. Apoptosis induced by box jellyfish (*Chiropsalmus quadrigatus*) toxin in glioma and vascular endothelial cell lines. Toxicon 40: 441–446.

Titelman, J., L. Gandon, A. Goarant & T. Nilsen, 2007. Intraguild predatory interactions between the jellyfish *Cyanea capillata* and *Aurelia aurita*. Marine Biology 152: 745–756.

Walker, M. J. A., 1977a. The cardiac actions of a toxin-containing material from the jellyfish, *Cyanea capillata*. Toxicon 15: 15–27.

Walker, M. J. A., 1977b. Pharmacological and biochemical properties of a toxin containing material from the jellyfish, *Cyanea capillata*. Toxicon 15: 3–14.

Walker, M. J. A., T. T. Martinez & D. V. Godin, 1977. Investigations into the cardiotoxicity of a toxin from the nematocysts of the jellyfish, *Cyanea capillata*. Toxicon 15: 339–346.

Wiersbitzky, S., G. Hegewald & B. Gibb, 1973. Glomerulotubular nephrosis in rabbits and injection of aqueous extracts from jellyfish (*Aurelia aurita*). Experimental Pathology 8: 232–235.